W9-BVS-060

changing the way the world learns

To get extra value from this book for no additional cost, go to:

http://www.thomson.com/wadsworth.html

thomson.com is the World Wide Web site for Wadsworth/ITP and is your direct source to dozens of on-line resources. *thomson.com* helps you find out about supplements, experiment with demonstration software, search for a job, and send e-mail to many of our authors. You can even preview new publications and exciting new technologies.

thomson.com: *It's where you'll find us in the future.*

The American Class Structure

In an Age of Growing Inequality

FIFTH EDITION

Dennis Gilbert
Hamilton College

WADSWORTH PUBLISHING COMPANY
I⊤P® An International Thomson Publishing Company

Belmont, CA• Albany, NY • Bonn • Boston • Cincinnati • Detroit • Johannesburg • London • Madrid
Melbourne • Mexico City• New York • Paris • Singapore • Tokyo • Toronto • Washington

Publisher: Denise Simon
Assistant Editor: Barbara Yien
Marketing Manager: Chaun Hightower
Project Editor: Jerilyn Emori
Print Buyer: Karen Hunt
Permissions Editor: Robert Kauser
Production: Robin Gold/Forbes Mill Press
Copy Editor: Robin Gold
Cover Designer: Stuart Paterson/Image House Inc.
Compositor: Forbes Mill Press/Wolf Creek Press
Printer: The Maple-Vail Book Manufacturing Group

Printed in the United States of America
  2 3 4 5 6 7 8 9 10

For more information, contact Wadsworth Publishing Company, 10 Davis Drive,
Belmont, CA 94002, or electronically at http://www.thomson.com/wadsworth.html

International Thomson Publishing Europe
Berkshire House 168-173
High Holborn
London, WC1V 7AA, England

International Thomson Editores
Campos Eliseos 385, Piso 7
Col. Polanco
11560 México D.F. México

Thomas Nelson Australia
102 Dodds Street
South Melbourne 3205
Victoria, Australia

International Thomson Publishing Asia
221 Henderson Road
#05-10 Henderson Building
Singapore 0315

Nelson Canada
1120 Birchmount Road
Scarborough, Ontario
Canada M1K 5G4

International Thomson Publishing Japan
Hirakawacho Kyowa Building, 3F
2-2-1 Hirakawacho
Chiyoda-ku, Tokyo 102, Japan

International Thomson Publishing GmbH
Königswinterer Strasse 418
53227 Bonn, Germany

International Thomson Publishing Southern Africa
Building 18, Constantia Park
240 Old Pretoria Road
Halfway House, 1685 South Africa

Library of Congress Cataloging-in-Publication Data

Gilbert, Dennis L.
    The American class structure in an age of growing inequality /
Dennis Gilbert. — 5th ed.
        p.   cm.
    Updated ed. of: The American class structure. 4th ed. 1993.
    Includes bibliographical references and index.
    ISBN 0-534-50520-1
    1. Social classes—United States. 2. Equality—United States.
3. Poverty—United States. 4. United States—Social conditions.
5. United States—Economic conditions.   I. Gilbert, Dennis L.
American class structure.   II. Title.
HN90.S6G54   1998
305.5'0973—dc21   97-40105

Contents

Chapter 9   Class Consciousness and Class Conflict   219

Chapter 10   The Poor, the Underclass, and Public Policy   251

Preface

I was 12 years old when the original version of *The American Class Structure* was being written in 1955. The author was Joseph Kahl, an unemployed Harvard Ph.D.., then living cheaply in Mexico. His book, which helped define the emerging field of social stratification, remained in print for twenty-five years. It earned this long run by presenting a lucid synthesis of the best research on the American class system. Each study was lovingly dissected by Kahl, who conveyed its flavor, assessed its strengths and weaknesses, summarized its most significant conclusions, and explained how they were reached.

The American Class Structure was not a theoretical book. Kahl created a simple conceptual schema with a short list of key variables drawn from the work of Karl Marx and Max Weber. He admitted that he had settled on this framework for the good and practical reason that it allowed him to draw together the results of disparate research reports. But the variables were interrelated, and Kahl believed that they tended to converge to create social classes in a pattern he called the American class structure. At the same time, he recognized that classes and class structure were abstractions from social reality—tendencies never fully realized in any situation but discernable when one stepped back from detail to think about underlying forces.

Sometime in the late-1970s, Kahl invited me to collaborate on a new version of *The American Class Structure*. He was then Professor of Sociology at Cornell, and I was his graduate student. The book we published in 1982 encompassed a body of stratification research that had grown enormously in sophistication and volume since the 1950s. *The American Class Structure: A New Synthesis* consisted almost entirely of new material but preserved the general framework of the original edition and its analyses of classic studies of the American class system. That edition and two subsequent

editions Kahl and I produced together, proved popular with a new generation of sociologists and sociology students. But when Wadsworth
Publishing asked us to start thinking about a fifth edition, Kahl, who had
by then retired to Chapel Hill, North Carolina, indicated that he would
rather spend his time listening to opera than reading page proofs again.
And since he would not be contributing to the fifth edition, he asked that
his name be taken off the cover.

Although there is now only one official author, the authorial "I" reverts
to"we" after this preface. Of course, most of this book is the product of a
long collaboration, and I am often at a loss to recall who wrote (or perhaps
rewrote) a particular passage. Retaining the "we" of previous editions
seemed perfectly natural. That said, I want to stress that I bear sole responsibility for every word included in this edition.

Readers of previous editions will find much that is familiar here, there
is also much that is new. Every chapter has been updated, with particular
emphasis on change in the class system. This edition gives expanded treatment to gender and race. It includes fresh material on jobs, wages, income,
wealth, poverty, residential patterns, social mobility, campaign contributions and many other subjects.

A new theme runs through the book: rising class inequalities. Data on
trends in earnings, income, wealth, jobs and related variables reveal a consistent pattern of growing inequality since the early 1970s (in sharp contrast
to the broadly shared prosperity of the 1950s and 1960s). The text repeatedly returns to a deceptively simple question: Why is this happening?

Like its predecessors, this edition of *The American Class Structure* is not
an encyclopedic survey of stratification research, nor is it an exercise in class
theory. It revolves around a short list of variables derived from classical theory, emphasizes selected empirical studies, and focuses on the socioeconomic core of the class system. Gender and race are treated in relation to
class, rather than as parallel dimensions of stratification. This approach reveals that the experience of class is inextricably bound up with gender and
race. For example, studies show that a married woman's sense of class identity reflects her husband's job, her own job, but also her attitude toward
gender roles. Residential segregation by class is increasing in the United
States, but so is segregation by race. One result has been the growth of affluent black neighborhoods.

In earlier prefaces, Kahl and I thanked many friends, colleagues and
students whose support made *The American Class Structure* a better book. I
want to add my thanks to those who contributed to this edition. Tim
Wickham-Crowley and two anonymous readers wrote thoughtful cri-
tiques of the last edition. Steve Rose (known to students of stratification
for his revealing income posters) was generous with his time, resources,
and ideas. Bill and Weslie Janeway shared their personal insights into the
class system. Bill Hadden, Philip Klinkner, Steve Rose, and Nina Serafino

read the manuscript in whole or part, offered valuable suggestions, and saved me from embarrassing errors. Chris Ingersoll designed the charts for Chapter 1. Without the help of research assistants Rebecca Pierce Bowmann and Julie Thomas and Hamilton Sociology Department secretaries Sally Carman and Carole Freeman, this long delayed book would have taken even longer to produce. Of course, I owe my largest debt of gratitude to Joe Kahl, a fine teacher, supportive colleague, and good friend.

Dennis Gilbert

The American Class Structure

in an Age of Growing Inequality

1

The Dimensions of Class

All communities divide themselves into the few and the many. The first are the rich and well-borne, the other the mass of the people... The people are turbulent and changing; they seldom judge or determine right...Give, therefore, to the first class a distinct, permanent share in the government. They will check the unsteadiness of the second, and as they cannot receive any advantage by a change, they therefore will ever maintain good government.

Alexander Hamilton (1780)

On the night the *Titanic* sank on her maiden voyage across the Atlantic in 1912, social class proved to be a key determinant of who survived and who perished. Among the women (who were given priority over men for places in the lifeboats), 3 percent of the first-class passengers drowned, compared with 16 percent of the second-class and 45 percent of the third-class passengers. Of the victims in first class, all but one had refused to abandon ship when given the opportunity. On the other hand, third-class passengers had been ordered to stay below deck, some of them at the point of a gun (Lord 1955: 107, cited in Hollingshead and Redlich 1958: 6).

The divergent fates of the *Titanic*'s passengers present a dramatic illustration of the connection between social class and what pioneer sociologist Max Weber called *life chances*. Weber invented the term to emphasize the extent to which our chances for the good things in life are shaped by class position.

Contemporary sociology has followed Weber's lead and found that the influence of social class on our lives is indeed pervasive. Table 1-1 gives a few examples. These data compare people at the bottom, middle, and top of the class structure. They show, among other things, that people at the bottom are more frequently the victims of violent crime, less likely to be in good health, and more likely to feel lonely. Those at the top are healthier, safer, and more likely to send their children to college; it is no wonder that they are, on average, happier with their lives and more conservative in their political outlooks.

Thoughtful observers have recognized the importance of social classes since the beginnings of Western philosophy. They knew that some individuals and families had more money or more influence or more prestige than their neighbors. The philosophers also realized that the differences were more than personal or even familial, for the pattern of inequalities tended to congeal into strata of families who shared similar positions. These social strata or classes divided society into a hierarchy; each stratum had interests or goals in common with equals but different from, and often conflicting with, those of groups above or below them.

Finally, it was noted that many of the political activities of people in society flowed from their class interests. As Hamilton said, the rich sought social stability to preserve their advantages, but the poor worked for social change that would bring them a larger share of the world's rewards.

This book is an analysis of the class structure of the United States today. We examine the distribution of income, prestige, power, and other stratification variables among the different classes in the country. We will point out how these variables react on one another; for instance, how a person's income affects beliefs about social policy or how one's job affects the choice of friends or spouse. And we will explore the question of movement from one class to another, recognizing that a society can have classes and still permit individuals to rise or fall among them.

To talk about a complex system and show how one part of it influences another, we must have a separate concept or word for each main aspect and

TABLE 1-1 Life Chances by Social Class[a]

	Lower Class	Middle Class	Upper-Middle and Upper Class
In excellent health[b]	28%	37%	53%
Victims of violent crime per 1000 population[c]	50	29	21
Psychological impairment index (average = 100)[d]	217	62	21
Feel lonely frequently or sometimes[e]	46%	35%	27%
Obesity in native-born women[f]	52%	43%	27%
Children, 18–24, in college[g]	15%	38%	54%
Dissatisfied with personal life[h]	22%	15%	5%
Favor liberal economic policies[i]	48%	38%	28%

[a]Classes defined by income

[b]Self-assessment, U.S. Public Health Service 1990: 164.

[c]U.S. Justice 1990: 239.

[d]Moderate to serious symptoms. Index calculated for this table from Strole et al. 1978: 308.

[e]DeStefano 1990: 33.

[f]Burnight and Marden 1967: 81.

[g]U.S. Census 1986a: 141.

[h]Gallup 1988: 234.

[i]In 1984 survey, favored increased federal spending on at least three of five social programs (food stamps, social security, Medicare, public schools, job creation). Center for Political Studies, "American National Election Study, 1984" (computed for this table).

a special method for estimating or measuring it. We begin by examining two key theories of stratification, to identify the major facets of the subject as a guide to concept formation. The writings of Karl Marx and Max Weber established an intellectual framework that strongly influenced subsequent scholars. (See the Suggested Readings at the end of the chapter for other important theorists.)

KARL MARX

Although the discussion of stratification goes back to ancient philosophy, modern attempts to formulate a systematic theory of class differences began with the work of Karl Marx in the nineteenth century; most subsequent theorizing has represented an attempt either to reformulate or to refute his ideas. Marx, who was born in the wake of the French Revolution and lived in the midst of the Industrial Revolution, emphasized the study of social class as the key to an understanding of the turbulent events of his time. His studies of economics, history, and philosophy convinced him that societies are mainly shaped by their economic organization and that social classes form the link between economic facts and social facts. He also concluded that fundamental social change is the product of conflict between classes.

Thus, in Marx's view, an understanding of classes is basic to comprehending how societies function and how they are transformed.

In Marx's work, social classes are defined by their distinctive relationships to the means of production. Taking this approach, capitalists, or the *bourgeoisie,* are a class consisting of the owners of the means of production, such as mines or factories. Likewise, workers, or the *proletariat,* are a class consisting of those who must sell their labor power to the owners of the means of production in order to earn a wage and stay alive. Marx maintained that in modern, capitalist society, each of these two basic classes tends toward an internal homogeneity that obliterates differences within them. Small business owners lose out in competition with big business owners, creating a small bourgeoisie of monopoly capitalists. In a parallel fashion, machines get more sophisticated and do the work that used to be done by skilled workers, so gradations within the proletariat fade in significance.

But notice that these are statements about trends, about long-run tendencies. At any given moment, distinctions within each class that stem from historical residues—even from markedly different earlier epochs—may influence the situation in important ways that shape behavior. Sometimes Marx called these subdivisions *fractions* of a class, and sometimes he seemed to consider them as momentarily separate classes. Generally his descriptions of contemporary situations in his writings as a journalist and pamphleteer show more complexity in economic and political groupings than do his writings as a theorist of history analyzing long-term trends.

Why did Marx look to production for the basis of social classes? In the most general sense, because he regarded production as the center of social life. He reasoned that people must produce to survive, and they must cooperate to produce. The individual's place in society, relationships to others, and outlook on life are shaped by his or her work experience. More specifically, those who occupy a similar role in production are likely to share economic and political interests that bring them into conflict with other participants in production. Capitalists, for instance, reap profit (in Marx's terms, *expropriate surplus*) by paying their workers less than the value of what they produce. Therefore, capitalists share an interest in holding wages down and resisting legislation that would enhance the power of unions to press their demands on employers.

From a Marxist perspective, the manner in which production takes place (that is, the application of technology to nature) and the class and property relationships that develop in the course of production are the most fundamental aspects of any society. Together they constitute what Marx called the *mode of production.* Societies with similar modes of production ought to be similar in other significant respects and should therefore be studied together. Marx's analysis of European history after the fall of Rome distinguished three modes of production, which he saw as successive stages of societal development: *feudalism,* the locally based agrarian society of the Middle Ages, in which a small landowning aristocracy in each district exploited the labor of a peasant majority; *capitalism*, the emerging commercial and industrial order of Marx's own lifetime, already international in scope and characterized by the dominance of the owners of industry over the mass of industrial workers; and *communism*, the technologically advanced, classless society of the future, in which all productive property would be held in common.

Unlike many later writers who believed that the level of technology by itself was the crucial determinant of social organization, Marx emphasized that modes of production entail patterns of both technology and social relations and that each can vary independently of the other. An agrarian society in which each producer cultivated land that he owned himself would not represent a feudal mode of production. Likewise, Marx viewed communism as a new mode of production, which could be built on the industrial technology already developed under the capitalist mode of production.

Marx regarded the mode of production as the main determinant of a society's *superstructure* of social and political institutions and ideas. He used the concept of superstructure to answer an old question: How do privileged minorities maintain their positions and contain the potential resistance of exploited majorities? His reply was that the class that controls the means of production typically controls the means of compulsion and persuasion—the superstructure. He observed that in feudal times, military and political power was monopolized by the landowners; with the rise of modern capitalism, political power was captured by the bourgeoisie when they gained control of the national government. In each case, the privileged class could use the power of the state to protect its own interests. For instance, in Marx's own time the judicial, legislative, and police authority of European governments dominated by the bourgeoisie was employed to crush the early labor movement, a pattern that was repeated a little later in the United States. As Marx expressed it in the *Communist Manifesto* (1848): "The executive of the modern State is but a committee for managing the common affairs of the whole bourgeoisie" (Marx 1979: 475).

But Marx did not believe that class systems rested on pure compulsion. He allowed for the persuasive influence of ideas. Here Marx made one of his most significant contributions to social science: the concept of *ideology*. Marx argued that human consciousness is a social product. It develops through our experience of cooperating with others to produce and to sustain social life. But social experience is not homogeneous, especially in a society that is divided into classes. The peasant does not have the same experience as the landlord and therefore develops a distinct outlook. One important feature of this differentiation of class outlooks is the tendency for members of each group to regard their own particular class interests as the true interests of the whole society. What makes this significant is that one class has superior capacity to impose its self-serving ideas on other classes. The class that dominates production, Marx argued, also controls the institutions that produce and disseminate ideas, such as schools, mass media, churches, and courts. As a result, the viewpoint of the dominant class pervades thinking in areas as diverse as the laws of family life and property, theories of political democracy, notions of economic rationality, and even conceptions of the afterlife. In Marx's words, "the ideas of the ruling class are in every epoch the ruling ideas" (Marx 1979: 172). In extreme situations, ideology can convince slaves that they ought to be obedient to their masters, or poor workers that their true reward will eventually come to them in heaven.

Marx maintained, then, that the ruling class had powerful political and ideological means to support the established order. Nonetheless, he regarded class societies as intrinsically unstable. In a famous passage from the *Communist Manifesto*, he observed

> The history of all hitherto existing society is the history of class struggles.
>
> Freeman and slave, patrician and plebeian, lord and serf, guildmaster and journeyman, in a word, oppressor and oppressed stood in constant opposition to one another, carried on an uninterrupted, now hidden, now open fight, a fight that each time ended either in a revolutionary reconstitution of society at large, or in the common ruin of the contending classes.
>
> In the earlier epochs of history, we find almost everywhere a complicated arrangement of society into various orders, a manifold gradation of social rank. In ancient Rome, we have patricians, knights, plebeians, slaves; in the Middle Ages, feudal lords, vassals, guild-masters, journeymen, apprentices, serfs; in almost all of these classes, again, subordinate gradations...
>
> Our epoch, the epoch of the bourgeoisie, possesses, however, this distinctive feature: It has simplified the class antagonisms. Society as a whole is more and more splitting up into two great hostile camps, into two great classes directly facing each other: Bourgeoisie and Proletariat. (Marx 1979: 473–474)

As these lines suggest, Marx saw class struggle as the basic source of social change. He coupled class conflict to economic change, arguing that the development of new means of production implied the emergence of new classes and class relationships. The most serious political conflicts develop when the interests of a rising class clash with those of an established ruling class. Class struggles of this sort can produce a "revolutionary reconstitution of society." Notice that each epoch creates within itself the growth of a new class that eventually seizes power and creates a new epoch; thus, change is explained by an internal dynamic that Marx called *the dialectic.*

Two eras of transformation through class conflict held particular fascination for Marx. One was the transition from feudalism to modern capitalism in Europe, a process in which he assigned the bourgeoisie (the urban capitalist class) "a most revolutionary part" (Marx 1979: 475). Into a previously stable agrarian society, the bourgeoisie introduced a stream of technological innovations, an accelerating expansion of production and trade, and radically new forms of labor relations. These changes were resisted by the feudal landlords, who felt their own interests threatened by those of the bourgeoisie. The result was a series of political conflicts (the French

Revolution was the most dramatic instance), through which the European bourgeoisie wrested political power from the landed aristocracy.

Marx believed that a second, analogous era of transformation was beginning during his own lifetime. The capitalist mode of production had created a new social class, the urban working class, or proletariat, with interests directly opposed to those of the rising bourgeoisie. This conflict of interests arose, not simply from the struggle over wages between capital and labor, but from the essential character of capitalist production and society. The capitalist economy was inherently unstable and subject to periodic depressions with massive unemployment. These economic crises heightened awareness of long-term trends separating rich from poor. Furthermore, capitalism's blind dependence on market mechanisms built on individual greed created an alienated existence for most members of society. Marx was convinced that only under communism, with the means of production communally controlled, could these conditions be overcome.

The situation of the proletarian majority made it capitalism's most deprived and alienated victim and therefore the potential spearhead of a communist revolution. However, in Marx's view, the objective situation of a class does not automatically cause its members to recognize their shared class interests and the need for militant class action—in short, to develop *class consciousness* leading toward political revolt. Some of Marx's most fruitful sociological work, to which we will return in Chapter 9, is devoted to precisely this problem. What intrinsic tendencies of capitalist society, Marx asked, are most likely to produce a class-conscious proletariat? Among the factors he isolated were the stark simplification of the class order in the course of capitalist development; the concentration of large masses of workers in the new industrial towns; the deprivations of working-class people, exacerbated by the inherent instability of the capitalist economy; and the political sophistication gained by the proletariat through participation in working-class organizations such as labor unions and mass parties.

What, in sum, can be said of Marx's contribution to stratification theory? His recognition of the economic basis of class systems was a crucial insight. His theory of ideology and his conception of the connection between social classes and political processes, although oversimple as stated, proved a fruitful starting point for modern research. As for his conception of change, a series of twentieth-century revolutions—including those in Mexico (1910), Russia (1917), and China (1949)—have established the significance of class conflict for radical social transformation. However, social revolutions have typically occurred in peasant societies during early stages of industrialization under foreign influence rather than in the advanced industrial countries where Marx anticipated them. In the advanced industrial countries, the proletariat used labor unions and mass political parties to defend its interests, thus rechanneling the forces of class conflict into the legal procedures of democratic politics.

A century after his death, it is apparent that Marx was a better sociologist than he was a prophet. He identified many of the central processes of capitalist society, but he was unable to foresee all the consequences of their unfolding, and his vision of a humane socialist future was certainly not realized in any communist country.

MAX WEBER

The great German sociologist Max Weber, who wrote in the early years of the twentieth century, was interested in many of the same problems that had fascinated Marx—among them, the origins of capitalism, the role of ideology, and the relationship between social structure and economic processes. Weber frequently benefited from Marx's work, even while reaching rather different conclusions. In the field of stratification, his special contributions were (1) to introduce a conceptual clarity that was often lacking in Marx's references to social classes and (2) to highlight those situations in which ideology, particularly religion, made an independent contribution to historical change.

Weber made a crucial distinction between two orders of stratification: *class* and *status*. *Class* had roughly the same meaning for both Weber and Marx. It refers to groupings of people according to their economic position. Class situation or membership, according to Weber, is defined by the individual's strength in economic markets (for example, the job market or securities markets), to the extent that these determine individual *life chances*. By life chances he meant the fundamental aspects of an individual's future possibilities that are shaped by class membership, from the infant's chances for decent nutrition to the adult's opportunities for worldly success.

Following Marx, Weber observed that the most important class distinction is between those who own property and those who do not. However, he noted that many significant distinctions can be made within these two categories. Among the propertied elite, for example, those who support themselves with stocks, bonds, and other securities (rentiers) are in a different class situation from those who live by owning and operating a business (entrepreneurs). The propertyless can be differentiated by the occupational skills that they bring to the marketplace: The life chances of an unskilled worker are vastly different from those of a well-trained engineer.

A *social class*, then, becomes a group of people who share the same economically shaped life chances. Notice that this way of defining a class does not imply that the individuals in it are necessarily aware of their common situation. It simply establishes a statistical category of people who are, from the point of view of the market (and the sociologist), similar to each other. Only under certain circumstances do they become aware of their common fate, begin to think of each other as equals, and develop institutions of joint action to further their shared interests.

Status, the second major order of stratification defined by Weber, is ranking by social prestige. In contrast with class, which is based on objective economic fact, status is a subjective phenomenon, a sentiment in people's minds. While the members of a class may have little sense of shared identity, the members of a status group generally think of themselves as a social community, with a common *lifestyle* (a familiar term we owe to Weber). In a classic essay on stratification, Weber (1946: 186–193) outlined these distinctions:

> In contrast to the purely economically determined "class situation," we wish to designate as "status situation" every typical component of the life fate of men that is determined by a specific, positive or negative, social estimation of honor...
>
> Status groups are normally communities. They are, however, often of an amorphous kind . . . . In content, status honor is normally expressed by the fact that above all else a specific style of life can be expected from all those who wish to belong to the circle. Linked with this expectation are restrictions on "social" intercourse (that is, intercourse which is not subservient to economic or any other of business's "functional" purposes). These restrictions may confine normal marriages to within the status circle and may lead to complete endogamous closure...
>
> Of course, material monopolies provide the most effective motives for the exclusiveness of a status group... With an increased inclosure of the status group, the conventional preferential opportunities for special employment grow into a legal monopoly of special offices for the members...
>
> With some over-simplification, one might thus say that "classes" are stratified according to their relations to the production and acquisition of goods; whereas "status groups" are stratified according to the principles of their consumption of goods as represented by special "styles of life."

In those passages, Weber specified many of the interrelations between class and status, between economy and society. Because of class position, a person earns a certain income. That income permits a certain lifestyle, and people soon make friends with others who live the same way. As they interact with one another, they begin to conceive of themselves as a special type of people. They restrict interaction with outsiders who seem too different (they may be too poor, too uneducated, too clumsy to live graciously enough for acceptance as worthy companions). Marriage partners are chosen from similar groups because once people follow a certain style of life, they find it difficult to be comfortable with people who live differently. Thus, the status group becomes an ingrown circle. It earns a position in the local community that entitles its members to social honor or prestige from inferiors.

Status groups develop the conventions or customs of a community. Through time they evolve appropriate ways of dressing, of eating, of living that are somewhat different from the ways of other groups. These ways get expressed as moral judgments reflecting abstract principles of value that separate "good" from "bad." The application of these principles to individuals establishes rankings of social honor or prestige. These distinctions often react back on the marketplace; to preserve their advantages, high-status groups attempt to monopolize those goods that symbolize their style of life—they pass consumption laws prohibiting the lower orders from wearing lace, or they band together to keep Jews or blacks out of prestigious country clubs or universities. (Weber regarded invidious distinctions among ethnic groups as a type of status stratification.)

A status order tends to restrict the freedom of the market, not only by its monopolization of certain types of consumption goods, but also by its monopolization of the opportunities to earn money. If they can get the power, status groups often restrict entry into the more lucrative professions or trades and access to credit. For example, entry into the electricians' union might be restricted to sons of current members. The local bank might be more willing to grant a loan to a member of the country club than to a social nobody. More generally, birth into a high-status family gives children advantages of social grace and personal contacts that eventually help careers.

Weber observed that, in theory, class and status are opposed principles. In its purest form, the class or economic order is universalistic and impersonal; it recognizes no social distinctions and judges solely on the basis of competitive skill or accumulated wealth. Status, in contrast, is based on particularistic distinctions: some people are "better" than others. Status groups want to restrict freedom of competition in both production and consumption.

But Weber recognized that, in practice, class and status are intertwined—at least in the long run. Historically, the status order is created by the class order; consumption, after all, is based on production. For example, although elite society might react against the status claims of the newly rich, it typically accepts their descendants if they have properly cultivated the conventions of the higher status group. On another level, the appearance of classes based on new sources of wealth—for instance, the emergence of an industrial bourgeoisie in Europe and America in the nineteenth century—signals a future restructuring of the status order as a whole.

Weber, like Marx, was interested in the relationship between stratification and political power. It would be accurate to say that for both men, stratification was essentially a political topic. But Weber was highly skeptical of the implication in Marx's work that all political phenomena could be traced back directly to class. For instance, Weber suggested that the institutions of the modern bureaucratic state exercise an influence on society that is not reducible to the control exercised by a single class. (In *The Eighteenth Brumaire*, Marx [1979: 594–617] himself reluctantly adopted this view for a special circumstance, but not Weber's corollary that a communist state might be grimly similar to a capitalist state, reflecting bureaucratic domination of society.)

Weber opposed the "pseudo-scientific operation" of Marxist writers of his day, who assumed an automatic link between class position and class consciousness (Weber 1946: 184). A common economic situation can and sometimes does lead to an awareness of shared class interests and a willingness to engage in militant class action, but it need not. Indeed, the very notion of class interest was highly ambiguous for Weber. Given his flexible conception of class, based on market-dependent life chances, he realized that modern society has several classes rather than just two and that they are continually changing. Furthermore, individuals can perceive their own situation and its convergence with that of other persons in a variety of ways that cannot be neatly divided into an accurate or "true" versus an inaccurate or "false" consciousness of shared interests, as the Marxists believed. He noted that the development of class consciousness is a complex process contingent on numerous factors (some of them already noted by Marx), including the "transparency" of the social arrangements that form the basis for divergent interests, the rate of social change, the dominant value system, the spread of radical interpretations of social reality that put dramatic labels on social experiences, and finally, the presence of leaders and associations, such as trade unions and political parties, capable of organizing class action into long-term conflict. Indeed, he investigated the different styles of party organization with such thoroughness that he almost seems to have considered them a semiautonomous influence in the total process.

Implicit in Weber's approach to stratification is the idea that status considerations can undermine the development of class consciousness and class struggle. For example, the politics of the American South has long been shaped by the tendency of poor whites to identify with richer whites rather than with poor blacks who share their economic position. Weber noted that political parties can develop around class, status, or other bases for conflict over power. The major American political parties are amorphous coalitions that have never been as clearly oriented toward the pursuit of class interests as, say, the working-class parties of Western Europe. None of this would have surprised Weber.

In sum, Weber accepted Marx's idea of the underlying economic basis of stratification in the long run. However, he identified another order of stratification by differentiating between class and status. He argued that the two interact with each other and with the political process in complex ways not fully recognized by Marx.

NINE VARIABLES

The scientific study of social stratification in the United States began in the 1920s and 1930s with the pioneering investigations of small-town class systems (which we review in Chapters 2 and 3). Since that time, sociologists have collected a mass of empirical evidence about the character of the American stratification system, both local and national. Unfortunately, the facts are described from numerous points of view and frequently conveyed in a highly technical manner. Our purpose in this book is to present the best examples of the research in a coherent form. To do so, we need a standardized language but a minimum of technical jargon.

We can combine the knowledge from the available empirical research by organizing it around a simple conceptual framework derived from our theorists, especially Weber, whose work suggests the difference between the economic, social, and political aspects of a stratification system. Under each of these rubrics, we can delineate a few basic variables that are significant for broad theories of stratification but still specific enough to be operationally defined and measured by the researcher. In the next few paragraphs, we will define these basic variables. In subsequent chapters, we will examine each one in detail, both as a concept and as a body of verified information about the United States.

The economic variables we will consider are *occupation, income,* and *wealth.* The first can be defined as a social role that describes the major work that a person does to earn a living. Modern societies are characterized by an elaborate and changing occupational hierarchy. Household *income* refers to the inflow of money over a given period of time, expressed as a rate (for example, $500 per week, $50,000 per year). Income is closely related to, but must be carefully distinguished from, *wealth,* which refers to assets held at a given point in time. The principal categories of wealth are monetary savings; personal property, such as homes or automobiles; and business assets, including commercial real estate, small enterprises, stocks, and bonds. Wealth can be viewed as an accumulation of past income. In certain forms, such as ownership of a business or of stocks and bonds, wealth becomes capital and is a source of new income. This possibility makes wealth especially important for students of stratification.

Our consideration of the status aspect of the American class system will emphasize *personal prestige, association,* and *socialization.* The first is the most obvious of our stratification variables; if we study a local community, we notice immediately that some people have higher *personal prestige* than others.[1] People have high prestige when neighbors in general have an attitude of respect toward them. Another word that is used for this attitude is deference, or the granting of social honor. Prestige is a sentiment in the minds of men and women, although they do not always know that it is there. The shrewd observer can often notice deference behavior that is not recognized by the participants, such as imitation of ideas or lifestyles. Consequently, it is necessary to study prestige in two ways: by asking people about their attitudes toward others and by watching their behavior.

People who share a given position in the class and status structures tend to have more personal contact or *association* with each other than with those in higher or lower positions. Such patterns of differential contact are significant because they promote similarities in behavior and opinion and a sense of community among the members of a class. Association (or interaction) is a variable that directs our attention to everyday social processes that

---

[1] The term *status,* employed by Weber, is a common synonym for *prestige.* We generally avoid this usage because *status* has other uses in sociology.

can be studied scientifically. By counting its frequency, measuring its duration, classifying its quality, and watching who initiates and who follows, we can draw systematic conclusions about association and can judge its consequences.

Socialization, as the term is employed in the social sciences, is the process through which an individual learns the skills, attitudes, and customs needed to participate in the life of the community. Although socialization takes place throughout the life cycle of the individual, we will be primarily interested in the socialization of children and teenagers. The early socialization experience of most members of a society is broadly similar; were it not, differences in expectations and behavior would be so great that social life would become impossible. However, there are significant variations in socialization among subgroupings within societies, especially complex societies. In the United States, research has uncovered class differences in socialization patterns, which can reinforce differences in values, channeling the young toward interactions with others of similar background and eventually toward assuming the class positions of their parents.[2]

Two variables related to the political aspect of stratification systems will be of interest to us: *power* and *class consciousness*. Weber defined *power* as the potential of individuals or groups to carry out their will even over the opposition of others (Weber 1946: 180). This classic definition implies that power is a significant dimension in quite varied social settings, from the power of parents over children to the power of the United States over the International Monetary Fund. We will restrict our use of the concept to broad political and economic contexts, dealing, for instance, with the power of the capitalist class over national economic priorities or the (declining) power of labor unions.

The degree to which people at a given level in the stratification system are aware of themselves as a distinctive group with shared political and economic interests is the measure of their *class consciousness*. In some circumstances, similar people do not have much contact with one another and think primarily in individualistic rather than in group terms. In other circumstances, they become highly group conscious, and then they are likely to organize political parties, trade unions, and other sorts of associations to advance their group interests. In general, Americans are thought to be less class conscious than Europeans; our traditions of equality lead some to deny that classes even exist.

Besides considering the eight variables just outlined, we will examine a final variable, *social mobility*, the extent to which people move up or down in the class system. This variable stands apart from the others because it implies an extended time dimension and because it cannot be uniquely assigned to the economic, social, or political aspect of stratification systems.

---

[2] Formal education can be treated as part of the broad socialization of individuals or as part of their direct preparation for occupational roles. We will deal with it both ways, so we do not list it as a separate variable.

We will focus on inter-generational mobility, the degree to which people ascend or descend relative to their parents. We will, at the same time, be determining the extent of "social succession"—that is, the proportion of people who simply inherit their parent's position. We will seek to answer two broad questions: What factors shape the structure of opportunities that make mobility possible? What characteristics of individuals account for differences in their ability to take advantage of mobility opportunities?

THE VARIABLES AS A SYSTEM

We have chosen these nine variables—occupation, income, wealth, personal prestige, association, socialization, power, class consciousness, and mobility—on pragmatic grounds. They constitute the set of variables that most efficiently organizes the existing empirical data on the American class system, and they are congruent with the thinking of the major theorists. Each variable, with the possible exception of power, can be measured by distinct and separate operations, and each can be used to stratify a given population. Thereafter, it is possible to study ways in which position on one dimension influences that on the others.

The nine variables do not form a closed system, all on the same level of abstraction, in the strict sense of scientific theory. However, they do constitute a useful conceptual scheme. By gathering data on each of these variables—as we will do for the United States—we can develop a thorough description of our class system. Moreover, because all the variables are mutually dependent, they provide a framework for thinking about the dynamics of class systems. For instance, numerous studies have demonstrated a connection between the stratification of occupations and each of the remaining eight variables. The occupational structure (that is, the distribution of people across occupations) of the advanced industrial nations is changing. An important question for students of stratification concerns the effect these changes are having on the other variables. Or, to pose a different sort of problem, we might look at our variables to help us understand how class systems—based, after all, on the unequal distribution of privilege—manage to persist. A few clues: The socialization of the less privileged appears to inculcate values supportive of the existing class order; mobility offers opportunities for dissatisfied individuals to change their class positions without changing the system; the privileged have disproportionate power over political and economic institutions.

Much is to be gained, then, by viewing these nine variables together. However, to say so is not to imply that they are of equal importance. They are not. We are convinced that societies that are similar in their technology and economic institutions have broadly similar class systems, that classes develop from economic roles, and that status communities emerge from stable distributions of economic privilege. In short, stratification systems are based on economic differences within societies. Therefore, among the variables defined here, those related to the economic aspect of stratification systems (occupation, income, and wealth) are fundamental. Nonetheless, the effects of these variables on the behavior of men and women are mediated through the other variables, which often exercise an independent or autonomous influence, especially in the short run. Thus, while recognizing the primacy of the economic variables, we believe that they can be studied most fruitfully as part of a more complex system.

WHAT ARE SOCIAL CLASSES?

A stratified society is one marked by inequality, by differences among people that are regarded as being higher or lower. The simplest form of inequality is based on the division of labor that always appears according to age and sex. Young children are everywhere subordinate to their elders; old people can have a high or a low position, depending on cultural values; women are often ranked below men.

Another form of inequality appears in every society (other than the smallest and most primitive ones) that ranks families or households rather than individuals.[3] A family shares many characteristics among its members that greatly affect their relationships with outsiders: the same house, the same income, and similar values. *If a large group of families is approximately equal in rank to each other and clearly differentiated from other families, we call these families a social class.* Social classes, defined in fashion, will be our primary focus in this book.

Our approach raises two questions: Why conceive of stratification in terms of discrete *classes*? And why think of classes as groupings of *families*? The first question arises because it is logically possible for a society to be stratified in a continuous gradation between high and low without any sharp lines of division. In reality, this is unlikely. The sources of a family's position are shared by many other similar families; there is only a limited number of types of occupations or of possible positions in the property system. One holds a routine position in a factory, office, or service setting; lives by manual skill or professional expertise; or manages people and money. People in similar positions have a tendency to mix with one another, to grow similar in their thinking and lifestyle. The similarities are shared within families and often inherited by children. In other words, the various stratification variables tend to converge and jell; they form a pattern, and this pattern creates social classes.

The pattern formed by the objective connections among the variables is heightened by the way people think about social matters because popular

---

[3] At times we will use the terms family and household interchangeably. When discussing income, we will often make the Census Bureau's more rigorous distinction between a group of related people residing together (family) and the broader concept that encompasses families, individuals residing alone, and unrelated individuals residing together (household).

thought creates stereotypes. Doctors are viewed as a homogeneous group, and distinctions among them tend to be ignored. Similarly, poor people tend to lump together all bosses, and rich people overlook the many distinctions that exist among those who labor for an hourly wage.

The second question arises because it is logically possible to study stratification of individuals rather than families. Why not define classes as discrete groups of *individuals* of equal rank? The simple answer, implied earlier, is that the members of a household live under the same roof, pool their resources, share a common economic fate, and tend, for all these reasons, to have a similar perspective on the world. This answer is not quite as persuasive as it was thirty or forty years ago. What made it seem self-evident in the past was that families were largely dependent on income produced by a "male head of household." The sociologist could place the family in the class system on the basis of his occupation, which tended to be a good predictor of the family's economic condition and its political outlook. Women, of course, were largely ignored in this conception of the class order, but it was arguably a reasonable approach to a world in which women's public economic role was quite circumscribed.

In the last few decades, women's economic and family roles have changed radically. Single women head a growing proportion of households. Married-couple families increasingly depend on two incomes. Where do we place a family in the class hierarchy if two spouses, both employed in working-class jobs, together produce a comfortable middle-class income? Suppose the husband is a factory worker and the wife is a teacher. Again, where do we place them? And if occupation is the key to political outlook, whose occupation counts here?

There are no fully satisfactory answers to such questions—the world is a complicated place. But there are, again, tendencies toward convergence and consistency. Family members (whatever disparities exist among them) still depend on common resources. They are viewed by outsiders as sharing the same position within the community. Husbands and wives typically have similar levels of education and, as a result, there is a correlation between the jobs held by working couples. Although married-couple families have grown increasingly dependent on wives' earnings, husbands are still the most important providers in the great majority of families. In the chapters that follow, we will repeatedly return to these issues, exploring the changing economic role of women in some detail. But we will continue to regard families as the basic unit of stratification analysis and define classes as groups of families or households.

At the same time, we will use the term "family" in the broadest possible sense to include households consisting of one person and larger domestic units "headed" by single females, single males, or couples (both heterosexual and homosexual). We will generally establish the class position of a family by the occupation of the family member who is the principal income earner. (In some cases, we will use household wealth or dependence on government payments to define class position).

In sum, we will interpret the stratification system through nine variables and discrete social classes composed of families. But we recognize that households may have inconsistent scores on the nine variables, that the lines dividing classes may be inconveniently fuzzy, and that the class placement of some families may be ambiguous. The reason for all this incoherence is not the inadequacy of the variables or definitions we employ, but the vague, fluid character of the stratification system itself. This book will emphasize the tendencies toward convergence, toward crystallization of the pattern, despite the many disturbing influences, often the result of social change, that keep the patterns from becoming as clear-cut in reality as in theory.

AN AMERICAN CLASS STRUCTURE

Some readers will have concluded by now that there is as much art as science in the study of social stratification—and they are probably right. We can make factual statements about, say, the distribution of income or patterns of association. But efforts to combine such information into broader statements about the class system run up against the inherent inconsistencies of social reality and are inevitably influenced by the viewpoint of the author.

We will, for example, be examining several general models of the class structure. Each tells us how many classes there are, how they can be distinguished from one another, and who belongs in each class. Some class models are more convincing than others because they make better use of the facts and illuminate matters that concern us. Some appear worthless. But there is really no way to establish that a particular model is "true" and another "false."

Our own model of the American class structure represents a synthesis of what we have learned writing this book. We will summarize it here and reconsider it in the last chapter. The model, illustrated in Figure 1-1, stratifies the population into six classes. Drawing from Marx, we distinguish a very small top class, whose income derives largely from return on assets, the *capitalist class*. But, drawing from Weber, we recognize multiple class distinctions among the largely nonpropertied majority. Below the capitalist class is an *upper-middle class* of well paid, university-trained managers and professionals. Next are the two largest classes, the *middle class* and the *working class*, distinguished from each other by the level of skill or knowledge and independence required on the job. They are followed by the *working poor*, people who work at low-skill, low wage, often insecure jobs. At the bottom is the *underclass*, with incomes heavily dependent on income from government programs.

Note that this model or map of the class system is based entirely on economic distinctions. We do not incorporate prestige differences, although we consider them important, because we believe they derive, in the long run, from economic differences. The model emphasizes *sources of income*: The top class draws income from capitalist property, the bottom class depends on

FIGURE 1-1 Gilbert-Kahl Model of the Class Structure

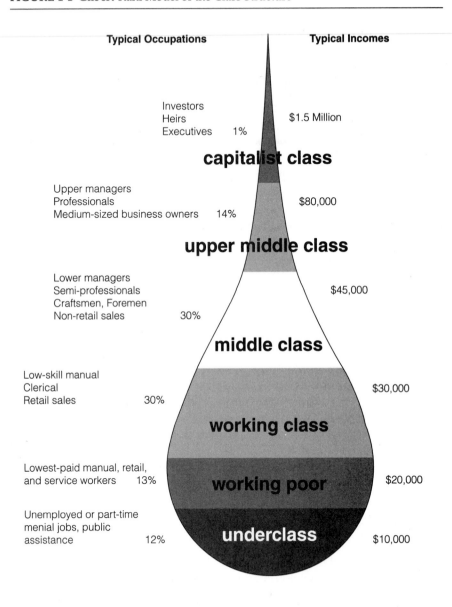

Typical Occupations   Typical Incomes

Investors
Heirs $1.5 Million
Executives   1%

capitalist class

Upper managers
Professionals $80,000
Medium-sized business owners   14%

upper middle class

Lower managers
Semi-professionals
Craftsmen, Foremen $45,000
Non-retail sales   30%

middle class

Low-skill manual
Clerical $30,000
Retail sales   30%

working class

Lowest-paid manual, retail,
and service workers   13% $20,000

working poor

Unemployed or part-time
menial jobs, public
assistance   12% $10,000

underclass

government payments, and the intermediate classes rely on earnings from jobs at differing occupational levels. We believe that the line dividing the top two classes from the classes below them has become the most important boundary in the class structure. The reason is that economic returns to capitalist property and the type of educational credentials possessed by the upper-middle class have grown rapidly in recent years. As we will see in Chapter 4, incomes at lower levels have tended toward stagnation.

IS THE AMERICAN CLASS STRUCTURE CHANGING?

We will return to this question repeatedly as we move from topic to topic. In particular, we will want to find out how the transformation of the U.S. economy in the last two or three decades has affected the class structure. In recent years, increasing class inequality has become a national political issue. Critics argue that the United States is becoming a less egalitarian, more rigidly stratified society. They say that poverty is increasing, the middle class is shrinking, social mobility is declining, and wealth is becoming more concentrated. We will examine data on wealth, income, jobs, mobility, poverty rates, political attitudes, and other factors to see whether the American class system is changing, and if so, how. Among the questions we will ask are these: Is the gap between rich and the rest of the population growing? Are opportunities to get ahead better or worse than they were in the past? Are neighborhoods becoming more segregated by social class? Is the balance of political power between classes changing?

The charts in Figure 1-2 preview some of our findings. They tell a story of a remarkable turn-around: Class inequalities, which fell in the 1950s and 1960s, rose steeply after the mid-1970s. This reversal is explicit in the four U-shaped curves. Individually, the charts in Figure 1-2 tell us the following about the years since 1975: (a) Wealth is increasingly concentrated in the hands of the richest 1 percent of households, (b) the income gap between the top 5 percent and the bottom 40 percent of families has widened, (c) the proportion of full-time workers earning poverty wages has increased, and (d) the percentage of Americans classified as poor has increased at the same time that (e) the proportion of prosperous families earning more than $75,000 (in inflation-adjusted dollars) has grown steadily.

We will take a second, more careful look at each of these charts in the appropriate chapters. For now we want to drive home the lesson of what has been called "the great U-turn" (Harrison and Bluestone 1988) and distinguish two periods in recent history. We will call the years from 1946 to approximately 1975 "the Age of Shared Prosperity" and the years from 1975 to the present "the Age of Growing Inequality."

FIGURE 1-2 From Shared Prosperity to Growing Inequality

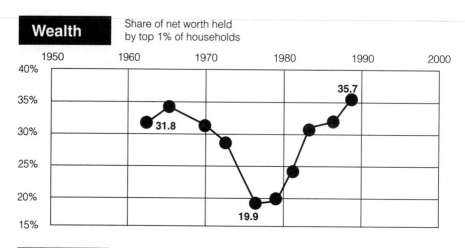

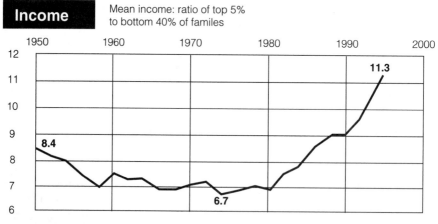

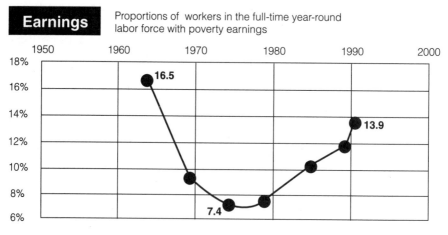

FIGURE 1-2 From Shared Prosperity to Growing Inequality (continued)

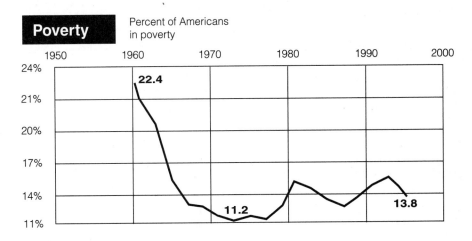

Poverty — Percent of Americans in poverty

22.4
11.2
13.8

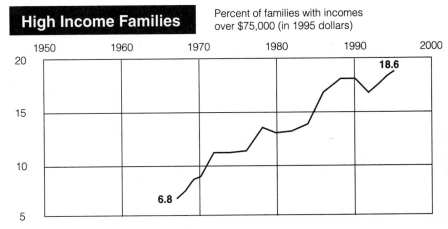

High Income Families — Percent of families with incomes over $75,000 (in 1995 dollars)

18.6
6.8

SOURCES: Wolff 1993; U.S. Census 1992, 1996d, 1996e.

SUGGESTED READINGS

Bendix, Reinhard, and Seymour Martin Lipset, eds. 1966. *Class, Status and Power: Social Stratification in Comparative Perspective.* 2nd ed. New York: Free Press.

A large collection of articles on important aspects of stratification in various countries. Part I covers basic theory, including Marx, Weber, and Davis and Moore's functionalist explanation of stratification.

Coser, Lewis. 1978. *Masters of Sociological Thought.* 2nd ed. New York: Harcourt Brace Jovanovich.

Contains short personal and intellectual biographies of Marx and Weber, which put them into the context of their times.

Crompton, Rosemary 1993. *Class and Stratification: An Introduction to Current Debates.* Cambridge, MA.: Polity Press.

Thoughtful guide to recent controversies.

Giddens, Anthony. 1973. *The Class Structure of the Advanced Societies.* New York: Harper & Row.

Attempts to rethink the main ideas of Marx and his critics to arrive at an acceptable modern interpretation of class structure.

Pakulski, Jan and Malcolm Waters 1996. *The Death of Class.* Thousand Oaks, CA. Sage.

Provocative book, arguing that the concept of social class no longer corresponds to social reality and should be abandoned by students of social inequality.

Grusky, David, ed. 1994. *Social Stratification: Class, Race and Gender in Sociological Perspective.* Boulder: Westview.

Anthology blending classic and postmodern readings.

Lenski, Gerhard. 1966. *Power and Privilege: A Theory of Social Stratification.* New York: McGraw-Hill.

Ambitious attempt to explain the development of stratification in each of several evolutionary stages, based on technology.

Marx, Karl. 1979. *The Marx-Engels Reader.* Edited by Robert C. Tucker. 2nd ed. New York: Norton.

Convenient collection of the writings of Marx and his partner, Friedrich Engels. Particularly relevant are "The German Ideology," Part I; "Wage Labour and Capital"; "Manifesto of the Communist Party"; and Engels' "Socialism: Utopian and Scientific."

Sernau, Scott, ed. 1996. *Social Stratification Courses: Syllabi and Instructional Materials.* Third Edition. Washington, D.C.: American Sociological Association.

Varied approaches to the study of stratification. Available from the ASA, 1722 N Street NW, Washington, DC 20036.

Weber, Max. 1946. *From Max Weber: Essays in Sociology.* Edited by H.H. Gerth and C. Wright Mills. New York: Oxford University Press.

A selection of Weber's most important sociological writings (except for his book The Protestant Ethic and the Spirit of Capitalism). *Especially relevant are: "Class, Status, Party"; "Bureaucracy"; "The Protestant Sects and the Spirit of Capitalism."*

2

Position and Prestige

Fame did not bring the social advancement which the Babbitts deserved. They were not asked to join the Tonawanda Country Club nor invited to the dances at the Union. Himself, Babbitt fretted, he "didn't care a fat hoot for all these highrollers, but the wife would kind of like to be among those present." He nervously awaited his university class-dinner and an evening of furious intimacy with such social leaders as Charles McKelvey the millionaire contractor, Max Kruger the banker, Irving Tate the tool-manufacturer, and Adelbert Dobson the fashionable interior decorator. Theoretically he was their friend, as he had been in college, and when he encountered them they still called him "Georgie," but he didn't seem to encounter them often . . .

Sinclair Lewis (1922: 190)

In the small group, in the local community, in the society as a whole, we notice that some people are looked up to, respected, considered people of consequence, and others are thought of as ordinary, unimportant, even lowly. Everywhere we see notables and nobodies.

Prestige is a sentiment in the minds of people that is expressed in interpersonal interaction: Deference behavior is expected by one party and granted by another. Obviously, this can occur only when there are values recognized by both parties that define the criteria of superiority (deference at pistol point is not the result of prestige). The parties need not be in perfect accord in their definition of the situation. Deference may be granted gladly or grudgingly. But as long as the subordinate recognizes that the superordinate *does* have the right to claim deference and feels constrained by group norms to grant it, a prestige difference exists.

Where inequalities in wealth and authority exist, the wealthy and the powerful can back their claims to prestige and deference with rewards and sanctions. People of lower prestige grant deference partly because they know that in the long run it is to their advantage to do so. Through symbolic submission to superiors, they gain favor in the eyes of those who are in a position to grant rewards—goods and services, promotion, protection, salvation, or perhaps just good will and a reassuring smile. The wealthy and the powerful can typically reinforce their claims to prestige with legitimating ideologies, propagated through schools, political institutions, mass media, churches, and other value-shaping institutions.

The most direct way to study prestige is to go into a local community where everyone knows everyone else and observe how they treat one another. That observation would be aided by some questions around the theme "What do you think of Jackson as compared with Albers?" As the observer moves from one sphere of local life to another, he or she will soon begin to see a pattern that groups individuals and families into clusters of people who are relatively homogeneous; we call them *social strata* or *prestige classes*. Within each, people feel similar and find it easy to communicate and share activities. Between strata, however, there can be some sense of distance and even tension: "They are not *our* sort of people."

The prestige hierarchy of a local community represents the synthesis of all the other stratification variables. It is the result of the evaluation by the people of the totality of the class structure. Such exhaustive mutual evaluations are only possible in small communities whose members have detailed knowledge of one another's lives and backgrounds. In other social contexts, people must depend on more limited and superficial clues to assign prestige to others, and the sociologists who study them must invent indirect methods of ranking. Nevertheless, an understanding of such local prestige structures is an excellent beginning for the study of the dynamics of stratification because if we can picture in detail the hierarchy of prestige, we can then look behind it for the factors that created it, both local circumstances and national conditions.

In this chapter, we will examine some classic investigations of prestige structures in small towns, a related study conducted more recently in two major cities, and a series of national surveys of the prestige associated with occupations. We will be particularly asking two questions: How do people create mental maps of the class system from their varied daily experiences of prestige distinctions? How can the sociologist turn the often vague and contradictory perceptions of respondents into coherent models of the class structure?

W. LLOYD WARNER: PRESTIGE CLASSES IN YANKEE CITY

The concern of American sociologists with the prestige aspect of stratification can be traced back to a series of community studies conducted by W. Lloyd Warner and his students and colleagues. Their research began in the early 1930s. Warner had already completed a three-year study of Australian aborigines and wanted to employ the techniques of social anthropology— the study of the "whole man" in the complete context of his sociocultural life—in the United States. Although Warner and his associates realized that in many ways, the metropolis was the typical social environment of modern society, they felt that the study of the great city as a whole from the viewpoint of social anthropology was too vast an undertaking for their techniques and resources, so they decided to start with a small community. They chose one "with a social organization which had developed over a long period of time under the domination of a single group with a coherent tradition" (Warner and Lunt 1941: 5). Their first study, of a New England town of 17,000 called "Yankee City," was followed by studies of even smaller communities in the South and Middle West (Warner et al. 1949b contains references to all of them).

Yankee City was once a famous seaport. It had a long history in New England commerce, having been a center of trade, fishing, and, more recently, manufacturing, especially of shoes and silverware. In many ways, its glory was in the past. In recent years, it had become merely a small city not too far from Boston, and many of its young people left for the more exciting life to be found in Boston and New York. Ethnically the town was relatively homogeneous but not perfectly so. Some families had been there for 300 years. Half its inhabitants had been born in the community, and another quarter came from other parts of New England and the United States. But the remaining quarter was from French Canada, Ireland, Italy, and Eastern Europe.

When Warner began the research, he made explicit to his staff the hypothesis that

> The fundamental structure of our society, that which ultimately controls and
> dominates the thinking and actions of our people, is economic, and that the
> most vital and far-reaching value systems which motivate Americans are to
> be ultimately traced to an economic order. Our first interviews tended to sus-
> tain this hypothesis. They were filled with references to "the big people with

money" and to "the little people who are poor." They assigned people high status by referring to them as bankers, large property owners, people of high salary, and professional men, or they placed people in a low status by calling them laborers, ditchdiggers, and low-wage earners. Other similar economic terms were used, all designating superior and inferior positions (Warner and Lunt 1941: 81).

However, after the research team had been in Yankee City for a while, they began to doubt that social standing could so easily be equated with economic position, for they found that some people were placed higher or lower in the prestige scale than their incomes would warrant. Furthermore,

Other evidences began to accumulate which made it difficult to accept a simple economic hypothesis. Several men were doctors; and while some of them enjoyed the highest social status in the community and were so evaluated in the interviews, others were ranked beneath them although some of the latter were often admitted to be better physicians. Such ranking was frequently unconsciously done and for this reason was often more reliable than a conscious estimate of a man's status . . .

We finally developed a class hypothesis which withstood the later test of a vast collection of data and of subsequent rigorous analysis. By class is meant two or more orders of people who are believed to be, and are accordingly ranked by the members of the community, in socially superior and inferior position. Members of a class tend to marry within their own order, but the values of the society permit marriage up and down. A class system also provides that children are born into the same status as their parents. A class society distributes rights and privileges, duties and obligations, unequally among its inferior and superior grades. A system of classes, unlike a system of castes, provides by its own values for movement up and down the social ladder. In common parlance, this is social climbing, or in technical terms, social mobility. The social system of Yankee City, we found, was dominated by a class order (Warner and Lunt 1941: 82–84).

Warner concluded from his interviews and observations, in other words, that prestige rank was the result of a combination of variables that included wealth, income, and occupation but also differential patterns of interaction and associated distinctions in social behavior and lifestyle.

One of the most important of the interaction patterns was determined by kinship. Not only were children assigned the status of their parents, but certain families had a prestige position that was not entirely explainable by their current wealth or income and seemed to flow from their ancestry.

When an individual had an equivalent rank on all the variables, his townsmen had no difficulty in deciding how much prestige to give him.[1] But when he had somewhat different scores on the several variables, there was difficulty in knowing exactly how to place him. This usually meant that the person was mobile and was changing his position on one variable at a

---

[1] Following Warner, we use the masculine pronoun because at the time of the study most women did not have jobs, and they were usually assigned the ranking of their husbands.

time; eventually he would be likely to get them all into line. Consequently, time was an important factor in stratification placement. For example, if a man who started as the son of a laborer became successful in business, he would be likely to move to a "better" neighborhood, join clubs of other business and professional men, and send his children to college. However, if he himself did not have a college education and polished manners, he would never be fully accepted as a social equal by the businessmen who had Harvard degrees. His son, however, might well gain the full acceptance denied the father.

After several years of study by more than a dozen researchers, during which time 99 percent of the families in town were classified, Warner declared that there were six groupings distinct enough to be called classes (Warner and Lunt 1941: 88):

Upper-Upper Class (1.4 percent). This group was the old-family elite, based on sufficient wealth to maintain a large house in the best neighborhood, but the wealth had to have been in the family for more than one generation. Generational continuity permitted proper training in basic values and established people as belonging to a lineage.

Lower-Upper Class (1.6 percent). This group was, on the average, slightly richer than the upper-uppers, but their money was newer, their manners were therefore not quite so polished, and their sense of lineage and security was less pronounced.

Upper-Middle Class (10.2 percent). Business and professional men and their families who were moderately successful but less affluent than the lower-uppers. Some education and polish were necessary for membership, but lineage was unimportant.

Lower-Middle Class (28.1 percent). The small businessmen, the school teachers, and the foremen in industry. This group tended to have morals that were close to those of Puritan Fundamentalism; they were churchgoers, lodge joiners, and flag wavers.

Upper-Lower Class (32.6 percent). The solid, respectable laboring people, who kept their houses clean and stayed out of trouble.

Lower-Lower Class 25.2 percent. The "lulus" or disrespectable and often slovenly people who dug clams and waited for public relief.

This proportionate distribution among the classes represents not only a long New England history, but also the special conditions of the Great Depression of the 1930s, which, for example, probably inflated the size of the lower-lower class.

Once the general system became clear to him, Warner said, he used a man's clique and association memberships as a shorthand index of prestige position. Thus there were certain small social clubs that were open only to upper-uppers, the Rotary was primarily upper-middle in membership, the fraternal lodges were lower-middle, and the craft unions were upper-lower. It seems that in cases of doubt, intimate clique interactions were the crucial test: A repeated invitation home to dinner appeared to be, for Warner, the best sign of prestige equality between persons who were not relatives.

PRESTIGE CLASS AS A CONCEPT

Warner maintained that the breaks between all these prestige classes were quite clear-cut, except for that between the lower-middle and the upper-lower. At that level, there was a blurring of distinctions that made placement of borderline families quite difficult. Of course, the placement of mobile families at all levels was difficult.

When he said that the distinctions between the classes were clear-cut, he meant in the minds of Yankee City residents. He saw his job of scientific observer as mainly one of staying around long enough to find out what the people really thought. In this, he ran into difficulties. Americans are uncomfortable with social inequality. Some people deny that it exists, while behaving as though it does.

> In the bright glow and warm presence of the American Dream all men are born free and equal. Everyone in the American Dream has the right, and often the duty, to try to succeed and to do his best to reach the top. Its two fundamental themes and propositions, that all of us are equal and that each of us has the right to the chance of reaching the top, are mutually contradictory, for if all men are equal there can be no top level to aim for, no bottom one to get away from (Warner et al. 1949b: 3).

Warner's recognition of this value conflict is reflected in his observation quoted earlier: "Ranking was frequently unconsciously done and . . . for this reason was often more reliable than a conscious estimate of a man's status. . . ." If rankings are unconscious, how do we know about them? The answer is, we observe words and behavior and then note the recurring patterns. Thus, sometimes class distinctions are explicitly described in the words of the people in the community, and sometimes they are abstractions created by the researcher to make sense out of the accumulated data.

The analyst has to create scientifically neat concepts or variables that approximate, as closely as possible, aspects of the real behavior of the subjects when they evaluate each other. Then the researcher has to formulate standard rules that show how to add up scores on each separate variable to arrive at the totality that will best predict the overall prestige judgment concerning a person that is made by others in the community. To be systematic and approach the goal of variables that make good statistics, the scientist is going to perform mental operations that are not identical to the automatic, half-conscious, and often contradictory thinking of the subjects. The aim of research is to predict, with reasonable accuracy, what most subjects will do in most circumstances. Furthermore, analysts must use variables that have theoretical meaning in other contexts, for only thus can they explain as well as describe.

Concepts are categories invented by social scientists to help explain what they have to say, but they are not invented out of pure imagination. They are based on careful observation, plus scientific reasons for abstracting from those observations a few simple factors that are worth studying in detail. In his more cautious moods, Warner admitted the degree of scientific abstraction that lay behind his six classes: "Structural and status analysts construct scientific representations (or "maps") which represent their knowledge of the structure and status interrelations which compose the community's social system" (Warner and Lunt 1941: 34). He also noted that some residents were more class-conscious than others and paid more attention to status distinctions (p. 69).

We can conclude that Warner's classes are ideal-type constructs that help him organize a vast amount of data on attitudes and behavior; they are not mere descriptions of the mental categories used by the inhabitants of Yankee City and other small towns. As such, they stand as hypotheses, and other observers have a right to test them against the data. The only sense in which these classes can be called real is to claim that they organize the data more usefully than any alternative set of hypotheses.

ARE THERE SIX CLASSES?

How can Warner's scheme be tested? How can we find out whether his six classes make a better fit to reality than five, or seven, or six that are cut differently from his? Much of the rest of the book will deal with aspects of this question because the full answer involves behavior evaluated in terms of values, interactions, class consciousness, and so on. Here we can give only a quick preview, emphasizing direct perception of prestige by local informants who rank people they know and live with.

Deep South, by Warner's colleagues Allison Davis, Burleigh R. Gardner, and Mary R. Gardner, presents an interesting chart illustrating the "social perspectives of the social classes," or the way the people at each level perceive the people at other levels. It is reproduced here as Figure 2–1. The chart suggests that people did not recognize six classes in Old City, a southern town of 10,000 inhabitants, but either four or five, though the scientists said that six classes existed in the white population. The people in the town made finer distinctions between persons close to themselves than between those who were far away. However, all the distinctions that were made coincided, even though they might have had different names. Thus, the lower-lowers made a distinction between "society" and the "way-high-ups but not society," but they did not subdivide "society." According to the authors, the line between society and the way-high-ups but not society was precisely the same as the line drawn by upper-uppers between "nice, respectable people" and "good people but nobody" (upper-middle and lower-middle). That is, when names of specific families were mentioned, both lower-lowers and upper-uppers placed them in the same one of these two groups.

FIGURE 2-1 The Social Perspectives of the Social Classes

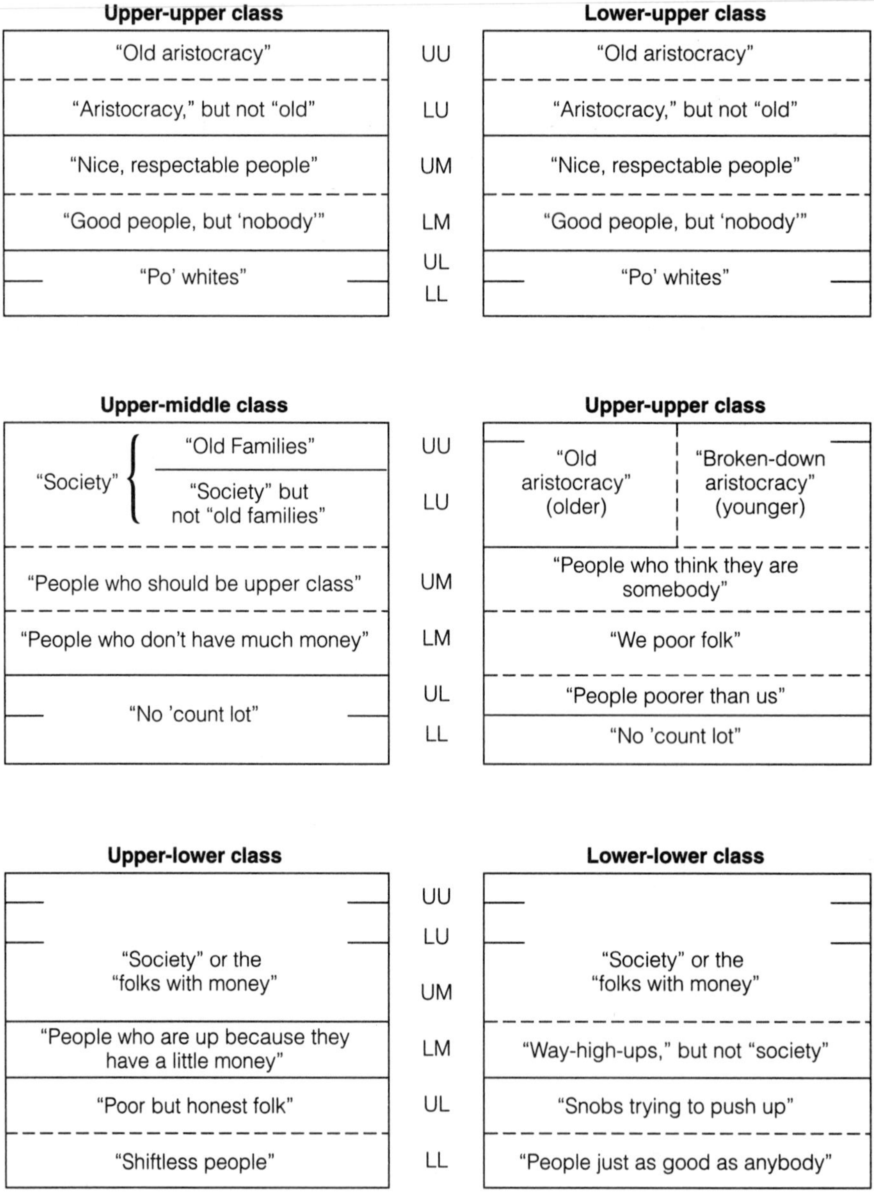

Upper-upper class

"Old aristocracy"	UU
"Aristocracy," but not "old"	LU
"Nice, respectable people"	UM
"Good people, but 'nobody'"	LM
— "Po' whites" —	UL LL

Lower-upper class

"Old aristocracy"	UU
"Aristocracy," but not "old"	LU
"Nice, respectable people"	UM
"Good people, but 'nobody'"	LM
— "Po' whites" —	UL LL

Upper-middle class

"Society" {	"Old Families"	UU
	"Society" but not "old families"	LU
"People who should be upper class"		UM
"People who don't have much money"		LM
— "No 'count lot" —		UL LL

Upper-upper class

— "Old aristocracy" (older)	"Broken-down aristocracy" (younger) —	UU LU
"People who think they are somebody"		UM
"We poor folk"		LM
"People poorer than us"		UL
"No 'count lot"		LL

Upper-lower class

— —	UU
— —	LU
"Society" or the "folks with money"	UM
"People who are up because they have a little money"	LM
"Poor but honest folk"	UL
"Shiftless people"	LL

Lower-lower class

— —	UU
— —	LU
"Society" or the "folks with money"	UM
"Way-high-ups," but not "society"	LM
"Snobs trying to push up"	UL
"People just as good as anybody"	LL

SOURCE: Reprinted from page 65 of *Deep South: A Social-Anthropological Study of Caste and Class,* by Allison Davis, Burleigh B. Gardner, and Mary R. Gardner. This chart and accompanying text from pages 71–73 of the same work are reprinted by permission of The University of Chicago Press. Copyright 1941 by The University of Chicago.

However, the upper-uppers made further subdivisions not recognized by the lower-lowers into "old aristocracy," "new aristocracy," and "nice, respectable people."

Davis and the Gardners wrote in *Deep South* (1941: 71–73):

> While members of all class groups recognize classes above and below them, or both, the greater the social distance from the other classes the less clearly are fine distinctions made. Although an individual recognizes most clearly the existence of groups immediately above and below his own, he is usually not aware of the social distance actually maintained between his own and these adjacent groups. Thus, in all cases except that of members of the upper-lower class the individual sees only a minimum of social distance between his class and the adjacent classes. This is illustrated by the dotted line in Figure 2–1. Almost all other class divisions, however, are visualized as definite lines of cleavage in society with a large amount of social distance between them.
>
> In general, too, individuals visualize class groups above them less clearly than those below them; they tend to minimize the social differentiations between themselves and those above . . . In view of this situation it is not surprising that individuals in the two upper strata make the finest gradations in the stratification of the whole society and that class distinctions are made with decreasing precision as social position becomes lower.
>
> Not only does the perspective on social stratification vary for different class levels, but the bases of class distinction in the society are variously interpreted by the different groups. People tend to agree as to where people are but not upon why they are there. Upper-class individuals, especially upper-uppers, think of class divisions largely in terms of time—one has a particular social position because his family has "always had" that position. Members of the middle class interpret their position in terms of wealth and time and tend to make moral evaluations of what "should be . . ." Lower-class people, on the other hand, view the whole stratification of the society as a hierarchy of wealth . . .
>
> The identity of a social class does not depend on uniformity in any one or two, or a dozen, specific kinds of behavior but in a complex pattern or network of interrelated characteristics and attitudes. Among the members of any one class there is no strict uniformity in any specific type of behavior but rather a range and a "modal average." One finds a range in income, occupation, educational level, and types of social participation. The "ideal type" may be defined, however, for any given class—the class configuration—from which any given individual may vary in one or more particulars.

How do investigators discover these "modal averages" of behavior: How do they find the classes? They first observe differing patterns of general behavior or style of life—the high-society crowd who hang around the country club versus the little shopkeepers who belong to the Elks. Then researchers pay particular attention to the names of specific families who belong to each group. If different informants use labels with varying shades of moral evaluation to describe the different groups, the analyst can equate the labels by interpretation: The "high-society crowd" of one informant and the "old aristocracy" of another turn out to be identical, because *both* the behavior described and the individuals who are said to belong are the same.

That all people do not recognize six levels is not so important, as long as the breaks that they do make all fit together in a consistent way. Naturally, people make the finest distinctions regarding those whom they know best and tend to merge others into broader categories. The analysts can take this into consideration and, if they wish, can subdivide a group according to the views of those in and immediately adjacent to it. The problem is like that of accommodating for perspective when making a map from aerial photographs. (Incidentally, these differences in perspective are not mere wrinkles to be ironed out by appropriate techniques—they are social facts well worth studying in themselves. If we knew more about them, we would know more about the dynamics of social perception that underlie all prestige judgments.) The practical problems in social mapping reduce to two: Do all observers put Albers above Jackson in the hierarchy, and if they distinguish between their ranks at all, do they all divide Jackson's group from Albers' at the same place? These are the questions of *ranking consistency* and of *cutting consistency*.

CLASS STRUCTURE OF THE METROPOLIS

Several decades after the appearance of the original Yankee City report, two of Warner's former students, Richard Coleman and Lee Rainwater, published the results of their study of prestige classes in two metropolitan areas: Boston and Kansas City. Working in metropolitan areas, Coleman and Rainwater could not duplicate the detailed ethnographic investigation that Warner conducted in Yankee City. Nonetheless, their book *Social Standing in America* (1978) shows the influence of their mentor, to whose memory the book is dedicated.

Social Standing in America was an ambitious undertaking, involving 900 interviews in the two cities. The statistical procedures employed were designed to provide representative samples of adults in Greater Boston and Greater Kansas City. Interviews were standardized and followed a fixed schedule of questions in the style of a social survey, but many questions were open-ended, allowing respondents to describe the class system in their own terms.

The hierarchy of prestige classes that Coleman and Rainwater stitched together from their analysis of the interviews is rather complex, so we offer a simplified, schematic version in Table 2–1. (Note that the annual incomes were recorded in 1971 dollars. Multiplying these amounts by four will give roughly equivalent values in inflated dollars of the late-1990s). Inspection of the table shows that the basic structure of the hierarchy is parallel to the one found by Warner in Yankee City. For instance, in both studies, the upper-upper and lower-upper classes correspond to a distinction between established families and "new money," although the distinction might be noticed

TABLE 2–1 Coleman and Rainwater's Metropolitan Class Structure

Class	Typical Occupations or Source of Income	Typical Education	Annual Income, 1971	% of Families
I. Upper Americans				
Upper-upper Old rich, aristocratic family name	Inherited wealth	Ivy League college degree; often postgraduate	Over $60,000	2%
Lower-upper Success elite	Top professionals; senior corporate executives	Good colleges; often postgraduate	Over $60,000	
Upper-middle Professional and managerial	Middle professionals and managers	College degree; often postgraduate	$20,000 to $60,000	19
II. Middle Americans				
Middle class	Lower-level managers; small-business owners; lower-status professionals (pharmacists, teachers); sales and clerical	High school plus some college	$10,000 to $20,000	31
Working class	Higher blue-collar (craftsman, truck drivers); lowest-paid sales and clerical	High school diploma for younger persons	$7,500 to $15,000	35
III. Lower Americans				
Semipoor	Unskilled labor and service	Part high school	$4,500 to $6,000	13
The bottom	Often unemployed; welfare	Primary school	Less than $4,500	

NOTE: Percentages adjusted for undersampling of "Lower Americans," acknowledged by authors.
SOURCE: Coleman and Rainwater 1978.

only by those who are themselves close to the top. In Boston, the upper-class respondents spoke of the former as "the tip-top—as close to an aristocracy as you'll find in America . . . Yankee families that go way back; the WASPs who were here first . . . the bluebloods with inherited income—they live on stocks and bonds" (p. 150). The same respondents described the lower-uppers as "a mix of highly successful executives, doctors, and lawyers with incomes of $60,000 at least, many $100,000 and more . . . They have help in the house, fancy cars, frequent and expensive vacations, and at least two houses . . . They're not considered top society because they don't have the right background—they're newer money, with less tradition in their lifestyle" (p. 151).

   In contrast with those at the top, Coleman and Rainwater delineated a bottom class characterized by dependence on irregular, marginal employment or public relief, often shifting from one to the other. As in Yankee City,

Boston and Kansas City families in this class were regarded as less than respectable and described in terms suggesting that they were physically and morally "unclean." However, many of Coleman and Rainwater's informants made a distinction between families on the very bottom and a class of semipoor families who worked more regularly and were slightly more orderly in their lifestyles.

As portrayed by Coleman and Rainwater, then, "upper America" and "lower America" neatly parallel corresponding prestige groupings in Yankee City. The same would appear to be true of "middle America," where Coleman and Rainwater's middle class and working class are equivalent to Warner's lower-middle and upper-lower classes. However, it was in the middle range of the class structure that they had the hardest time organizing the views of Kansas City and Boston respondents into prestige categories. In judging prestige, city respondents at this level gave almost exclusive emphasis to income and standard of living and paid relatively little attention to other stratification variables, such as occupation and association, that had seemed important to Warner.

Coleman and Rainwater reported that their middle American respondents recognized three levels among themselves, often called "people at the comfortable standard of living," "people just getting along," and "people who aren't lower class but are having a real hard time" (pp. 158–159). But the two sociologists found these categories inadequate and insisted on a more traditional distinction between middle class and working class (each of which, they suggest, can each be subdivided along income lines). The distinction they are making is essentially between white collar and blue collar workers. They do place the lowest-paid white-collar workers in the working-class category. But there are no blue-collar workers, even the best paid, in their middle class (See Table 2-1).

How do Coleman and Rainwater justify substituting their own judgment here for their respondents' judgments? They argue that lifestyle and associational differences that emerged in the interviews show that the middle-class/working-class distinction is more fundamental than any income distinction. For instance, among families at the same "comfortable" income level, they noted important differences in consumption patterns. The middle-class families at this level were likely to spend more on living room and dining room furniture and less on the television sets, stereos, and refrigerators and other appliances that were attractive to working-class families with similar incomes. The working-class families owned larger and more expensive automobiles and more trucks, campers, and vans. Moreover, income equality in middle America does not appear to produce social equality; patterns of friendship, organizational membership, and neighborhood location parallel lifestyle distinctions (pp. 182–183). Some respondents implicitly recognized these differences. A "comfortable" working-class man observed (p. 184):

> I'm working class because that's my business; I work with my hands. I make good money, so I am higher in the laboring force than many people I know.

But birds of a feather flock together. My friends are all hard-working people . . . We would feel out of place with higher-ups.

The wife of a white-collar man was more explicit (p. 184):

I consider myself middle class. My husband works for a construction company in the office. Many of the construction workers make a lot more than he does. But when we have parties at my husband's company, the ones with less education feel out of place and not at ease with the ones with more education. I think of them as working class.

The difficulties Coleman and Rainwater encountered in delineating the prestige hierarchy of middle America raise general, and by now familiar, questions about the methods by which the sociologist defines the structure of prestige classes in a community. They described their approach to prestige measurement as close to Warner's. However, the metropolitan context of their research imposed an important limit on their ability to replicate his methodology. Warner's basic method involved matching the standing of one family against another on the basis of what others said about them and whom they associated with (Warner, et al. 1949b). Because the respondents in Boston and Kansas City were unlikely to know or associate with one another, this approach was impractical. Synthesizing individual judgments of the class system was problematic for Coleman and Rainwater because their data consisted of verbal statements about general symbols rather than details about particular others in the community.

Coleman and Rainwater were clear that their version of the prestige hierarchy was not a mirror image of the class structure as understood in the community but rather a simplified model defined by the researchers. "Ultimately, to number and name the American social classes is a task for the social scientists. The status structure is too complex to be comprehended fully by average persons from their inevitably narrow vantage point" (1978: 120). They approached this "task" in two stages. First they attempted "to piece together hundreds of 'narrow' views on the subject and construct from them a single picture that is a summation of public impressions about the social class hierarchy" (p. 120). Subsequently this "image-based picture" was elaborated by combining it with data on behavior to produce "a picture that, in our view, is more nearly the truth about classes in America" (p. 120). As in Warner's work, the outcome is significantly dependent on the "clinical" judgment of the investigators.

The basic vehicle for the initial step was an open-ended question in the survey instrument: "How many classes would you say there are [in America], and what names would you give to them (or what names have you heard other people use)? What else could you say to describe each of the classes you've mentioned?" (p. 120). The most frequent answer to the first part of the question was that there were three classes. However, this response was given by only one-third of the informants. Other answers included two, four, five, six, seven, and nine classes or a flat refusal to say. One Bostonian replied, "You have too many classes for me to count and name . . . Hell! There may be fifteen or thirty. Anyway, it doesn't matter a damn to me" (p. 121). Even respondents who suggested the same number of classes differed in the way they described and named the classes, especially when the number was greater than three.

The disparate character of responses to both parts of the question suggests, as Coleman and Rainwater admit, that there is no "clear public consensus as to the name and number of social classes in the United States" (p. 20), at least in large cities. However, the variations among images of the class system that emerged in Boston and Kansas City were in some ways systematic. As in the earlier studies by the Warner group, the researchers found that respondents tended to construct more class boundaries close to their own position in the prestige hierarchy and fewer at a distance from themselves. Furthermore, when they asked people for reactions to a series of hypothetical class names, they found that the names "were 'cheapened' in their assumed reference by respondents of successively lower-status" (p. 127). Thus, upper Americans understood middle-middle class to refer to those earning $15,000 or more a year in the early 1970s, typically college graduates. But middle Americans associated the label with incomes of $12,000 to $13,000 and lower Americans assumed incomes of $9,000 to $10,000 (p. 127).

Coleman and Rainwater noted that the simplest division, that into a two-class model, emphasized the distinction between "us" and "them" and was usually offered by people slightly below average in their own position who expressed both "awe and animus" toward those above them in the hierarchy with more money (p. 121); this perception is linked to a conflict model of society. Citizens who thought in terms of a three-class model also generally used income as the basic criterion for dividing up the social system, but they added another class below themselves at the bottom—people who were poor, who lived in the slums on welfare—and gained some sense of pride by having someone to be above. Models that contained four or more classes usually used criteria other than money to make additional distinctions and often came from respondents in higher positions. They talked of education, the way people lived, family background, and other factors that modified the influence of money by itself. These respondents implicitly used an individualistic and competitive instead of a conflict model of society. (For a theoretical elaboration of this difference, see Ossowski 1963; for a revealing study in England, see Bott 1954 and 1964.)

Coleman and Rainwater imposed order on the many conceptions of the class system proposed by their respondents. As they viewed the problem, "All these seeming inconsistencies and variations should be thought of not as confusion but as a rich lode of imagery about the status system; represented therein are the best efforts of usually inarticulate people to express categorically and concretely the underlying continuum of status" (1978: 123). The trick was to find the criteria for cutting points in the continuum, explicit or implicit in the comments of respondents, which indicate the class groupings that were most salient for them. This could be done without assuming that Bostonians and Kansas Citians agreed completely among themselves on the number of classes or the proper names for them. Approaching the matter in this way, Coleman and Rainwater located thirteen logical dichotomies, of which only six "emerged with sufficient frequency and strength of supporting argument to be considered centrally significant" (p. 124). They were the following:

1. A source-of-income cut, placing those with inherited wealth socially above those with self-earned wealth.
2. A level-of-income cut, establishing "people who really have it made" financially above those who are "doing well but aren't really well-to-do."
3. A cut based on educational credentials and associated occupational accomplishments—"the degrees versus the non-degrees"—ranking the former above the latter.
4. A cut based on publicly perceived income level and associated standard of living—"people who have a comfortable salary and all the necessities plus a few luxuries" versus families who are "just getting along."
5. Another cut based on income and standard of living—"people getting by who have a decent house and the husband has a decent-paying job" versus "people who have to work at jobs that don't pay well . . . Their housing is not slum, but it's undesirable."
6. A cut based on source of income, this time private earnings versus public charity—"the poor who are working and largely self-supporting" versus "the welfare class."

These six cuts combine to suggest a seven-class model similar to the one outlined in Table 2–1. It is worth noting that, for the most part, the classes delineated by these cuts are not simply *quantitatively* distinguished from each other, as they might be on a single scale, such as income. The cuts imply *qualitative* differences among classes. For example, the first cut separates new money from old money; the third cut divides those with from those without college degrees. A full application of all the cuts allowed Coleman and Rainwater to further subdivide the seven classes shown in Table 2-1, but we need not follow them that far.

How does the Coleman-Rainwater metropolitan model differ from the Gilbert-Kahl model outlined in Chapter 1? See Table 2-2. There are some variations in class labels, but the main differences stem from the underlying bases of the models. Our model is based on economic position, in particular occupation and sources of income. Coleman and Rainwater, like their mentor Warner, have created a prestige model, largely based on public perceptions of the class order. For this reason, they make the distinction between old money and new money—in effect splitting our capitalist class in half—and lean toward the traditional blue-collar/white collar distinction to define the middle and working classes. Despite these differences, the two maps of the class system are broadly similar. This may, in some degree,

TABLE 2-2 Three Class Models Compared

Gilbert & Kahl *(National, 1990s)*		*Coleman & Rainwater* *(Metropolitan, Early 1970s)*		*Warner* *(Small City, 1930s)*	
Capitalist	1%	Upper Upper	2%	Upper Upper ⎫	3%
		Lower Upper		Lower Upper ⎬	
Upper Middle	14	Upper Middle	19	Upper Middle	10
Middle	60	Middle	31	Lower Middle	28
Working		Working	35	Upper Lower	33
Working Poor	25	Semipoor	13	Lower Lower	25
Underclass		The Bottom			
Total	100%		100%		100%

reflect their common debt to a tradition of sociological thinking about class. But the real key to their similarities is that in Kansas City and Boston, as well as Yankee City, prestige is largely, though not quite wholly, derived from economic position.

PRESTIGE OF OCCUPATIONS

Warner, Coleman and Rainwater, and many other investigators have stressed the importance of occupation for the prestige evaluations Americans make of one another. Especially in metropolitan settings, where people do not have a detailed knowledge of one another's income, family background, lifestyle, associations, and so forth, they are forced to fall back on a few shorthand indicators of personal prestige, such as occupation. They know, of course, that occupation is a fair indicator of two other sources of prestige: income and education. Physicians are typically affluent; not many janitors hold college degrees. They may also associate particular lifestyles and patterns of interaction with occupations, such as the general distinction between blue-collar and white-collar workers. These expectations account for the emphasis they place on occupation in making prestige assessments. For the sociologist engaged in a large-scale research operation, occupation is especially useful: It is more visible than income, and it can be studied with census data as well as social surveys and qualitative field studies. Furthermore, because census data are available for earlier periods, we can use occupation as an indicator in historical research.

There have been numerous studies of occupational prestige, going back at least fifty years, but the best known are national polls conducted under the auspices of the National Opinion Research Center (NORC) at the University of Chicago, beginning with a 1947 survey (NORC 1953) supervised by Cecil North and Paul Hatt. In this section, we will examine preliminary results from a 1989 NORC survey of 1,500 people and review some of the published analyses of earlier NORC studies of occupational prestige (Reiss 1961; Hodge, Treiman, and Rossi 1966; Nakao and Treas 1990).

Keiko Nakao and Judith Treas, who conducted the 1989 survey, describe their procedure as follows:

> Each respondent was asked to evaluate 110 occupations according to their "social standing" and to sort out small cards bearing the occupational titles onto a nine-rung ladder of social standing (from "1" for the lowest to "9" for the highest) (pp. 1–2; emphasis added).

The individual ratings from this procedure were averaged and converted into occupational prestige scores, ranging, theoretically, from 0 to 100. In practice, however, there was enough skepticism about the high-prestige jobs and sufficient respect for the low-prestige jobs that the scores rarely went above 80 or below 20.

Table 2–3 presents a sample of the occupational prestige scores from the 1989 study. The results at the top and bottom of the scale are consistent with what we have seen in the Warner group's community and metropolitan studies (which, of course, are based on much more than occupation). The highest-ranking occupations are professional and managerial (physician, professor, plant manager, hospital administrator), ordered by the level of expertise or administrative responsibility they entail. Virtually all assume a college education or better.

The lowest are unskilled, manual jobs, such as garbage collector, janitor, and filling station attendant. Between these extremes are the less demanding office or sales positions (bookkeeper, secretary, cashier in a supermarket) and the skilled manual jobs (electrician, plumber, welder). But note that there is no clear distinction between white-collar and blue-collar jobs (in Warner's terms, lower-middle and upper-lower). The electrician ranks above the bookkeeper, the carpenter above the traveling salesman, and the truck driver above the store clerk. Obviously, when faced with this sort of task, respondents are interested in something more than just where someone works or the color of a shirt collar (Glenn and Alston 1968).

When interviewees in the first NORC survey were asked the main factor they had weighed in making their ratings, the most frequent replies were pay, service to humanity, education, and social prestige, but none of these criteria was volunteered by more than 18 percent of the sample (NORC 1953: 418). Whatever the bases of their judgments, the surveys demonstrate that respondents did have a scale in mind on which they could place occupations with a rough consensus. Although there were significant differences among individuals in their relative ratings of occupations, sociologists were more impressed with the great consistency of the average ratings that were given to occupations by relevant subgroups of the population. The correlations between the ratings made by random pairs of

TABLE 2-3 National Opinion on Prestige of Occupations

Occupation	NORC Score
Physician	86
Department head in a state government	76
Lawyer	75
College professor	74
Chemist	73
Hospital administrator	69
Registered nurse	66
Accountant	65
Public grade school teacher	64
General manager of a manufacturing plant	62
Policeman	59
Superintendent of a construction job	57
Airplane mechanic	53
Farm owner-operator	53
Electrician	51
Manager of a supermarket	48
Secretary	46
Bookkeeper	46
Insurance agent	46
Plumber	45
Bank teller	43
Welder	42
Post office clerk	42
Travel agent	41
Barber	36
File clerk	36
Assembly-line worker	35
Housekeeper in a private home	34
House painter	34
Cook in a restaurant	34
Cashier in a supermarket	33
Bus driver	32
Logger	31
Salesperson in furniture store	31
Carpenter's helper	30
Salesperson in shoe store	28
Garbage collector	28
Bartender	25
Cleaning woman in private home	23
Farm worker	23
Janitor	22
Telephone solicitor	22
Filling station attendant	21
Table clearer in restaurant	21

SOURCE: Nakao and Treas 1990.

individual raters ran about 0.6, but the correlations between the average ratings made by the prosperous and the poor, people in high- and low-prestige
occupations, blacks and whites, men and women, residents of the Northeast
and the South, and city and country dwellers were all 0.95 or above. Even
those who proposed different criteria for judging occupations did not differ
in the way they ranked occupations (Reiss 1961: 189, 193; Treiman 1977:

60–74; Nakao and Treas 1990: 12). In other words, differences were largely ideosyncratic, reflecting personal views, rather than systemic variations by class, race, gender, or geography.

What systematic differences there were can be summed up in two principles: (1) People tended to raise in rank their own and closely related occupations, and (2) people agreed with each other more concerning occupations that were well known. For example, in 1947 many respondents did not know how to place "nuclear physicist" (answers in response to a question about this included "assistant to a physic" and "he's a spy"); yet they had no trouble deciding about a "scientist."

However, even in less esoteric fields brief occupational titles of the sort employed in the NORC surveys are somewhat ambiguous. Confronted by "lawyer," the respondent does not know if the reference is to a small-town general-practice lawyer or a partner in a major Wall Street firm. Of course, we can never capture with a survey the richness of detail that Warner reports from a community study because surveys force us to depend on a few simple categories. On the other hand, there is no substitute for the systematic knowledge a survey can provide. It is especially useful for making broad comparisons between different communities or different historical periods, but we must always remember that we are using social symbols about general types of jobs, leaving out a lot of concrete detail.

The consistency of prestige ratings across social subgroups is matched by their stability over time. The correlation between scores in the 1947 and 1963 NORC surveys was virtually perfect (0.99). The correlation for occupations rated in both the 1963 and the 1989 surveys was reportedly at a similarly high level. Moreover, retrospective comparisons between the NORC studies and several earlier U.S. studies suggested that there had been virtually no changes in the rankings of occupational prestige since 1925 (Hodge, Treiman, and Rossi 1966).

OCCUPATIONS AND SOCIAL CLASSES

The NORC surveys reveal how the public places occupations on a continuum, but the surveys leave us in the dark about how people might cut that continuum into occupational classes. The problem is obviously similar to the one we raised earlier concerning the grouping of families in Yankee City or Boston into prestige classes. Some evidence on this question was gathered in a national survey of approximately 2,000 adults conducted by the University of Michigan's Survey Research Center in 1975 (Jackman 1979). The SRC questionnaire asked respondents to place a series of occupations into one of five class categories: poor, working class, middle class, upper-middle class, and upper class. This question and several others in the SRC study replicate items in a 1946 survey conducted by Richard Centers, whose influential work on class consciousness and political attitudes we will examine in Chapter 9 (Centers 1949). The results from the 1975 survey are presented in Table 2-4.

TABLE 2-4 Social Class Assignment of Occupations

Occupations	Poor	Working	Middle	Upper Middle	Upper	Don't Know	Total
Corporation directors and presidents	—	1	3	22	*72*	2	100
Doctors and lawyers	—*	2	4	36	*57*	2	100
Executives and managers	—*	3	14	*56*	26	2	100
Supervisors in offices and stores	—*	17	*56*	23	3	1	100
Small businessmen	2	18	*61*	18	1	1	100
Schoolteachers and social workers	—*	19	*60*	17	3	1	100
Foremen in factories	1	40	*47*	10	1	2	100
Plumbers and carpenters	1	*44*	40	12	2	1	100
Workers in offices and stores	2	*60*	33	3	1	1	100
Assembly-line factory workers	5	*75*	16	2	1	2	100
Janitors	25	*67*	6	1	—*	1	100
Migrant farm workers	73	22	2	1	1	2	100

Social Class Assigned (percent)

NOTE: * = Less than 1. Numbers in italics represent modal category. Sample size = 1,850.
SOURCE: Jackman 1979: 449.

The data indicate that people do not have a difficult time associating occupations with social classes (there are few "don't knows") and that there is considerable popular agreement about where occupations should be placed in the five-class system that was suggested by the interviewer. In virtually all cases, the most frequent class assignment (indicated in italics) accounted for more than 50 percent of the responses.

As we can anticipate by now, class placement of occupations is easiest at the top and bottom of the scale. There is, for example, strong agreement that high-ranking corporate officers are upper class and janitors and assembly-line workers are working class. The greatest ambiguity centers on the middle-ranking occupations, especially skilled manual workers and foremen. Note also that the distinction between working class and middle class does not neatly correspond to the difference between blue-collar and white-collar jobs. Here, as when ranking occupations in the NORC surveys, people are more discriminating than that. The majority place the lower white-collar positions in offices and stores in the working class.

The 1975 data differ little from the original results that Centers obtained in the 1940s, except for minor variations caused by the exact wording of the question. Both studies support the idea that the public perceives more than a rank order of occupational titles: It groups those titles into class categories, that have widespread meaning, even though the agreement about the placement of particular occupations is less than perfect.

CONCLUSION: PERCEPTION OF RANK AND STRATA

Over the years, investigators have tried various devices to systematize the creation of groups or strata or classes out of an undifferentiated rank order. They generally find that the rank order is seen by the public with more clarity than are subdivisions into groups—there is, to recall our earlier language, greater *ranking consistency* than *cutting consistency* among respondents. Some people divide the world into conflicting groups of "us" versus "them," and they are likely to use labels such as working class versus middle class. Others stress the openness of our society and the chance for people to keep climbing a notch or two, and they are likely either to deny class divisions or to make many distinctions based on minor differences. These findings tell us that perceptions of the class order are intimately linked with degrees and styles of class consciousness, which we will explore further in Chapter 9.

When a sociologist presents respondents with either a list of the neighboring families in a small town or a list of occupations and requests that they arrange them in categories of equivalent prestige, the stimulus presented is partly structured and partly ambiguous. Over time, the respondents will have run into varied situations that indicate prestige differences, but these situations have not been so clear and so consistent that they completely determine the subject's perceptions. Their experiences of prestige differences on the job, at a family gathering, in church, at a high school reunion, and in their own neighborhoods are likely to be quite varied and even contradictory. Consequently, the respondent in the survey will have a general idea of a prestige hierarchy, but not a perfectly sharp one with fixed labels for each level plus standard status symbols to indicate who belongs where. This ambiguity allows (in fact, almost requires) the respondent to project personal views into the data and shape them slightly to suit individual desires and prejudices. Therefore, people at different levels give somewhat different answers for two reasons: Their outer experience has been different, and their inner frames of reference are different.[2] Both stimuli and perceiver vary from one respondent to another.

There is no such thing as an "objective" status structure that can be viewed by the completely neutral observer. Prestige is embedded in attitudes about the relationships between persons, positions, and symbols, and it varies according to the perspective from which they are viewed. The neutral observer has to find a way of summarizing the agreements and disagreements but should not foolishly conclude that the result is a reality that is more "real" than the subjective perceptions of the respondents (and the behavior they determine).

---

[2] There is, strictly speaking, another cause of varied and inconsistent answers: People do not pay full attention to a task of this type, and a lot of random error creeps in.

From the many studies we have reviewed, three conclusions stand out and are confirmed by every one of them. (1) In American society, there is a prestige hierarchy of both persons and occupations. (2) This hierarchy or rank order is divided by most citizens into a few categories or classes, but there exists considerable ambiguity about just how to define and differentiate them. (3) There is more agreement about rank order than about the criteria used in making ranking decisions, and more agreement about ranks than about divisions or strata. There is enough consensus to allow us to arrange a rank order that permits further operations to seek the variables that correlate with it and to begin to study the causes of that covariation. We then find that the covariation between occupational prestige and personal prestige is higher than for most pairs of variables that sociologists are interested in. Either occupational prestige largely causes personal prestige, or else they both flow from the same underlying causes.

Some tentative principles seem to explain the differences in perceptions of both ranking and grouping, and these appear in a number of the different studies, though the data are not completely consistent. The principles can be summarized as in the list that follows; they interact with and sometimes offset the effects of one another.

1. People perceive a rank order.
2. People tend to enhance their own position:
   a. By raising their own position relative to others.
   b. By varying the size of their own group. Here the evidence is not consistent. Apparently there is a tendency to narrow the group when thinking of individual persons about whom invidious distinctions can be made (especially those lower on the scale), and enlarge it when thinking of general categories of persons who are closely similar to one another.
   c. By perceiving separate but equal groups, thus accepting difference but denying hierarchy.
3. People agree more about the extremes than about the middle of the prestige range. This is probably a result both of clarity of stimulus and of aspects of perceptual organization. There are more people in the middle, and they are less publicized. Also, errors at the extremes can be made in only one direction.
4. People agree more about the top of the range than the bottom and make more distinctions about the top than about the bottom. This probably reflects stimulus reality, for those at the top are more conspicuous and more differentiated.
5. People lump together into large groups those who are furthest from them.
6. The better persons and occupations are known, the more agreement there is concerning them.

7. People have more consensus about the relative rank of persons or occupations that are closely related to each other in some functional way, such as the hierarchy of doctors, nurses, and orderlies.
8. People at the top are more consistent with one another, and make more divisions into groups, than do those on the bottom.
9. People in the middle or at the bottom are more likely to conceive of class differences in financial terms, while those at the top are more conscious of prestige distinctions from lineage and style of life.
10. Mobility is a source of ambiguity in perception of the prestige order. People find it difficult to "place" mobile individuals. Perception of high rates of mobility leads to the conclusion that class boundaries are amorphous or nonexistent.

These ten principles connecting social facts with the way people perceive those facts are sufficient to explain why there is no straightforward answer to the question that is asked so often: How many social classes are there in America? The moment we try to answer the question with data that come from the views of ordinary citizens, we are confronted with ambiguities and contradictions. The best we can do is to piece together a simplified but coherent map of prestige classes out of these inconsistent materials—as Coleman and Rainwater did for Boston and Kansas City. An alternative, taken by the authors of this book and many theorists before them, is to develop a model that does not claim to be based on *perceptions* of the class order. These models are typically based on economic distinctions rather than prestige rankings. But they also must also be fashioned out of inconsistent materials—drawn from history, economics, and demography. Both approaches involve as much art as science. We can expect the analyst to know the facts and make convincing use of them to develop a map of the class system. The map may be more or less effective as a device to interpret social life. But in the final analysis, the class system is not like the solar system, an objective reality we can hope to discover. Our conception of it will always be provisional.

SUGGESTED READINGS

Aldrich, Nelson W., Jr. 1988. *Old Money: The Mythology of America's Upper Class.* New York: Vintage.

Old money and not-so-old money. The ethos of the national upper class explained from within.

Boston, Thomas D. 1988. *Race, Class and Conservatism.* Cambridge, MA.: Unwin and Hyman

Black class structure and the debates surrounding it.

Bourdieu, Pierre. 1984. *Distinction: A Social Critique of the Judgement of Taste.* Cambridge, MA.: Harvard University.

Class cultures and their functions in defining and reproducing class differences.

Curtis, Richard F., and Elton F. Jackson. 1977. *Inequality in American Communities.* New York: Academic.

> *Compares several communities on various indicators of stratification. Authors conceive stratification as a continuum rather than as a structure of classes.*

Ossowski, Stanislaw. 1963. *Class Structure in the Social Consciousness.* New York: Free Press.

> *A systematic account of competing basic conceptions of class structure in the history of Western social thought.*

Ostrander, Susan 1984. *Women of the Upper Class.* Philadelphia: Temple University Press

> *The lives of upper-class women and their role in the maintenance of their class.*

Veblen, Thorstein. 1934. *The Theory of the Leisure Class.* New York: Modern Library. (First published in 1899.)

> *This book introduced the idea that "conspicuous consumption" was the way to buy prestige in competitive America.*

Warner, Lloyd W., et al. 1973. *Yankee City.* New Haven: Yale University Press. *Handy abridgment of the entire series of volumes on Yankee City.*

3

Social Class, Occupation and Social Change

No business which depends for existence on paying less than living wages to its workers has any right to continue in this country.

Franklin Delano Roosevelt

A post-industrial society, being primarily a technical society, awards place less on the basis of inheritance or property… than on education and skills.

Daniel Bell (1976)

In 1924, sociologists Robert S. Lynd and Helen Merrell Lynd went to Indiana to describe a "typical" American town. For more than a year, the couple studied the everyday life of a community of 35,000 they called "Middletown." To establish a historical baseline for their work, the Lynds reconstructed life in 1890, when Middletown had only 11,000 people and was going through the first stages of industrialization. They eventually returned to the city and wrote a second book about it, telling of its growth to 47,000 by 1935 and its reactions to the days of boom and bust that had followed the first research. Thus, we have three points of observation: 1890, 1924, and 1935.

The Lynds' two books, *Middletown* (1929) and *Middletown in Transition* (1937) are classic studies of American life.[1] They are especially valuable to us because they illustrate, in the miniature scale of a small town, how the American class structure was transformed by industrialization. Their narrative links social change, occupation, and social class—our key concerns in this chapter—and provides a historical backdrop against which we can understand the more recent social transformation associated with the emergence of a "postindustrial society."

MIDDLETOWN: 1890 AND 1924

In their first book, the Lynds reported that the daily life in Middletown was conditioned by a fundamental social distinction: the division of the population into working class and business class. Generally, members of the first class (70 percent of the population) supported themselves by working with *things*, while members of the second class (30 percent) made a living in activities oriented toward *people*. Being born on one side or the other of this cleavage was "the most significant single cultural factor" in the lives of Middletown's inhabitants, influencing matters as varied who they married, when they got up in the morning, how they dressed, and whether they drove Fords or Buicks and worshiped with the Holy Rollers or the Presbyterians.

One central theme of the first volume was that, from 1890 to 1924, there were basic changes in the work pattern of both the business and working classes—changes, incidentally, that resulted in a wider gap between them. These changes flowed from three causes: larger population, more machinery, and increasing emphasis on money.

Middletown in 1890 was a market town that was just beginning to turn to manufacturing. The work habits and values of its people were extensions of the traditions of their farmer parents. Those farmers were people who had conquered a wilderness. There had been land for all who would work

---

[1] For reassessments of Middletown, he book and the town, see Caplow 1980 and Caplow and Chadwick 1979.

it, and from such plenty, there emerged a society that lacked gradations of rank and privilege, a society that stressed individual initiative and progress, family solidarity, simplicity of manners and style of life, equality among neighbors.

As trading and handicraft manufacturing succeeded farming as the base of livelihood, the old traditions could easily continue. A man earned whatever his own efforts deserved. True, there developed a gradation in income that extended from unskilled through skilled laborers to bosses (who were often former craftsmen) and a few professional people. This gradation, however, was not sharply divided into levels; a person often moved through several steps in a few years, and it was understood that the system was open to everybody in fair and equal competition. Income and prestige were direct outcomes of competence at work, and everybody could understand and agree that as competence increased with age and experience, it brought the right and necessity of teaching and directing the work of people who were less skillful, and thereby it earned a higher income.

The spread of machine production began to change all this. The machine undermined the value of the traditional knowledge and skills of the master craftsman. It demanded speed and endurance. After a few weeks' experience with the machine, a 19-year-old boy could outproduce his 42-year-old father. The old apprentice system was gradually abandoned, and, with its passing, the traditional distinction between skilled and unskilled workers was blurred.

There were basic changes among the business class as well. The old businessman was a small merchant or manufacturer, whose capital consisted mostly of his personal savings. He had started as a worker and had become a businessman. His relations with both employees and customers were personal, even intimate. The new businessman operated with bank credit, had too many employees and customers to know them personally, and had ties with other businessmen all over the country. True, the small grocer and his kind still existed in 1924, but the major part of production and exchange in Middletown was passing into the hands of the new businessman (who was sometimes a branch manager of a national corporation).

Money had become the significant link between people. In the old days, each family was more self-sufficient; families processed most of their own food and clothing (from purchased raw materials, of course); they entertained themselves at home and with the neighbors; when they did buy things, they paid cash. By 1924, "credit was coming rapidly to pervade and underlie more and more of the whole institutional structure within which Middletown earned its living" (1929: 45). Businessmen were much more dependent on the banks and individuals on the credit agencies and merchants who allowed them to buy on credit. More articles of use were bought and fewer were homemade; more activities had changed from family and neighborhood affairs to commercial propositions.

As money became central to daily life, Middletown's workers and businessmen became more concerned with accumulating cash and less interested in the intrinsic satisfactions of their work. Within the business class, friendships, social activities, church membership, and political principles were increasingly viewed from the instrumental standpoint of their contribution to one's business success. In turn, social status in these other realms became dependent on financial status. Newcomers to Middletown were judged by material externals—where they live, what kind of car they drive. "It's perfectly natural," a leading citizen told the Lynds. "You see, they know money, and they don't know you" (1929: 80-81).

By 1924, Middletown was becoming too large and its productive system too complex and mechanized for community prestige to flow automatically from skill at work. People did not understand just what the activities of others were; a grocer knew little about glass blowing or automobile-parts manufacturing, and a glass blower knew little about financing a grocery store on credit. Concurrently, the money nexus was becoming increasingly important as more spheres of life became parts of the commercial market. The result was that people began to use money as a sign of accomplishment, a common denominator for prestige.

The Lynds did not find it necessary to divide the population any further than business class and working class when they wrote about Middletown in the 1920s. True, they recognized some gradation within each group, but they said that the working class was becoming more homogeneous through time as machines degraded skill, and the business class remained a small and basically undifferentiated group. They felt that no other distinction in Middletown was as important as this one.

Underlying this system was the belief that, in general, authority and income on the job and style of life and prestige off the job were automatic results of relatively free competition among individuals. This notion was brought to Middletown from the farms and villages; it was the American heritage of the frontier. Although some people were beginning to question this belief in 1924, the vast majority of both workers and businessmen clung to it. Free enterprise was not a mere theoretical discussion by economists of the workings of the market; it was a quasi-religious belief of the people.

MIDDLETOWN REVISITED

When the Lynds returned for a restudy in 1935, Middletown had grown larger. Its industrial base was more mechanized, more centralized into larger units, and more subservient to national corporations like General Motors, which owned plants in Middletown. The city had passed through the boom of the late 1920s and the devastating crash of the early 1930s. But in some ways Middletown was surprisingly unchanged. In the midst of a national depression, the people of Middletown retained their essential optimism and their faith in unregulated capitalism. They believed that prosperity was "normal" and setbacks were "merely temporary" (1937:13).

The Lynds found that the trend toward mechanization that was dominant in Middletown from 1890 to 1924 had continued. But machines were becoming so efficient that a productive labor force that was only slightly larger could produce vastly more goods. As a result, new employment was flowing into the service sectors of the economy and the occupational structure was changing. Using the census figures for 1920 and 1930, the Lynds reported that the proportional increase of workers in the production industries was 26 percent, whereas the increase in the service industries was 66 percent. The service-sector labor force was becoming increasingly professionalized. There were more schoolteachers, more social workers, more dental hygienists. The new efficiency of the productive machinery could support more services of a "luxury" type.

Within the factories, there were changes in the composition of the work force. The trend toward semiskilled machine tenders continued—proportionally more workers were in that category, fewer either unskilled or highly skilled. But the large numbers of half-trained workers employed on a complex production line needed a few very well-trained people to invent the machines and to direct and coordinate their work. These technicians, engineers, and managers were usually trained for their jobs in school and college, rather than being promoted from the ranks. For an increase of only 10 percent in the number of production workers who labored for an hourly wage, there was an increase of 31 percent in managers and officials, 128 percent in technical engineers, and 900 percent in chemists, assayers, and metallurgists. At the same time, the number of independent owner-manufacturers dropped 11 percent (1937: 68).

Technological change was transforming Middletown. More machine tenders were now gathered into fewer but bigger factories. The gap between worker and manager had widened. And the managers had changed. They were not so often self-made owner-businessmen; they were more likely to be members of "the new middle class" of professionals and administrators, trained in college, who worked for a salary. No longer could the complex system of Middletown be adequately described by the simple division into business class and working class.

The Lynds described what they saw in 1935 with a schema of six groups or classes emerging from tendencies in the occupational system. Interestingly enough, their model, based on occupational niches, is almost the same as the Warner's based on prestige, consumption style, and interaction networks. To facilitate comparisons with Warner and other class models, we have added class labels to the Lynds' descriptions, as follows:

1. *An Emergent Upper Class* developing out of Middletown's old middle class. Includes wealthy manufacturers, managers of local units of national corporations, and some wealthy "dependents," including one or two highly successful lawyers.
2. *An Upper-Middle Class*, consisting of smaller manufacturers, merchants, and professionals, as well as the higher paid employees of the

city's major enterprises. The Lynds describe the two groups in this class as socially united but often divergent in economic interests. They mix easily with members of the top class.

3. *A Lower-Middle Class* of small retailers, employed semi-professionals, clerks, salesmen, civil servants. Not quite the social peers of the second class.

4. *An Upper-Working Class*, the "aristocracy of local labor," including foremen, skilled craftsmen, and machinists.

5. *A Lower-Working Class*, encompassing most semi-skilled and unskilled wage workers, such as machine operatives, truckmen, and laborers.

6. *A Marginal Class*, irregularly employed and poorly housed, consisting of poor whites from nearby border states and the city's few blacks.

The Lynds noted that the "businessfolk" of the top three classes tended to identify with one another and against the three working classes. At the same time, they observed that Middletown's nascent class system was largely invisible to its inhabitants. Proud of their democratic traditions, they sustained the belief that any social cleavages in the community were "based not upon class differences but rather upon the entirely spontaneous and completely individual and personal predilections of the 12,500 families who compose its population" (1937:460–461).

The processes by which occupation and income are transformed into style-of-life symbols, interaction networks, and personal prestige, and political power are sharply illustrated by the Lynds account of the emergence into prominence of the "X family." Four brothers founded the family fortune before the turn of the century by starting a small glass-manufacturing plant on a capital of $7,000. The legends of Middletown contain many tales of the simplicity and humbleness of these business pioneers. When one of the brothers died in 1925, a newspaper recalled that "'he always worked on a level with his employees. He never asked a man to do something he would not do himself'" (1937:75).

These men built an industrial enterprise that became world famous. They grew rich, they contributed vast sums to local charities and development projects, and they became politically powerful. The family was inevitably drawn into community affairs. The brothers and their sons put up money to keep the banks from failing in the depression, thereby gaining control. Believers in high profits and low wages, they led the fight against unionization. They bought a substantial interest in the local newspaper and served on the board of the university. The X clan did not run Middletown by itself. There were other powerful business interests and the people of Middletown still voted. But, concluded the Lynds, "the business class in Middletown runs the city. The nucleus of business-class control is the X family" (1937: 77).

Not until the second generation did the X family gain leadership in consumption affairs:

> Around the families of the four now grown-up sons and two sons-in-law
> of the X clan, with their model farms, fine houses, riding clubs, and air-
> planes, has developed a younger set that is somewhat more coherent, ex-
> clusive, and self-consciously upper-class (1937:96)

These families were set off from lesser business folk by their residence in an exclusive neighborhood and elite patterns of leisure, symbolized by the annual horse show. The men of the older generation had always been preoccupied with business and, by and large, disinclined to cultivate an opulent lifestyle. The younger generation, born to great wealth, also worked hard, but "played expensively" at its own sports, with the members of a well-defined social set.

Great wealth has to be used. Whether it is plowed back into business, exchanged for elegance in leisure, used to patronize health, the arts, and education, or devoted to the training of a new generation who will be taught how to spend it gracefully, it enhances the power and prestige of those who own it. The second generation of owners is bound to be different: They cannot have the motives of ambitious individuals born in poverty. They are born to the wealth that brings power, the income that brings luxury, and the values of those who have been reared to expect both; these values have been polished by attendance at universities with national prestige, rather than at the local college. Such people will be different from first-generation entrepreneurs of the same age—a difference that will be less important during the day at business than during the evening at play. They will be bound to seek the company of others like themselves. They will, in other words, create and become an upper class with links to the national elite.

INDUSTRIALIZATION AND THE TRANSFORMATION OF THE NATIONAL CLASS STRUCTURE

The changes the Lynds chronicled in Middletown reflect the sweeping transformation of the United States into an industrial society. Although the process was under way in the 1840s and 1850s, it was consolidated after the Civil War. Between 1870 and 1929, the national population tripled, but the output of the American economy grew tenfold, largely as the result of the colossal gains in productivity made possible by new industrial technologies. By 1900, the United States was the world's leading industrial nation (Ross 1968; Hacker 1970).

Closely associated with this achievement were other developments of critical significance for the American class system, most of which we have seen played out on a small scale in Middletown. For instance, Americans changed from predominantly rural residents and farmers to city dwellers employed at urban occupations. In 1870, approximately three-quarters of the population lived in rural places (under 2,500 inhabitants), and more than half the labor force was employed in farming. By 1930, most Americans lived in towns and cities, and nearly 80 percent of the labor force was engaged in nonfarm occupations (Ross 1968: 26). During the same period, occupational tasks were becoming increasingly subdivided and specialized. Thus, the steelworker and the metallurgist were among a multitude of successors to the blacksmith.

As the industrial economy expanded, the country's labor requirements grew faster than the native population. This labor deficit was met by encouraging immigration from abroad, a policy so successful that by 1920, 22 percent of the industrial labor force was foreign born.

Finally, the period saw a leap in the scale of economic organization. Large corporations emerged as the dominant force in the American economy. Industrial technology requires large enterprises, but it was the drive to control national markets and generate monopoly profits that produced such giants as Standard Oil, United States Steel, E.I. du Pont de Nemours, and General Electric, through a wave of corporate mergers at the turn of the century (Ross 1968: 40–41).

In the next few pages, we will try to show how the class structure was affected by these changes. First we will look at three broad class groupings (upper class, working class, and middle class) that were decisively transformed by industrialization. Then we will examine how the occupational structure was altered. Thus, we will be looking at the same phenomena from two different perspectives.

THE NATIONAL UPPER CLASS

Our account of the upper class draws on the work of sociologist C. Wright Mills, whose best known work, *The Power Elite* (1956), is examined in greater detail in Chapter 8. Writing of the history of the upper class in that book, Mills observed that the United States had never had a national aristocracy of the type known in Europe. Without a feudal past or a single national center of wealth, power, and prestige like Paris or London, an enduring "pedigreed" class never developed. But before industrialization, there had been a series of relatively stable and compact regional upper classes across America:

> Up the Hudson, there were patroons, proud of their origins, and in Virginia, the planters. In every New England town, there were Puritan shipowners and early industrialists, and in St. Louis, fashionable descendants of French Creoles living off real estate. In Denver, Colorado, there were wealthy gold and silver miners. And in New York City, as Dixon Wecter has put it, there was a class made up of coupon-clippers, sportsmen living off their fathers' accumulation, and a stratum [of families]... trying to renounce their commercial origins as quickly as possible (Mills 1956: 48).

The prestige of these regional upper classes rested on old family fortunes, some on the eastern seaboard reaching back to colonial times. By gradually assimilating new moneyed families, the established families were able to keep prestige and wealth in close correspondence. But this system was overwhelmed in the post–Civil War era.

Mills noted that before the Civil War, there were very few millionaires. George Washington, among the richest men of this time, left an estate of $530,000 in 1799. In 1840, there were only thirty-nine millionaires in New York City and the entire state of Massachusetts. However, the postwar expansion of the industrial economy created unprecedented opportunities for rapid accumulation of new wealth. By 1892, a survey revealed more than 4,000 millionaires in the country (Mills 1956: 101–102). The wealth of the Lynds' "X" family and the fortunes associated with names such as Rockefeller, Woolworth, Carnegie, Morgan, and Vanderbilt originated in this period.

The post–Civil War fortunes destroyed the neat coincidence of family lineage, wealth, and prestige. The established families initially resisted the social pretensions of the new rich, but this was not easy because the size and national scope of the new wealth dwarfed the older family fortunes. Not by chance, this era produced one of the most famous attempts to define membership in upper-class society, Ward McAllister's list of the top 400. McAllister, who established himself as an arbiter of New York society on the basis of his close ties to the prestigious Mrs. John Jacob Astor, decided in the 1880s that there were "only about four hundred people in fashionable New York Society. If you go outside that number, you strike people who are either not at ease in a ballroom or else make other people not at ease" (Mills 1956: 54). The list was a defensive attempt to preserve the prestige of the old families against the pretensions of the social climbers. In 1892, McAllister finally published his list, which actually contained only 300 names and was basically a directory of the best-known pre–Civil War families. Only nine of the richest men of the day (those with fortunes in excess of $30 million) were among the 300 (Mills 1956: 54).

McAllister's effort to specify the makeup of upper-class society was obsolete even before it was formulated. By 1883, Mrs. Astor had already overcome her disdain for Mrs. Vanderbilt's new railroad money and accepted an invitation from her to a fancy-dress ball. Gestures of this sort, Mills argued, have prevented the emergence of an American aristocracy. "Always in America, society based on descent has been either bypassed or bought-out by the new and vulgar rich" (Mills 1956: 52). In the late nineteenth and early twentieth centuries, the family fortunes built after the Civil War were merged with the old society to create a new prestige class, national in scope such as the new economy and centered in major cities such as New York, Philadelphia, Chicago, and San Francisco. (This class comprises the "upper-uppers" and some of the "lower-uppers" that Coleman and Rainwater identified in Kansas City and Boston.) The progress of this merger can be seen in the *Social Register,* an elite directory that first appeared during this era and proved to be the most successful and enduring effort to specify the socially elect. The first *Social Register,* published in New York in 1887, contained 881 families in a careful mix of new and old. Within a few years, registers were issued in other major cities. During the period 1890–1920, the rate of admissions to the *Social Register* was high, reflecting rapid assimilation of the new rich; since then, annual admissions have declined to more modest levels. By 1940, more than three-quarters of the families covered by Ferdinand Lundberg's (1968) study of America's great fortunes (most accumulated between the Civil War and World War I) were listed in the *Social Register* (Baltzell 1958: 20).

THE INDUSTRIAL WORKING CLASS

The economic transformation that forced a restructuring of the upper class also recast the bottom of the American class structure. Limited before the Civil War to the mills of New England, factory production spread across the country and created a modern industrial working class. In 1860, there were 1.5 million industrial wage earners in the country; in 1900, there were 5.5 million (Brooks 1971: 39). By 1930, 17 million workers were employed (largely in manual occupations) in the sectors most directly affected by industrialization: manufacturing, mining, and utilities and transportation.

Many of the new industrial workers were immigrants, actively recruited from abroad by business and government and drawn by the economic opportunities offered by an expanding economy. In the decades immediately after the Civil War, the immigrants were typically English, Irish, German, or Scandinavian, people historically and culturally similar to the native-born population. After 1890, however, the character of immigration changed, and most of the immigrants of the late nineteenth and early twentieth centuries came from southern and eastern Europe. These "new immigrants," as they have been called, were Italians, Jews, and Slavs, and by the time they arrived, the best jobs in the industrial economy were controlled by others. The new immigrants were therefore relegated to the least attractive manual-labor jobs, while native-born Americans, including the offspring of the earlier immigrants, dominated skilled occupations and supervisory positions (Hacker 1970: 88–92; Link and Cotton 1973: 13–15). Thus, a close association grew up between stratification and ethnicity, with the highest positions being occupied by the members of the most culturally Americanized groups. This system was reinforced by active discrimination against many immigrants and their children in schools and politics (Brooks 1971: 76–77).

The working conditions and wages offered America's expanding working class were dismal indeed, especially in the jobs available to recent immigrants. A contemporary who visited Italian and Jewish sweatshops in New York City described people crowded into rooms

> Small as boat's cabins; crouched over their work in a fetid air which an
> iron stove made still more stifling and in what dirt; hunger hollowed
> faces; shoulders narrowed with consumption, girls of fifteen as old as
> grandmothers, who had never eaten a bit of meat in their lives (Allen
> 1952: 52–53).

In Chicago, another observer described the early-morning procession of
industrial workers:

> Heavy brooding men, tired anxious women, thinly dressed, unkempt lit-
> tle girls, and frail, joyless lads passed along, half awake, not uttering a
> word as they hurried to the factories (Allen 1952: 52–53).

A Commission on Industrial Relations, appointed by President Wilson,
concluded in 1915 that "a large part of our industrial population are...living
in conditions of actual poverty...It is certain that at least one third and pos-
sibly one half of the families of wage earners employed in manufacturing
and mining earn in the course of the year less than enough to support them
in anything like a comfortable and decent condition" (Link and Cotton
1973: 38). The commission discovered that the children of the poor were
dying at three times the rate of middle-class children and that 12 to 20 percent of all children in six major cities were underfed and undernourished.
In 1897, the average workweek in industry was nearly sixty hours. Long
hours and the indifference of employers to safety conditions produced an
appalling record of industrial accidents. An incomplete survey showed that
at least half a million workers were killed, crippled, or seriously injured on
the job in 1907 (Link and Cotton 1973: 38).

In the early years of the new century, some improvement began in these
conditions. State legislatures, over the determined opposition of upper-class
employers, started to pass legislation controlling safety conditions, providing compensation for injured workers, prohibiting child labor, and regulating the labor of women. As industrial productivity increased, hours of work
declined and wages increased. However, the life of the average worker remained unenviable. A survey of industrial accidents in 1913 (more reliable
than the 1907 study) found that 25,000 persons had been killed and 700,000
seriously injured at work (Link and Cotton 1973: 38). Moreover, the benefits
of increased pay and reduced hours were very unevenly distributed among
workers; for the majority of workers who were unskilled and not unionized, the gains were slight, especially relative to growing industrial productivity (Ross 1968: 37–40).

The contrast between the conditions of the working class and the opulent
lifestyles of the new rich convinced many Americans that class divisions were
becoming sharper in America. This impression was reinforced by the evidence of class conflict between the emerging industrial working class and its
upper-class employers. The period from the Civil War to World War I saw the
most violent labor confrontations in American history. For example, in 1877,
when railroads cut the wages of their already overworked (fifteen to eighteen
hours per day) and underpaid workers, a wave of strikes hit the nation's rail

system. State militia and federal troops (called out by officials sympathetic to the railroads to "maintain order") provoked bloody confrontations in which scores of people were killed or injured. Millions of dollars of damage was done to company property. In 1892, at Homestead, Pennsylvania, a private army of 300 Pinkerton detectives hired by management fired on striking employees of the Carnegie Steel Corporation (later United States Steel). In the ensuing battle, three Pinkertons and seven workers were killed. A few days later, the company prevailed on the governor to send state militia to take over Homestead, which was under the (peaceful) control of a strike committee. Multiple indictments against strike leaders, charging crimes from murder to "treason against the state of Pennsylvania" broke the strike, and the union itself, by depleting the union treasury with legal expenses. However, not all strikes ended in management victories. When textile workers in Lawrence, Massachusetts, struck in 1912, martial law was immediately declared, and three dozen strikers were arrested and summarily sentenced to prison. After several violent incidents, including an attack by police on a group of women and children from striking families, the strikers gained public sympathy and won concessions from employers (Brooks 1971: 50–55, 84–92, 118–121).

In the wake of the 1877 rail strikes, E.L. Godkin, the prestigious editor of *The Nation*, wrote, "We have had an uprising not against political oppression or unpopular government but against society itself." These events should serve to disabuse Americans of the illusion that "this was the one country in which there was no proletariat, no dangerous class." The problem, as Godkin saw it, was that the United States had been invaded by people of alien "blood" (immigrants, that is) who had no respect for "American ideals." The solution? Reinforcement of the army and militia and a reduction of loose talk "about the laborer and his rights" (Litwack 1962: 53–56).

If such violent incidents were exceptional, they were nonetheless symptomatic of the era. Workers who sought decent wages and working conditions had to contend with powerful companies, which could generally count on the support of state and federal governments, the courts, and, as Godkin's remarks suggest, the press. Only gradually did the federal government begin to assume a more balanced attitude toward labor conflicts. As late as the 1930s, the legal right of workers to form unions was still in question.

In the late nineteenth and early twentieth centuries, the United States came as close as it ever has to Marx's conception of a capitalist society. Readers of the *Communist Manifesto* would recognize the emergence of bourgeoisie and proletariat in an era of rapid industrial development, the creation of a political and ideological superstructure beholden to the bourgeoisie, and evidence of violent class conflict. However, events did not take the course an orthodox Marxist might have anticipated. Despite Godkin's observation about an "uprising against society," American workers proved to be more interested in earning reasonable wages than in creating a socialist society, and they used both their unions and their votes to help achieve their goals. In addition, two tendencies prevented stark polarization between two classes, a development that Marx regarded as a prerequisite to proletarian revolution. One was differentiation *within* the working class among ethnic groups, among the races, and among workers at different skill and wage levels, which kept the entire working class from developing a shared, combative class consciousness. The other was the expansion of the middle class (as we have seen in Middletown) in response to the special needs of a sophisticated industrial economy, which offered the chance of advancement to workers or their children. Added to this were long-term gains in the standard of living of most groups that appeared to absorb their attention more than did the differences between rich and poor. We will return to these themes in Chapter 9.

THE NEW MIDDLE CLASS

For an understanding of the transformation of the middle class in industrializing America, we turn again to the work of C. Wright Mills. In *White Collar* (1951), drawing on the earlier work of Lewis Corey (1935; 1953), Mills distinguished between two groupings he called "old middle class" and "new middle class." The former is composed of small entrepreneurs: farmers, shopkeepers, self-employed professionals, and the like. The hallmark of this class is its independence, based on the entrepreneur's ownership of the property with which he or she works. The new middle class is composed of the salaried white-collar people whose emergence the Lynds chronicled in Middletown—a heterogeneous grouping of office workers, salespeople, and salaried professionals and managers. Early in the nineteenth century, the old middle class comprised as much as 80 percent of the total population, but as the country's industrial society matured, the working class became the majority and the new middle class displaced the old. Mills used labor force statistics to portray the shift in the composition of the middle classes (see Table 3–1).

The changing balance within the middle class, Mills observed, represented the decline of *property* and the rise of *occupation* as the principal basis of stratification. In the countryside, many small farmers were forced off the land or became tenants as agricultural prices fell relative to the prices of the urban products that farmers buy. Behind declining prices were the domination of national economic policy by the new industrialists and financiers of the Northeast, intensified competition from foreign agriculture, and, ironically enough, significant advances in agricultural technology, which enabled fewer farmers to feed an expanding urban population.

In the cities, many businesspeople shared the financial plight of the small farmers. However, in this case the modest relative decline in the proportion of business owners in the labor force masked what was actually happening: Each year, millions of small business enterprises collapsed, only to be replaced by millions of new enterprises, most of them doomed to the

TABLE 3-1 The Middle Classes

	1870 (percent)	1940 (percent)
Old middle class	85	44
Farmers	62	23
Businessmen	21	19
Free professionals	2	2
New middle class	15	56
Managers	2	6
Salaried professionals	4	14
Salespeople	7	14
Office workers	2	22
Total	100	100

SOURCE: Mills 1951:65.

same fate. In the developing post–Civil War economy, the small-business owner was relegated to the highly competitive retail sector of the economy, while much of the rest of the economy came under the domination of large corporations.

Just as the decline of the old middle class was evident in the fate of the farmer and small-business owner, the emergence of the new middle class was tied to the triumph of the corporation. As the operations of corporations grew in productive efficiency, financial magnitude, and geographic scope, a decreasing proportion of the work force was required in the actual production of things, but an increasing proportion was needed to manage, design, sell, and keep account of production. Moreover the growth of the corporate economy imposed new tasks on government. The expansion of corporate and public bureaucracies meant rapid growth of the new middle class.

However, as the proportion of white-collar employees grew, their relative standing in the class system declined. Mills observed that, in the past, white-collar people had strong claims on superior prestige. In contrast with blue-collar workers, they wore street clothes at work, engaged in occupational tasks that were more mental than muscular, and were likely to have direct contact with the entrepreneur, from whom they could, in effect, "borrow" prestige. White-collar workers typically had a high-school education. Only a small proportion of the population stayed in school that long. Nearly all were white and native-born Americans who spoke good English. White-collar workers were paid substantially more than manual workers.

Virtually all these traditional bases of white-collar prestige were weakened over time. The very growth of white-collar employment made white-collar status less special. As we will show later in this chapter, the character of work in the expanding white-collar bureaucracies increasingly approximated the experience of the assembly line. Contacts between most white-collar employees and ranking executives became infrequent. As the

American-born sons and daughters of immigrants entered the labor market, the significance of nativity was reduced. The spread of education broke the new middle class's exclusive claim on the high school diploma. At the same time, the wage gap between blue-collar and white-collar workers began to close. In 1890, the average annual income of white-collar workers was approximately twice that of urban manual workers. By the mid-1930s, it was 1.6 times, and by the late 1940s, 1.2 times (Mills 1951: 72–73). Of course, well-paid professionals and managers, the top strata of the new middle class, earned salaries well above blue-collar wages. But the situation of the great mass of white-collar employees was approaching that of the working class, and the line between the two strata steadily grew more indistinct.

NATIONAL OCCUPATIONAL SYSTEM

The Lynds' chronicle of Middletown and the parallel national accounts by Mills and others suggest that the post–Civil War transformation of the American class system can be viewed as a result of a series of shifts in the distribution of occupations. We know, for instance, that there was a proportional decline in the number of farmers and a proportional increase in the number of factory workers and salespeople in the work force. Approaching social change this way has the advantage of being precise—we can count the number of people employed in different occupations at various points in time. It is also directly relevant to the technological and organizational shifts that have reshaped the economy. Fortunately, there is an excellent source of data on occupational change: the national census conducted by the federal government every ten years. In 1820, the Bureau of the Census began to ask questions about occupations, although systematic and comparable series of statistics do not reach back that far. In the early 1940s, an official of the Census Bureau, Dr. Alba M. Edwards, delved into the bureau's dusty archives and reorganized and standardized the material on occupations, going back to 1870 (Edwards 1943).

To produce a series of comparable historical statistics, Edwards needed to find a way of classifying thousands of specific occupations into a few niches, each one with homogeneous "social-economic status." He decided against using a single indicator of social standing. Instead he lumped together factors such as the nature of the work, the skill and training involved in it, the income it brought, and common opinion about its prestige. He made his decisions on the basis of general knowledge at the time, without systematic tests.

The occupational hierarchy that Edwards proposed has been repeatedly modified in detail by the Census Bureau, but it is still recognizable in hundreds of government publications on occupations and on factors related to occupation, from fertility to income distribution. Moreover, versions of the Edwards schema have been adopted by many nongovernment researchers.

TABLE 3–2 Major Occupational Groups, with Examples

Professionals
Public school teachers
Doctors
Computer systems analysts
Registered nurses
Librarians
Engineers

Managers and administrators
Sales managers
Public administrators
Bar managers
Corporate executives
Proprietors of small retail establishments
Accountants

Technicians (often classified with professionals)
Laboratory technicians
Draftspersons
Computer programmers

Sales workers
Retail-store clerks
Insurance agents
Manufacturer's sales representatives

Clerical workers
Bank tellers
Bookkeepers
Secretaries
Cashiers
Data entry operators

Craft, precision production, and repair workers
Carpenters
Masons
Foremen in construction or manufacturing
Machinists
Telephone repair workers
Auto mechanics

Operatives
Assembly-line workers in manufacturing
Butchers
Gas station attendants
Truck drivers

Service workers
Waiters and waitresses
Police and firefighters
Child care workers
Maid and janitors
Domestic servants

Laborers (excludes farm)
Unskilled construction workers
Freight and stock handlers
Garbage collectors

Farm workers
Farmers (owner-operator)
Farm managers
Unpaid farm-family laborers
Migrant farm laborers

TABLE 3-3 Occupational Groups: Earnings, Education, and Prestige

Occupational Group	Median Earnings 1995		Median Years of Education	Median Prestige Score
	Men	**Women**		
Professionals	$47,340	$33,300	16.2	82
Managers, Admins.	$46,530	$30,640	14.6	75
Technicians	$36,040	$26,800	14.2	—
Sales	$35,060	$20,280	13.5	67
Craftsmen	$30,421	$21,343	11.9	68
Clerical	$27,420	$21,141	13.0	65
Operatives	$25,370	$16,740	11.3	58
Service	$21,330	$14,480	11.8	47
Laborers	$18,860	$14,860	11.2	46
Farm	$17,350	$11,880	11.0	61/50

NOTES: Earnings for full-time year-round workers. First prestige figure refers to Professionals and Technicians. Farm prestige figures refer to farm owners and farm workers, respectively.
SOURCES: U.S. Census 1996e, U.S. Labor 1990b, and Reiss 1961:68.

The major occupational categories in the schema, along with examples of specific occupations typical of each category, are given in Table 3-2. Because this system of classification is so widely used, it is worth studying the examples to get a sense of the range of each occupational grouping.

Edwards maintained that his classification was the most practical means for making a *scale* of occupations that would increase in prestige, education, and income as one ascended step by step. We can test this claim using current government data on income and education, together with occupational prestige scores drawn from the NORC survey. These statistics are presented in Table 3-3 for the original Edwards categories, plus the new group of service workers.

Ordered as they are in the table (which is not quite the way they are usually presented), the major occupational categories are fairly consistent with the income, education, and prestige rankings. Note that we have placed craftsmen above clericals, blurring the usual line between white- and blue-collar workers but achieving greater ranking consistency.

The median statistics employed in the table hide an inevitable problem with occupational scales of this type. Some of the categories are quite broad and include very dissimilar occupations. This creates difficulties for stratification analysis, especially for the upper strata. For example, the professional grouping includes doctors and engineers, who rank high on all three dimensions. However, the most numerous professionals by far are public school teachers, who are characterized by more modest earnings and prestige. (The wide gap in median earnings between professional men and women reflects, in part, just such differences in the mix of professions represented.) Another example is the managerial category, which is large enough

TABLE 3–4 Occupational Structure of the United States: 1870 to 2005

| | Percent of the Labor Force | | | | | | | |
	1870	1900	1930	1950	1972	1972*	1990	2005
Occupation:								
Professionals and technicians	3	4	7	9	14	13	17	17
Managers, officials, and proprietors	6	6	7	9	10	9	13	11
Sales workers	{4	5	6	7	7	10	12	11
Clerical workers		3	9	12	17	16	15	17
Craftsmen and foremen	9	11	13	14	13	13	12	11
Operatives	10	13	16	20	17	16	11	9
Laborers, except farm	9	13	11	7	5	6	4	4
Service workers	6	9	10	10	13	13	13	16
Farmers and farm laborers	53	38	21	12	3	5	3	3
Total (rounded)	100	100	100	100	100	100	100	100
Number in labor force (millions)	12.9	29.0	48.7	59.0	81.7	82.1	117.9	136.2
Percent of labor force female	15	18	22	28	38	38	45	47

*The last three columns, based on the 1980 Census classification of occupations, are not directly comparable with the first five, which are consistent with the 1970 Census classification.

SOURCES: 1870 from Edwards 1943; 1900–1950 from U.S. Census 1975; 1972 from U.S. Census 1973 and U.S. Labor 1984; 1990 from U.S. Labor 1991; and 2005 from U.S. Labor 1995.

to encompass the manager of a fast-food restaurant and the president of a multinational corporation.

The Edwards schema, even in revised form, is less than perfect as a stratification measure. However, it is doubtful that Edwards or his successors at the Census Bureau could have done much better, given the difficulty of reducing the multitude of occupations represented in the work force (the government's *Dictionary of Occupational Titles* lists more than 20,000 of them) to a simple set of categories. The original purpose of Edwards' scheme was to organize occupational data to make broad historical comparisons, and it serves well for that purpose.

THE TRANSFORMATIONS OF THE AMERICAN OCCUPATIONAL STRUCTURE

Using Edwards' compilation and more recent census data, we have charted the evolution of the American occupational structure since 1870. The results are assembled in Table 3-4, along with projections of the future from the Labor Department's "occupational outlook" experts.

The occupational upheaval that first transformed Middletown is reflected here: Since 1870, farm occupations have declined from more than half the labor force to 3 percent in recent years. As the percentage of agricultural

FIGURE 3-1 Occupational Distribution of the Labor Force

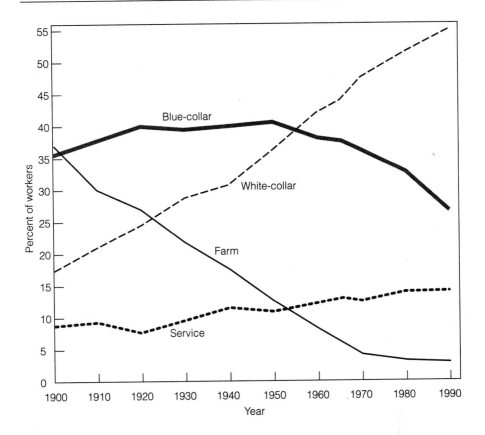

SOURCES: U.S. Census 1976: 387; U.S. Labor 1990b:183–188.

workers declined, the proportion of operatives, the grouping that includes most factory workers, expanded until it became the single largest occupational category in 1950. However, there is a small but significant surprise here. After 1950, the proportion of operatives began to decline. This happened despite the steady expansion of manufacturing production. The other two major blue-collar categories, craftsmen and nonfarm laborers, have followed a similar curved trajectory, although the proportional decrease in laborers began much earlier.

While the agricultural categories have continuously contracted and the blue-collar groupings first expanded and then contracted, white-collar employment has grown steadily, as it did in Middletown. (See Figure 3-1) By 1980, white-collar workers constituted more than half the labor force. Two classifications within that group have grown especially rapidly in recent decades—professional/technical and clerical—both having increased their proportion fivefold.

The relatively modest growth of a third white-collar grouping, the managerial occupations, obscures the radical transformation of this classification. The decline of the independent entrepreneur and complementary increase of salaried managers, noted by Mills in the period 1870–1940, has not abated. The proportion of managers who were self-employed dropped from more than half in the 1950s to 9 percent in the 1980s. Currently, only 8 percent of all workers are classified as self-employed. The United States has truly become a nation of employees. (Levitan and Taggart 1976: 67; U.S. Census 1984a: 395; U.S. Census 1995a: 408).

The service worker category has also changed in composition as it has grown. At the beginning of the century, most of the workers falling under this rubric were domestic servants (or in the more neutral current language of the Census Bureau, "private household workers"). Today the largest and most rapidly expanding service occupations include hospital attendants, waiters and waitresses, janitors, and security guards. While making a hospital bed or washing windows in an office building may be no more satisfying than performing similar tasks in a private home, the change does imply a more democratic set of class relationships. Of course, there are still some privileged American households where domestic chores are done by hired servants. In 1994, the Census Bureau found a little short of 1 million private household workers, enough people to fill a good-sized city (U.S. Census 1995a: 413).[2]

FROM AGRICULTURAL TO POSTINDUSTRIAL SOCIETY

There are several keys to the long-term patterns of occupational change, outlined in Table 3-4 and Figure 3-1. Earlier we noted the growth of large bureaucratic organizations associated with the modern corporation and the expansion of government. This tendency increases the demand for managers; certain professionals, such as accountants; and, above all, clerical workers. The phenomenal expansion of clerical employment is also linked to the increasing participation of women in the labor force, as noted earlier.

However, the fundamental cause of changes in the occupational structure since 1870 has been *the transformation of the United States from an agricultural to an industrial and finally to a postindustrial society.* In 1870, agriculture was still the mainstay of the economy, and most working Americans were farmers. By 1900, the United States had become an industrial society, with most workers employed in manufacturing and related sectors of the economy. After 1970,

---

[2] The current figure is probably an underestimate. Because some domestics are undocumented aliens and many employers do not pay legally mandated payroll taxes, private household employment is often concealed.

TABLE 3-5 From Agricultural to Postindustrial Society

Periods and Predominant Employment Sectors

Agricultural Society	*Industrial Society*	*Postindustrial Society*
1776–1900	**1900–1970**	**1970 –**
Agriculture	Manufacturing	Retail & wholesale trade
Fisheries	Mining	Finance & insurance
Forestry	Construction	Real Estate
	Transportation	Health Care
	Public Utilities	Education
	Communications	Government
		Legal Services
		Business Services
		Social Services
		Lodging & Restaurant
		Personal Services

SOURCE: Based on employment data from U.S. Census 1975.

NOTE: Postindustrial includes some smaller service sectors not listed here.

the United States could be described as a postindustrial society, with most jobs in service-producing rather than goods-producing sectors of the economy.[3] These three steps are outlined in Table 3-5.

Both economic shifts were conditioned by technological change. The transition to an industrial society was predicated on new technology, which substituted machine power for animal and human muscle. The emergence of a postindustrial society depended on continuing technological advances, which steadily increased productive efficiency and enabled a shrinking proportion of the labor force to meet the material needs of the rest. This dependence on technology, of course, has increased the demand for scientific and technical personnel. Engineering, for instance, is one of the three largest professional occupations (U.S. Labor 1996b).

As the proportion of the labor force employed in producing goods has declined, employment has expanded in the diverse service-producing sectors, such as retailing, finance, law, hotels, health care and education. The net result of these changes has been to move workers out of the sectors that require large numbers of blue-collar production workers and into those that depend more on white-collar or service workers.

Early assessments of postindustrial society ranged from sunny optimism to dark pessimism.[4] Interpreters on the sunny side anticipated a world of prosperity, opportunity, and social mobility. The postindustrial economy would require fewer low-skilled manual workers but growing

---

[3] We have abandoned the usual practice of classifying transportation, utilities, and communications as service-producing sectors because they are historically and occupationally closer to the industrial economy.

[4] On the origins of the term *postindustrial*, see Bell 1976:13-40.

armies of engineers, technicians, computer specialists, financial managers, accountants, and other highly trained professionals and managers. One writer foresaw the "white-collarization of America" (Wattenberg 1974:26). Another expected a "status upheaval" (Bell 1976:134).

Interpreters on the dark side, in contrast, predicted shrinking opportunities for most Americans and growing social inequality. The sons and daughters of factory workers were more likely to become janitors and food service workers than engineers and computer systems analysts. These writers did not associate "white-collarization" with opportunity but with the declining prestige, shrinking rewards, and increasingly menial character of office work (Braverman 1974; Glenn 1975). Writers on both sides of the debate saw a parallel between the earlier industrial revolution and the shift toward a postindustrial society. Both anticipated another round of sweeping social transformation in the wake of economic change.

The changes we have examined in the occupational structure (Table 3-4) provide support for both sunny and dark side interpretations of postindustrial society. As the sunny-siders anticipated, the proportion of professionally and technically trained workers has increased rapidly in recent decades and will, according to the Labor Department projections, continue to grow. But the proportion of service workers has also been expanding. The Labor Department's more detailed projections for specific occupations also suggest a mixed picture. Table 3-6 lists the fifty occupations that Labor Department analysts expect to add the largest number of new jobs between 1994 and 2005. Together these occupations account for nearly 70 percent of the new jobs during this period. We have divided the list into three broad education/skill levels and ordered occupations by the number of new jobs they are expected to produce.

What conclusions can be drawn from these detailed projections? Clearly, job growth is tilted toward the service-producing sectors characteristic of a postindustrial economy, such as health care, restaurants, and education. The sunny side interpreters of postindustrial society would be impressed with projected job growth in such sophisticated fields as computer technology, accounting, and finance. But the dark side analysts would be quick to point out the low skilled—the janitors, hospital orderlies, and kitchen workers—have equal weight in the job projections. Moreover, there are relatively few new mid-level positions that require skills short of a college education. Notably absent from this list are skilled blue-collar jobs; in fact, such occupations are prominent among those with the largest job losses, according to the same Labor Department projections (U.S. Labor 1995: 82)

These trends suggest that the postindustrial economy is producing a pattern of *occupational polarization*, marked by good jobs for well-educated managers and professionals, but limited opportunities for other workers. The postindustrial occupational structure may be one of the keys to the increasing social inequality of recent years—a theme to which we will return later in this chapter.

TABLE 3–6 The Fifty Largest-Growing Jobs, 1994–2005

Occupation	Number of New Jobs (thousands)	Percent
Professional and managerial	**4,981**	**40.6**
Registered Nurses	473	
General Managers & Top Executives	466	
Computer Systems Analysts	445	
Teachers, secondary school	386	
Marketing & Sales Worker Supervisors	380	
Clerical Supervisors & Managers	261	
Teachers, elementary	220	
Teachers, special education	206	
Food Service & Lodging Managers	192	
Social Workers	187	
Lawyers`	183	
Financial Managers	182	
Computer Engineers	177	
College & university faculty	150	
Teachers, preschool & kindergarten	140	
Computer scientists	134	
Residential counselors	126	
Human services workers	125	
Accountants & auditors	121	
Physicians	120	
Marketing, advertising & PR managers	114	
Coaches & Physical Trainers	98	
Science & engineering managers	95	
Technical and skilled	**2,497**	**20.4**
Secretaries	390	
Receptionists & Information Clerks	318	
Truckdrivers light & heavy	271	
Maintenance repairers, general utility	231	
Cooks, except short order/fast food	206	
Licensed practical nurses	197	
Police & detectives	166	
Correction officers	158	
Automotive mechanics	126	
Medical assistants	121	
Police patrol officers	112	
Cooks, short order/& fast food	109	
Engineering & science technicians	92	
Low skill	**4,781**	**39.0**
Cashiers	562	
Janitors & Cleaners	559	
Salespersons, retail	532	
Waiters & Waitresses	479	
Home Health Aides	428	
Guards	415	
Nursing aides, orderlies, & attendants	387	
Teacher aides & educational assistants	364	
Child care workers	248	
Personal & home care aides	212	
Food preparation workers (except cooks)	187	
Hand packers & packagers	160	
Amusement & recreation attendants	139	
Counter & rental clerks	109	
Total	**12,259**	**100**

WOMEN WORKERS IN POSTINDUSTRIAL SOCIETY

A dramatic increase in the proportion of female workers has accompanied the development of postindustrial society. At the beginning of the century, women constituted only a tiny proportion of the labor force. As late as 1950, only 28 percent of all workers were women, but in recent decades women's labor force participation has accelerated, while the participation of men has begun to decline. Women will soon constitute half the labor force. (Table 3-4; Bianchi 1995; Blau 1984: 302; Wetzel 1995).

Women workers have long been segregated into a relatively small number of "pink-collar" occupations—fields that are almost entirely female. Secretaries, cashiers, hairdressers, and elementary school teachers are among pink-collar workers. Pink-collar jobs typically offer lower pay, less prestige, and slimmer opportunities for advancement than other positions requiring similar levels of education and training. But women workers are increasingly moving out of the pink-collar ghetto. (Blau 1984; Howe 1977).

In the twentieth century, women were drawn into the labor market by the ballooning demand for clerical workers and later for service workers. Beginning in the early 1970s, growing numbers of wives joined the labor force to bolster family incomes that were being eroded by the stagnating wages and the rising unemployment rates of male workers. Increasingly, couples recognized that middle-class and upper-middle class living standards required two incomes.

The pull of economic forces was reinforced by the push of social forces: A declining birth rate, a rising divorce rate, and the increasing proportion of births to single mothers all led women to seek work outside the home. By 1970, the implicit marital compact (husband as permanent provider, wife as permanent housekeeper-caregiver) that had tied men and women to traditional family roles was collapsing. Attitudes toward working wives and mothers were shifting—encouraged perhaps by the women's movement, but also compelled by the new economic and marital uncertainties.

These changes were particularly felt by the women of the baby boom generation—mothers of today's college students. Compared with their own mothers and grandmothers, the boomers were much closer to men in their educational achievement, late to marry, and more likely to be childless or to delay childbearing. As a result of these factors and others outlined earlier, boomer women were more likely to join the labor force, more inclined to work full-time and continuously, and more like men in their occupations. In the 1970s and 1980s, as the boomers replaced their less career-oriented mothers in the workplace, they closed a good part of the economic gap between men and women.

For example, in 1979, a woman employed full-time, year-round, earned, on average, 60 cents for every dollar earned by a similarly employed man. This figure had remained stubbornly unchanged for many years. By 1992, the relative pay rate had risen to 71 cents—and even higher for the baby

TABLE 3–7 Occupational Distribution by Sex, 1995

	Percent of Workers	
	Males	Females
Managers	14.6	12.8
Professionals	12.7	16.7
Technicians	2.8	3.5
Sales Occupations	11.3	13.0
Clerical workers	5.6	25.4
Craftsmen	18.3	2.1
Operatives	14.3	6.0
Services workers	10.1	17.7
Laborers, except farm	6.0	1.7
Farm	4.3	1.3
Total	100.0	100.0
Number (millions)	67.4	57.5

SOURCE: U.S. Labor 1991.

boomers.[5] In 1970 (the year Hillary Clinton entered Yale Law School), women were 5 percent of law school graduates; by 1990, they were 41 percent. During the same period, the proportion of women among medical graduates climbed from 8 to 33 percent; and the proportion of females among people classified by the Census Bureau as "managers" grew from 18 to 40 percent (Bianchi 1995: 115, 127-9; U.S. Census 1996a, 1983; U.S. Labor 1991).

Yet women remain, on average, quite different from men in the marketplace. Even among baby boomers, the majority of women (unlike the majority of men) *do not* work full time, year round (U.S. Census 1996e: 38, 40). Their yearly earnings, although advancing, are therefore likely to lag behind men's. As Table 3-7 indicates, women's occupational profile is different from men's. Women are more likely to be employed as clerical or service workers and less likely to hold one of the skilled blue collar jobs. Fifty-six percent of working women are clerical, sales, or service workers (6 percent less than in 1970). Although women appear to be well represented in the general category of "professionals," more detailed occupational data indicate that 60 percent of female professionals work in just three fields: nursing, teaching, and social work. The high-prestige, high-income professions, including law, medicine, engineering, and architecture, continue to be dominated by men (U.S. Census 1984c: 166–175; U.S. Labor 1996b).

Currently, nearly half of working women hold pink-collar jobs—if we define a pink-collar occupation as one that is 75 percent female.[6] A more sophisticated way to measure occupational segregation is the "index of occupational

---

[5] This commonly used comparison refers to the median annual earnings of men and women.

[6] Based on the Labor Department's 1995 employment data for approximately 300 occupations (U.S. Labor 1996b).

dissimilarity," which is the percentage of workers who would have to switch jobs to eliminate gender segregation in occupation. This measure ranges from one hundred (absolute segregation) down to zero (equal distribution of men and women across occupations). Relatively stable in the 1950s and 1960s, the index fell from 68 in 1970 to 53 in 1990.[7] Thus, the pink-collar phenomenon endures, though it is slowly weakening.

Like the advance in women's relative wage noted earlier, the decline of occupational segregation marks the arrival of the more career-oriented baby boom women. To a lesser extent it also reflects structural change in the postindustrial economy. Declining employment in sex-segregated occupations drives the index down. The postindustrial decline in blue-collar employment and growth of more integrated white-collar occupations tends to reduce occupational segregation.

Both pink-collar employment and what might be called the Hillary Clinton phenomenon—ambitious, successful women married to ambitious successful men—are of growing importance for the class system. The reason, as we will see in Chapter 4, is that U.S. households are increasingly dependent on women's earnings.

TRANSFORMATION OF THE BLACK OCCUPATIONAL STRUCTURE

In his classic 1944 report on black America, Gunner Myrdal declared, "The economic situation of the Negroes in America is pathological. Except for a small minority enjoying upper or middle class status, the masses of American Negroes. . . are destitute" (Myrdal 1944 cited in Smith and Welch 1989: 519). Even as Myrdal wrote, the situation he described was changing. From 1940 to 1980, the weekly earnings of the average black worker (in inflation-adjusted "real" dollars) rose 400 percent. Because white wages were growing at a slower rate, the wages of black men advanced from 43 to 73 percent of the white male average.[8]

Table 3-8 reveals the sweeping transformation of the black occupational structure over the same period. In 1940, 80 percent of black workers were still concentrated in the four lowest occupational categories. By 1980, nearly 70 percent were in the upper six categories. The occupations of black workers had shifted even more rapidly than those of whites. As a result, the occupational distribution of blacks had moved much closer to that of whites. The change for black women was even more dramatic than this general picture for both sexes. As late as 1960, one-third of employed black women cleaned white people's houses. Only a small percentage worked at the white-collar jobs that were typical of white women. By the 1980s, about half of black female workers held white-collar positions (Farley 1984: 48–49).

---

[7] Based on 500 occupations covered in the decennial census.

[8] Figures refer to mean weekly wage (Smith and Welch 1989: 521).

TABLE 3-8 Occupational Structure of Black Workers, 1940–1980

| | *Percent of Workers* | | |
	1940	*1960*	*1980*
Professionals and technicians	3	5	10.9
Managers, administrators, and proprietors	1	2	4.5
Sales workers ⎫	2	8	2.7
Clerical workers ⎭			18.5
Craftsmen and foremen	3	7	9.7
Operatives	10	21	20.1
Laborers, except farm	14	14	7.4
Service workers	34	34	24.4
Farmers	15	3	
Farm laborers	17	5	1.7
Total	100	100	100

SOURCES: U.S. Census 1980a: 75: U.S. Census 1983: 417.

Black workers were affected by the same broad processes of socioeconomic change that were altering the world of white workers, but in ways that were peculiar to an oppressed minority. In the 1930s and 1940s, under the pressures of new industrial unions and the needs of economic mobilization for World War II, the discriminatory barriers that had kept blacks out of many industrial jobs began to fall. With the mechanization of southern agriculture in the 1940s and 1950s, large numbers of black sharecroppers and farm workers were forced off the land, typically into urban slums and low-wage urban employment.

The civil rights movement of the 1960s produced antidiscrimination legislation and affirmative action programs in government and private industry, which opened many new jobs to blacks. At the same time, blacks were closing the educational distance between themselves and whites. In 1940, the average educational gap between blacks and whites in their late 20s was more than three years of schooling. By the early 1980s, the gap for this same age group had been reduced to half a year (Farley 1984: 17–18). As a result, many young blacks were able to qualify for attractive white-collar jobs.

Despite the progress since 1940, a considerable occupational gap remains between blacks and whites. Compared with whites, blacks are still underrepresented at the top of the occupational structure and overrepresented at the bottom (Table 3–9). There are, to take an extreme example, about two white hospital orderlies for every white physician, but fifteen black orderlies for every black physician (U.S. Labor 1996b). As this example suggests, the benefits of change have been unevenly distributed among black Americans. In 1995, the proportion of black professionals and managers was five times what it had been in 1940, but "service occupations" remained the largest single black occupational category, as it had been 50 years earlier. Furthermore, in the early 1990s, many black workers had no job at all.

TABLE 3-9 Occupational Distribution by Race, 1989

	Percent of Workers	
	Black	White
Managers	9.3	14.5
Professionals	10.7	15.0
Technicians	2.8	3.2
Sales workers	8.9	12.6
Clerical workers	16.9	14.5
Craftsmen	8.1	11.2
Operatives	14.9	9.8
Service workers	21.7	12.4
Laborers, except farm	5.5	3.8
Farm	1.2	3.1
Total	100.0	100.0
Number (millions)	13.3	106.5

SOURCE: U.S. Labor 1990b: 183–188.

The net result of these diverse trends has been increasing class differentiation within the black population. Although large numbers of young, well-educated workers are moving into jobs that were open to few of their parents, the underclass of low-wage or unemployed workers appears to be growing. Income inequality among blacks has been increasing since the late-1960s (U.S. Census 1996a: B6-7). By 1980, sociologist William J. Wilson, an influential student of black America, was writing about a "deepening economic schism in the Black community" (Wilson 1980: 142).

Another disturbing development has been the stagnation of black progress compared with that of whites in recent years. The ratios of black to white wages, for both men and women, were the same in 1994 as they had been in 1970 (U.S. Census 1995a; 1980b).[9] African-Americans made their biggest relative gains in the booming post-World War II economy. But the occupational shifts associated with the postindustrial economy have had an especially negative effect on back workers. In relative terms, young black men actually slipped backwards in the 1980s, emerging further behind whites in earnings and employment rates. According to one careful study, groups most affected among the young were quite diverse, including high school drop-outs, college graduates, and factory workers in the Midwest (Bound and Freeman 1992).

WAGES IN THE AGE OF GROWING INEQUALITY

The shift from an industrial to a postindustrial economy roughly coincided with the transition, referred to in Chapter 1, from the Age of Shared

---

[9] Figures refer to median wages of full-time workers.

FIGURE 3-2 Earnings Index for Full-Time Workers

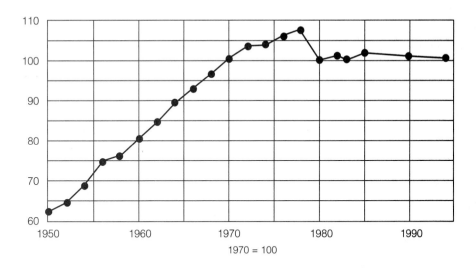

1970 = 100

NOTE: Real earnings, calculated from mean annual earnings (before 1970) or median weekly earnings (after 1970).

SOURCES: U.S. Census 1975, *Historical Statistics of the United States*, p. 164 (before 1970), and U.S. Census, *Statistical Abstract of the United States*, various editions (after 1970).

Prosperity to the Age of Growing Inequality. We have labeled the quarter century following World War II the Age of Shared Prosperity because during these years the annual earnings of American workers at all levels grew at a healthy pace. But, as Figure 3-2 indicates, this pattern of wage growth came to an abrupt halt in the 1970s (Danziger and Gottschalk 1995; Levy and Murnane 1992; U.S. Labor 1994).

We call the period since the early 1970s the Age of Growing Inequality. Three key tendencies characterize job earnings during these year: (1) On average, men's earnings (in inflation adjusted "real" dollars) have stagnated, in sharp contrast with the growth of preceding decades, (2) women's earnings have risen steadily, and (3) both the distribution of men's earnings and distribution of women's earnings have become increasingly unequal. As Figure 3-3 reveals, men employed full time were earning, on average, about the same in 1994 as they had in 1970. Women were making significantly more in 1994 than in 1970. These contrasting trends narrowed the gender gap in earnings.

Over the same period the disparity between high-wage and low-wage workers widened. In 1970, a worker in the 90th percentile of male wage-earners earned about 6.50 times more than a worker in the 10th percentile. By 1991, he earned 8 times more. Similarly measured, the gap

FIGURE 3-3 Median Weekly Earnings, by Gender

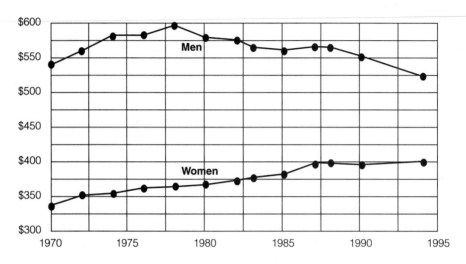

NOTE: Full-time workers in 1994 dollars.

SOURCE: U.S. Census, *Statistical Abstract of the United States,* various editions.

between low and high wage women also grew, though less dramatically.[10]

Men's earnings have, quite literally, been polarizing: Paychecks are sinking at the bottom *and* rising at the top of the labor market. An especially discouraging feature of the new age has been the growing number of men working for what can be considered poverty wages. In 1990, 14 percent of all men who worked full-time, year round did not make enough to maintain a family of four above the federal poverty line ($12,200 in 1990).[11] Figure 3-4, one of the U-shaped curves we previewed in Chapter 1, shows that the proportion of men in this sad situation dropped in the 1960s only to rise again in the 1980s. On the other hand, the proportion of men earning more than $40,000 also increased during the 1980s.[12]

The most rapid gains came at the very top of the job ladder, among the CEOs of major corporations. From 1980 to 1992, the real earnings of CEOs

---

[10] These figures refer to weekly wage and salary for all employed workers. Data limited to full-time workers reveal a similar pattern of change. See Danziger and Gottshalk 1995: 115, 129 and U.S. Labor 1994: 74.

[11] Of course, many of these men were not supporting families or had working wives who could supplement their modest incomes. And many lived in households smaller than four. The federal poverty line is explained in Chapter 10.

[12] According to tabulations by Levy and Murnane (1992: 1333-1334) the proportion of prime age men earning more than $40,000 (in 1988 dollars) grew from approximately 15 to 19 percent, while the proportion earning less than $20,000 increased from 32 to 38 between 1979 and 1988. The figures refer to part- and full-time workers, age 25 to 54.

FIGURE 3-4 Low-Wage Workers

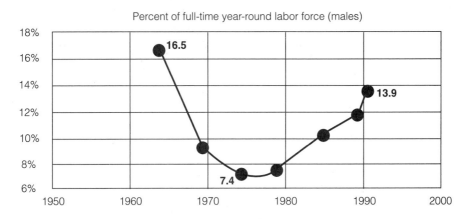

Percent of full-time year-round labor force (males)

SOURCE: U.S. Census 1992.

TABLE 3-10. Relative Pay of Chief Executive Officers

| | CEO Earnings as Multiple of | | |
Year	Factory Workers	Teachers	Engineers
1960	41	38	19
1970	79	64	37
1980	42	39	22
1992	157	112	66

SOURCE: *Business Week*, April 26, 1993.

rose by 360 percent to an average of $3.8 million—a remarkable increase in a period of stagnant wages. In 1992, CEOs made 157 times as much as factory workers, 112 times as much as teachers, and 66 times as much as engineers. As Table 3-10 indicates, this was a considerable leap over the ratios prevailing at the beginning of the period.

GROWING INEQUALITY OF WAGES: WHY?[13]

The contrast between the Age of Shared Prosperity and the Age of Growing Inequality is epitomized by the fate of a young man, recently graduated from high school, looking for work. A generation ago he might have found

[13] This section draws on the following sources: Burtless 1990, 1995; Danziger and Gottschalk 1995; Frank and Cook 1995; Freeman 1996; Gittleman 1994; Harrison and Bluestone 1988; Kodrzycki 1996; Krugman and Lawrence 1994; Levy and Murnane 1992; Sachs and Shatz 1994.

a blue-collar job in manufacturing or construction that would enable him to support a family at a reasonable standard. Today, with the same qualifications, the young graduate would likely find a service or retail position that pays much less in real terms; he would have a tough time supporting a family on his own. Since the late-1970s, the earnings gap between high school and college educated workers has widened. So have the gaps between skilled and unskilled and between younger and older workers. Obviously, the tide is running against our high school grad.

In an era of growing inequality, the less educated, the less skilled, and the less experienced have been the biggest relative losers. But the phenomenon of wage polarization does not end with them. At the same time that high school graduates are falling behind college graduates, earnings inequality is increasing *among* high school grads and *among* college grads. In fact, such *within group* inequality in earnings is rising among the members of virtually any definable group in the labor force: doctors, lawyers, waiters, carpenters, men, women, older workers, younger workers, workers in manufacturing, workers in service industries—even workers in specific firms.

Why is this happening? Four quite plausible explanations have received particular attention in the literature:

1. *Economic Restructuring.* As we have seen, in a postindustrial society, employment tends to shift out of manufacturing and into service-producing sectors of the economy (see Table 3-5). The expanding sectors typically provide ample opportunities for well-educated managers and professionals, but fewer good-paying jobs, like those once available in the goods-producing sectors, for workers with limited education and skills.

2. *Expanding International Trade.* In recent years, U.S. participation in international trade has increased. Trade creates economic winners and losers at home. Imports of cheap consumer goods produced by low-wage workers in developing countries undercut opportunities for the least skilled American workers. At the same time, increased demand from abroad for high-tech American products such as computers or commercial aircraft creates high-paying jobs for American workers with advanced skills and education. The accompanying expansion of capital markets does the same.

3. *Technological Change.* New technologies, in particular those associated with computers and modern communications, also tend to displace unskilled workers and create increased demand for workers with relevant skills and education. Automation cuts into the blue-collar industrial labor force. The remaining production jobs are likely to require higher skills than they would have in the past.

4. *Weakened Wage Setting Institutions.* Since the 1970s, "the invisible hand of the market" has gained influence over wages at the expense of institutions, such as labor unions, "internal labor markets," and minimum wage legislation, that have shielded workers from market

forces. Fewer workers are represented by labor unions and covered by collective bargaining agreements, which tend to boost the earnings of unskilled workers and limit wage differentials. Fewer corporations draw on internal labor markets (promoting existing employees rather than hiring outside the firm), a practice which also constrains wage differences. Over time, federal law has allowed inflation to erode the real value of the minimum wage, an important protection for the working poor. (These developments are treated in Chapters 9 and 10).[14]

There is good evidence for the influence of these factors—although the relative importance of each has been the subject of considerable debate. By now, most thoughtful writers on the subject have concluded that multiple causes are at work. The very pervasiveness of the phenomenon of growing inequality in earnings undercuts simple explanations. For example, the loss of manufacturing jobs has certainly played a role, but it cannot account for the rising *within-group* inequality among service sector workers or, for that matter, in the manufacturing sector itself.

We conclude this chapter by examining two intriguing interpretations of recent trends: *The Great U-Turn*, by Bennett Harrison and Barry Bluestone, and *The Winner Take-All Society*, by Robert Frank and Philip Cook. The first looks at developments affecting workers in the bottom half of the labor market, while the second looks at the very top.

HARRISON AND BLUESTONE: NEW CORPORATE STRATEGIES

American corporations were in serious trouble in the early 1970s, according to Harrison and Bluestone. Labor's share of national income was up. The business share was down. Profits had been falling continuously since the mid-1960s. Corporate executives were contending with a demanding workforce, abrupt increases in the price of energy, rising taxation, growing government regulation, and accelerating inflation. But the central problem facing American business was rising competition from abroad. Producers in Europe, Japan, and newly industrializing countries such as South Korea, and Taiwan were claiming large shares of the U.S. market for products from shoes and textiles to steel and automobiles.

According to Harrison and Bluestone, American capitalists responded to this challenge on three fronts: They adopted new strategies to reduce labor costs, they pursued short-term profit through speculative financial dealings,

---

[14]A fifth category of explanations might be added to this list: labor-market demography. The labor market participation of baby boomers, immigrants (legal and illegal), new college graduates, and women are said to have contributed to wage polarization. Labor supply is inevitably a factor in wage determination, but evidence that *change* in the supply of workers from these groups has significantly contributed to change in the distribution of earnings is not strong.

and they sought favorable government policy in areas from taxes to workplace safety rules. The authors suggest that all three have contributed to increasing inequality, but they particularly stress corporate labor strategies.

Corporations shrank their labor costs, say Harrison and Bluestone, by becoming lean and mean—that is, by reducing the number of people they employ and cutting the compensation of those who remain. Massive layoffs by major corporations, so-called "downsizing," were commonplace in the 1980s and have remained so in the 1990s. Industrial corporations took advantage of cheap labor abroad by opening plants in developing countries. Another strategy was "outsourcing," buying components or finished products from low-wage companies at home and abroad. To reduce the cost of labor, companies imposed wage freezes, reduced benefits, and adopted "two-tier" wage systems providing lower pay scales for new hires. After downsizing their permanent workforces, corporations increased their use of part-time, temporary and "leased" employees, who typically worked at lower wages, often without benefits. Corporations also found ways to reduce, if not escape, the influence of labor unions. Often unionized plants were closed and replaced by non-union plants in regions of the United States or foreign countries unfriendly to union activity. The very threat of such action could be used to intimidate union members and their organizations.

Harrison and Bluestone's account incorporates most of the explanatory factors we listed earlier. In an analysis relevant to the first factor, they measured the effect of replacing manufacturing jobs with service-sector jobs. They found that about one-fifth of the increase in wage inequality can be attributed to these job shifts and four-fifths of the increase to the changing wage distribution *within* sectors of the economy. In other words, the postindustrial shift toward services is important, but not as important as the growing wage inequality affecting sectors across the economy. Their description of corporate strategies to hold down wage costs helps explain these broader developments. On the other hand, the authors have little or nothing to say about the role of technology in reducing opportunities for less educated workers.

Unlike many writers on such topics, Harrison and Bluestone do not simply attribute change to impersonal economic forces. They see corporate leaders making conscious decisions that have adverse consequences for their workers. They argue that there were alternatives: Instead of laying off workers, squeezing wages, and focusing on short-term profit, corporations could have saved good jobs by investing in new technology, modernizing their plants, and retraining their employees.

FRANK AND COOK: WINNER TAKE ALL

What do Michael Jordon, Luciano Pavarotti, Tom Hanks, Danielle Steel, and Michael Eisner have in common? According to economists Frank and Cook, they all compete successfully in "winner-take-all markets"—more precisely, in fields in which rewards are heavily concentrated in the hands of a few top performers. The key to their extraordinary success is relative rather than absolute ability: They are a little better than the competition. For example, there are probably hundreds of tenors in the world who are almost as talented as Luciano Pavarotti, but they will never play major opera houses, get big recording contracts, or appear on television. Like the hundreds of actors who are almost as talented as Tom Hanks, these tenors will be lucky indeed if they can manage to support themselves as performers.

Frank and Cook assert that the winner-take-all phenomenon, well established in the entertainment industry and in professional sports is rapidly spreading to other fields, such as business, law, journalism, medicine, and academia. They point to Michael Eisner, CEO of Disney Corporation. In the 1990s, his annual compensation reached into the hundreds of millions of dollars. Although Eisner is an extreme case, corporate CEOs now routinely earn millions yearly. A widening chasm separates their compensation from the rewards available to other employees—even ranking executives of the same firms. For reasons we do not have space to detail here, the authors are convinced that winner-take-all competition has negative consequences for individuals and the economy as a whole.

The key question for Frank and Cook is why are winner-take-all markets proliferating? Some writers point to the cozy relationships between executives like Eisner and the corporate boards that set their compensation. Others emphasize the emergence of "a culture of greed" in the 1980s which became increasingly tolerant of excessive rewards for those at the top. Frank and Cook stress technological and economic factors. They note that mass production, large-scale organization, and vast markets favor large relative rewards at the top. For example, few authors can claim the $12 million Danielle Steel gets for one of her (less than profound) novels. But a publisher who commits much larger sums to producing and promoting mass-market fiction might be ill-advised to sign a writer almost as popular. Corporations like Disney operate on a scale that, even twenty years ago, would have seemed extraordinary. With so much money riding on every decision, stockholders are not inclined to hire someone almost as able as Eisner for a lot less.

Improvements in communications and transportation, along with growing international trade, expand the arena in which winner-take-all markets can flourish. Yet they are by no means universal. Frank and Cook cite evidence that the large relative rewards flowing to CEOs in the United States are exceptional. In 1990, American CEOs earned 150 times the salary of the average worker, while German CEOs made 21 times the average German worker, and Japanese CEOs made 16 times the average Japanese worker (Frank and Cook 1995: 70, from Crystal 1991). They suggest that the reason for this enormous disparity is the open competition for the top slots in American corporations. In Germany and Japan—as in the United States until recently—executives typically spend their entire careers with a single firm and are promoted from within. Their corporate employers are not compelled to bid against one another for CEOs.

Much of Frank and Cook's argument rests on the idea that barriers to market competition have been falling. Corporations, sports teams, TV-networks, universities, and other well-financed organizations are much more willing to raid one another's talent than they have been in the past. Federal deregulation of commercial aviation, trucking, banking, communications, the securities industry, and other sectors increases competition between firms and raises the bidding for top managers and professionals. Falling barriers to international trade and investment probably have the same effect.

Frank and Cook illuminate some of the market forces that are polarizing earnings, but their central concept of winner-take-all remains problematic. The notion works well for singers and CEOs, but how much does it tell us about the more general phenomenon of rising inequality? The authors attempt to extend the idea to people they call "minor-league superstars"—approximately one million successful doctors, dentists, lawyers, stockbrokers, accountants, and others who earned more than $120,000 in 1989. They show that inequality of earnings among people in these occupations grew in the 1980s. But winner-take-all assumes that a few top players suck up a large share of the available rewards, leaving little for the less talented. It is difficult to imagine that this is the case among people in large, varied professions like law and accounting.

Another problem: Frank and Cook focus on occupational earnings, as we have often done in this chapter. But, as we will learn in the next chapter where we broaden our perspective on inequality, those with the highest incomes typically depend more on investments than on jobs or professions.

CONCLUSION

This chapter has emphasized the evolution of the occupational structure and its implications for the class system. We have seen the country transformed from an *agricultural society* to an *industrial society*, built around the urban, goods-producing sectors of the economy, and finally to a *postindustrial society*, dominated by service-producing sectors from fast-food to health care (These shifts are summarized in Table 3-5). In the transition to industrial society, the farm population dropped precipitously, while in the cities, an industrial proletariat grew and a stratum of white-collar office workers appeared. During these years, a national capitalist class emerged. As industrial society gave way to postindustrial society, farm employment was reduced to numerical insignificance, the industrial proletariat contracted, the white-collar sector swelled and diversified, and a new stratum of service-oriented menial workers emerged. Both industrial and postindustrial society saw the displacement of independent entrepreneurs by salaried employees.

Postindustrial society drew what we (half-facetiously) described as sunny and dark-side interpretations. The sunny interpretations highlight opportunities for well-trained professionals, technicians and managers in the postindustrial economy. The dark interpretations emphasize the loss of good-paying blue-collar jobs associated with the decline of manufacturing. For the sunny-siders the emerging service-producing sectors mean new jobs for hospital administrators, medical technicians, accountants, financial analysts, hotel managers and computer specialists. For the dark-siders, the service-producing sectors mean employment in lowly service occupations for fast food workers, janitors, and hospital orderlies. Ironically, the occupational trend data and occupational projections we examined provide support for both positions.

These dual trends also support the distinction, introduced in Chapter 1 and employed in this chapter, between the Age of Shared Prosperity (1946 to 1975) and the current Age of Growing Inequality (1975–). Although the transition to the Age of Growing Inequality roughly coincides with the transition to postindustrial society, we do not mean to identify one with the other. The developments we associate with postindustrial society have probably contributed to rising inequality, but they may be less important than other sources of inequality. On the other hand, we cannot say that inequality will inevitably continue to grow under postindustrial circumstances. By the same token, the years of industrial society (1900–1970) were not always years of shared prosperity.

The distribution of job earnings gives us a fairly precise way to measure the growing inequalities associated with economic change. The data reveal increased wage polarization, especially among men. At the top of the distribution, earnings have climbed rapidly—in the case of corporate CEOs, spectacularly. At the bottom, earnings have stagnated or even fallen.

The last sections of the chapter explored the possible reasons for the rising wage inequality, including economic restructuring, expanding trade, changing technology, and weakened wage setting institutions. Of these broad explanatory factors, economic restructuring and technology are the most clearly associated with the emergence of postindustrial society. From Harrison and Bluestone (*The Great U-Turn*) we learned how corporate strategies have contributed to the declining fortunes of blue collar workers. Frank and Cook (*The Winner-Take-All Society*) showed how increased market competition has reinforced the concentration of earnings at the very top.

"Earnings" refers to the part of income from jobs, professional practices, and other self-employment. In the next chapter we will look all sources of income, and instead of focusing on individual workers, we will consider entire households. From this broadened perspective we will reexamine the question of growing inequality.

SUGGESTED READINGS

Bell, Daniel. 1976. *The Coming of Post-Industrial Society.* New York: Basic Books.

> *Broad and optimistic interpretation of postindustrial society, emphasizing changes in technology, economic organization, and occupational structure.*

Braverman, Harry. 1974. *Labor and Monopoly Capital.* New York: Monthly Review Press.

> *Key work on the division of labor and the transformation of work in capitalist industrial societies. Challenge to Bell.*

Danziger, Sheldon and Peter Gottschalk 1995. *American Unequal.* Cambridge, MA.: Harvard University.

> *Growing inequality, the economy, and public policy.*

Farley, Reynolds, ed. 1995. *State of the Union: America in the 1990s.* 2 volumes. New York: Russell Sage Foundation.

> *Valuable set of essays on long-term trends in areas including employment, occupation, earnings, income, education, economic differences between men and women.*

Harrison, Bennett, and Barry Bluestone. 1988. *The Great U-Turn: The Corporate Restructuring and Polarizing of America.* New York: Basic Books.

> *The transformation of the American economy in the 1980s and its distributive impact.*

Jaynes, Gerald David, and Robin M. Williams, Jr., eds. 1989. *A Common Destiny: Blacks and American Society.* Washington, D.C.: National Academy Press.

> *A comprehensive study of black Americans by an outstanding panel of social scientists. Emphasis on change over the last fifty years.*

Terkel, Studs. 1974. *Working: People Talk About What They Do All Day and How They Feel About It.* New York: Avon.

> *A provocative series of interviews with people, from ranking executives to unskilled laborers.*

Wilson, William J. 1980. *The Declining Significance of Race.* 2nd ed. Chicago: University of Chicago Press.

> *Argues that changes in occupational structure may be more important for blacks than traditional forms of prejudice.*

4

Wealth and Income

The life I've always wanted
I guess I'll never have
I'll be working for somebody else
Until I'm in my grave
I'll be dreaming of a life ease
And mountains
Oh mountains o' things
To have a big expensive car
Drag my furs on the ground
And have a maid that I can tell to bring me anything.

Tracy Chapman, Mountains of Things

Procrustes, a giant of Greek mythology, had the bizarre habit of altering the stature of his house guests to fit the length of the available bed by either stretching them or chopping off their legs. In the next few pages, we will apply Procrustes' approach to the study of income. We have put together an imaginary parade in which the heights of the marchers are made proportional to their incomes (Pen 1971: 48–59). The parade is a convenient way of gaining an overview of the distribution of income, our first concern in this chapter. Later we will consider the distribution of wealth, and the trend toward greater inequality in the distribution of wealth and income.

In Chapter 1, we defined income as monetary gain over a period of time. Job earnings, which we examined in the last chapter, are one source of income, but as we will see, there are several other important sources.

THE INCOME PARADE

The income procession is organized as follows: All the 100 million households counted by the Census Bureau will be represented in the parade.[1] In good Procrustean fashion, the marchers will be stretched or trimmed in proportion to the household's total income. Those representing households with the average annual income ($43,000 in 1994), will be of average height. By this standard, a marcher from an $80,000 household would be almost 11 feet tall. A marcher from a $20,000 household would be a little under three feet tall.

Because we want a quick impression of the distribution of income, we will make the entire procession pass by our reviewing stand at a uniform pace, in exactly one hour. This will be rough on the marchers, but it has a particular advantage for us. At any moment, we will be able to tell how much of the parade has gone and how much is to come just by looking at our watches. Let's begin the parade with the shortest marchers, the income pygmies, and work up to the giants. (For an overview of the parade see Figure 4-1)

The procession opens on an odd note. In the first few seconds we see nothing except a few wisps of hair moving across the horizon. It seems that the first people are marching in a deep ditch. They do not appear above ground because they are business people who have suffered income losses during the past year and have had to borrow from the bank or use up their own capital to cover them. Given the high failure rate of small businesses, we should not be surprised at this sight, however peculiar.

---

[1] In fact, the U.S. Census survey data we have used to create the parade assume 98.990 million households. In this section we will freely round off all figures with the hope of gaining clarity from simplicity. The parade is based on data from U.S. Census 1996a; U.S. Labor 1996a; Rose 1992; and unpublished tabulations provided by the Census Bureau and Labor Department. Income figures refer to 1994.

FIGURE 4-1 The Income Parade

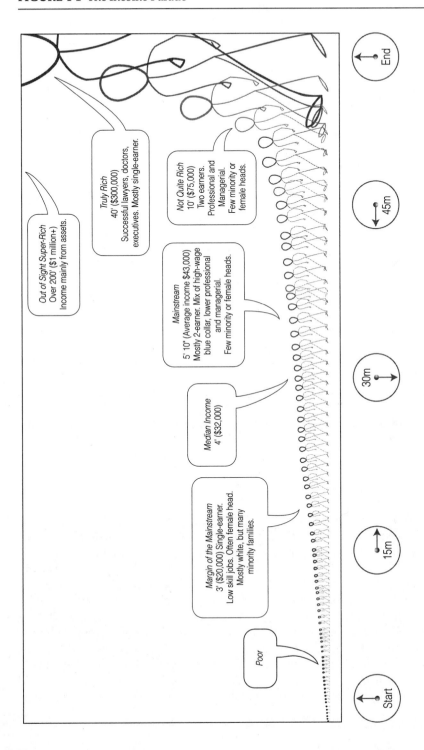

Five Minutes: Poor. After they pass, we see tiny people, the size of a match or a cigarette. Five minutes into the parade, the marchers are still only one foot tall because they receive about $7,500 a year. All who have gone by so far (and many who are to come, depending on the size of their families) are "poor" by the federal government's official poverty standard. There is a notable overrepresentation of women among these dwarfs. Nearly one-fifth are female heads of families who are single, divorced, or separated. Another fifth are elderly women living alone. The majority of marchers at this early point in the parade depend on government income-transfer programs such as public assistance, social security, or veterans benefits. Many of the marchers are malnourished because their food needs were not reduced when we shrank them to their present size. They could get government food stamps, but many of those who qualify do not receive stamps, either because they are too proud to accept them or because they are ignorant of their rights. There are also some eligible people who do not collect cash assistance to which they are entitled.

Blacks and Hispanics show up in disproportionate numbers in the first part of the parade. Nevertheless, the majority of the dwarfs who lead the procession are white and "Anglo." Actually, the single most encompassing characteristic of the dwarf marchers is that they are not employed, because they are old or disabled, temporarily out of work, studying, home with children, or not especially interested in working. Yet there is a substantial minority that does work, and among them are many people who work full-time, all year long, without exceeding dwarf height; they simply are not paid much for their labor.

Twenty Minutes: On the Margin of the Mainstream. As the procession moves on, the marchers get taller, but only very gradually, despite the breakneck pace we have imposed. After twenty minutes—one-third of the parade has passed—we are still seeing three-foot midgets who live on $20,000 a year. As their not-quite-normal height suggests, the marchers are on the lower margin of the broad mainstream—above the official poverty line ($15,000 for a family of 4) but well below the average household income.

We see many members of female-headed households at this point in the parade. There are many retired people. But the typical household has one wage earner, who works at a low-skilled blue-collar, clerical or service job.

What sort of lifestyle do these midgets buy with their money? The answer depends in part on household size. Families of three or four lead austere, sometimes precarious lives. Smaller households may enjoy greater comfort and security. The average housing expenditure is a meager $400 per month. However, retired people at this income level are frequently homeowners, free of rent and mortgage obligations. Most households at this point in the parade own cars, typically older models purchased used. Food, housing, utilities, transportation, and taxes consume most of the household budget, so there is little money available for other necessities. For example, the average member of a four-person household would have only $225 a year to spend on clothing. Because these households have little or no savings, even a few weeks of unemployment or an unexpected bill can threaten their standard of living.

Thirty to Forty Minutes: In the Mainstream. At half past the hour, we catch sight of little people, about four feet eight inches, who, by definition, receive the median income of $32,000. (The median is the amount dividing the bottom 50 percent from the top 50 percent in any distribution.) Seven minutes later, we notice marchers whom we can look in the eyes, assuming, of course, that we ourselves receive the mathematical average (mean) income of $43,000 and are therefore of average height. A careful count reveals that minorities are only slightly underrepresented in this part of the parade. But the proportion of female-headed households has dropped abruptly to below average levels.

Many of these average-sized marchers have lower-level managerial or professional positions. Others hold better-paying blue-collar jobs. But most of the households represented here depend on the earnings of two or more workers. This is especially true of the black and Hispanic families marching in this part of the parade.

These average marchers live substantially better than the midgets we saw fifteen minutes ago. They are more likely to own homes and drive late-model cars. They manage to cover the basics of food, housing, utilities, and transportation with about half of their incomes, leaving more for other necessities and some luxuries, such as family vacations. Nonetheless, they often feel financially pressed and are no more likely to have money left over at the end of the year than are the midgets.

Fifty to Fifty-Five minutes: Not Quite Rich. Nearing the end of the parade now, we see marchers who would fascinate an NBA scout: They are 10- to 12-foot hulks. The hulks are not quite rich, but their $75,000 to $85,000 incomes allow them to live more gracefully and comfortably than the smaller people who went before. There is money for fashionable clothing, new cars, quality furniture, and some domestic help at home.

The hulks typically hold professional and managerial jobs. But few households attain hulk status with one good job. They are even more likely than the average-sized people to depend on the earnings of two working spouses. Female-headed families are rarely found here. Minority marchers have not vanished from the parade, but their ranks have thinned out in the last 10 minutes.

The Final Minutes: Rich and Super-Rich. The procession has only 3 minutes to run. Yet some of the most extraordinary moments lie ahead. It took nearly twenty minutes before we saw three-foot midgets and another thirty-five minutes before the twelve-foot hulks appeared, but now we are looking at 15-foot giants. With household incomes of $110,000, they would be considered rich by most of the smaller marchers. In about two minutes, we will be looking at fifty-foot Goliaths, seconds after that, 200-foot King

Kongs, and then towering leviathans, thousands of feet tall. If we look down the line at the people who have yet to pass, it appears as if a steep mountain peak is advancing on our reviewing stand.

In these last minutes of the procession, the character of the marchers is changing rapidly. Instead of depending on two salaries, as did most of the giants we saw a moment ago, Goliath households do not typically include a working wife. The Goliaths, for whom $500,000 is a respectable income, are generally highly successful professionals (most often lawyers and doctors), certain sales people (for example, stock brokers), corporate executives, or the owners of prosperous small enterprises. Another shift comes with the arrival of the King Kongs, who would consider $1 million a modest living. About 70,000 households have incomes in excess of this amount; they rush past us in the last seconds of the parade.[2] Many hold important positions; for instance, the earnings of the top officers of the biggest corporations place them in the ranks of these mammoth marchers. However, the greater part of income at this point in the parade does *not* come from jobs. These marchers own substantial business enterprises, commercial real estate, and valuable portfolios of stocks and bonds. Such income-producing assets, rather than salaries, account for their colossal incomes and overpowering stature.

How do these monster marchers spend their money? A typical urban-based family with a $500,000 income owns two homes—a $1 to $2 million dollar apartment in the city and a substantial week-end house in the country. In addition to housing expenses the family's annual budget includes $50,000 for domestics (including a nanny, if needed), $30,000 for private schools, and $50,000 for daily expenses, including food. This budget might sound modest to software entrepreneur Bill Gates who recently built a $30 million house in Seattle or CNN founder Ted Turner, who bought an $80 million ranch in New Mexico for himself and wife Jane Fonda. Turner, like many others in his class, likes to be surrounded by open spaces. The New Mexico ranch, one of seven large western properties he has acquired, covers 578,000 acres, room enough for 22 lakes, 30 miles of fishing streams and more than 8,000 elk (*Forbes* Oct. 14, 1996 and *New York Times* Nov., 20, 1996).

LESSONS FROM THE PARADE

This chapter will elaborate some of the themes introduced by the procession. But before going on, we should pause to review what we have just seen and list the general lessons that can be drawn from the parade.

---

[2] IRS data on 1994 tax returns, reported in the *Wall Street Journal*, June 12, 1996.

1. **Many Midgets, Few Giants.** Our most general impression, confirmed by Figure 4-1, is an extremely gradual increase in income levels until a break point late in the procession. At half past the hour, we were still looking at runts four feet eight inches tall. The slow climb continued until the final minutes of the parade, when heights rose precipitously as the small numbers of Americans with very high incomes, and finally colossal incomes, strode by.

2. **Living Standards.** The parade tells us something about the relative welfare of different segments of the population. It took about twenty minutes before we caught sight of the $20,000 midgets and fifty minutes before the $75,000 hulks appeared. The parade was in its last seconds when we saw the $500,000 giants. We saw that many families at the $20,000 level could not afford to own a home, most families at the $75,000 level were quite comfortably housed, and families earning $500,000 were likely to own two luxury residences.

3. **Job(s).** The number of workers in each household was one of the main determinants of its place in line. At the beginning of the parade, we noted that many households had no job income. The midget households that followed typically had one wage earner. Among families in the $75,000 hulk range, two working spouses was the rule. But at the highest levels of income, working wives were less common and less crucial for a family's living standard.

4. **Sources of Income.** Jobs are the main source of income for most households. However, during the parade, we noted shifts in the relative importance of different income sources. For many of the early marchers, government transfer payments, such as Social Security, public assistance, educational grants, and veterans benefits, were crucial. In the broad middle of the parade, households depended on wage or salary income from jobs or, less commonly, on entrepreneurial income from small businesses and professional practices. In the final moments of the procession, we saw marchers who are supported by their wealth in the form of income-producing assets, such as stocks, bonds, and rental property.

5. **Occupation,** the marchers showed us, is a key determinant of household income, but not the overpowering factor we might have anticipated. From the reviewing stand, we saw low-skilled blue-collar, clerical, and service workers gradually give way to more skilled workers and then to managers and professionals. But there was considerable overlapping of occupational categories. We saw managers relatively early in the parade, and a few blue-collar workers marched with the giants. One reason for this is that a two-income blue-collar household can easily out-earn a low-level manager who does not have a working spouse. Another is that occupational pay scales overlap, even for very different occupations. For example, among men working full time, the top 20 percent of mechanics make more than the bottom third of accountants (U.S. Census 1993:153–154).

6. **Women's Shifting Role** was one of the defining features of the parade. At the beginning of the parade we saw many older women and female heads of families. Among married couple families some twenty minutes into the parade, wives without jobs were typical. But among the more prosperous households toward the end of the parade, working wives were the rule. Only in the final moments of the parade did women's employment rates and the significance of their earnings decline.

7. **Minorities.** Blacks and Hispanics were at a disadvantage in the parade. They were overrepresented among the early marchers, often by female heads of households. On the other hand, given their traditional position in American society, we were not prepared for their strong representation in the middle of the parade. Later in this chapter, we will take a closer look at the experiences of women and minority households.

8. **Income and the Class Structure.** The parade suggests that relationship between the distribution of income and the class structure is clear at the extremes but somewhat blurred in the middle. We can think about the problem in terms of the class model we introduced in Chapter 1. The little people at the very beginning of the parade, who have no employment income or work at very low-wage jobs, correspond to our underclass and working poor. The towering marchers we saw in the last two minutes of the parade represent the top of the upper-middle class and the capitalist class. But in the middle of the procession, we found factory workers marching with middle-class managers. And we saw dual-income working-class marchers looking down on single-income upper-middle class marchers. In sum, the class structure as we have defined it (or, for that matter, as Coleman and Rainwater defined based on social prestige in Boston and Kansas City) does not exactly match the distribution of household income. The mismatch is greatest in the middle of the parade.

THE DISTRIBUTION OF INCOME

Table 4-1, based on the annual Census Bureau survey confirms our broad impression of the income parade: A large chunk of the population survives on modest annual incomes (below $25,000); an affluent few enjoy high incomes (more than $75,000); about half the population could be described as middle income ($25,000 to $75,000).

Ethnicity and family structure create variants on this basic pattern. The income distributions of white, black and Hispanic families are compared in Table 4-2. Note the high proportion of minority families in the broad middle of the income distribution: about 40 percent of minority families have incomes between $25,000 and $75,000. The difference between these relatively

TABLE 4-1 Income by Household Type, 1994

Income	Percent		
	All Households	Family Households	One-Person Households
Under $5,000	4.1	2.8	8.1
$5,000–$15,000	18.6	12.0	38.7
$15,000–$25,000	16.7	14.9	21.5
$25,000–$35,000	14.2	14.5	13.1
$35,000–$50,000	16.3	18.2	10.0
$50,000–$75,000	16.5	20.2	5.6
$75,000–$100,000	7.0	9.0	1.4
Over $100,000	6.6	8.5	1.5
Total	100.0	100.0	100.0
Number (in thousands)	98,990	69,305	24,732
Median Income	$32,264	$39,390	$17,324

NOTE: Households include families, individuals, and unrelated persons sharing housing.
SOURCE: U.S. Census, 1996a.

TABLE 4-2 Family Income by Race and Hispanic Origin, 1994

Income	Percent		
	White	Black	Hispanic
Under $15,000	13.0	32.1	30.4
$15,000–$25,000	14.6	18.4	21.0
$25,000–$50,000	33.0	28.3	29.8
$50,000–$75,000	20.9	13.3	11.7
Over $75,000	18.4	7.9	7.2
Total	100.0	100.0	100.0
Number (in thousands)	58,444	8,093	6,202
Median Income	$40,884	$24,698	$24,318

NOTE: Persons of Hispanic origin can be of any race.
SOURCE: U.S. Census, 1996a.

comfortable middle-income families and the families struggling on lower incomes is often the difference between families headed by married-couples and single females.

The bar graph (Figure 4-2) compares the median incomes of married couple and female-headed families, by ethnicity. (The median income, as we noted earlier, is the midpoint in the distribution, dividing the top 50 from the bottom 50 percent.) The graph highlights the economic plight of female-headed families. It also reveals that the income gap between female and couple-headed families is greater than that between majority and minority families. Of course, female-headed families are, as the figures beneath the

FIGURE 4-2 Median Family Income by Ethnicity and Family Type, 1994

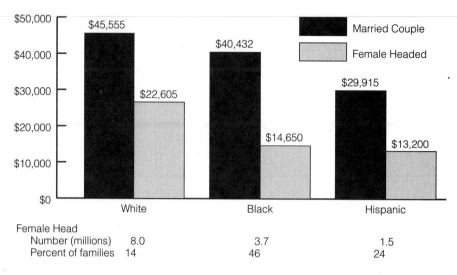

	White	Black	Hispanic
Female Head			
Number (millions)	8.0	3.7	1.5
Percent of families	14	46	24

NOTE: Hispanic may be of any race.
SOURCE: U.S. Census 1996a.

graph reveal, more prevalent among minority households—reflecting in part the economic strains to which they are subjected. Nonetheless most female-headed families are white.

In this section and the income parade, we have singled out female-headed families. What about male-headed families—that is, families headed by single men?[3] They are less significant for our analysis because there are relatively few male-headed families (less than 5 percent of all families), and their median income ($28,000) is close to that for married-couple families without working wives. Female-headed families, on the other hand, constitute almost 20 percent of families. Their median income ($18,000) is well below that of couples without working wives (U.S. Census 1996a).

SOURCES OF INCOME

In the income parade we noted shifts in the predominant sources of income. Table 4-3 traces this tendency. The top panel of the table is largely based on the annual Census Bureau survey, the best source for lower income households. The bottom panel is based on federal income tax returns, the best

---

[3] This category does not include single men living alone or in nonfamily households.

TABLE 4-3 Sources of Income

A. By Income Quartile

Percent

Income Quintile	Wage and Salary	Small Business	Capitalist	Government Transfers	Other	Total
Bottom	51.1	0.0	3.5	39.0	6.4	100
Second	66.0	3.8	5.7	17.7	6.8	100
Third	72.9	3.4	7.4	9.9	6.4	100
Fourth	77.0	3.4	8.3	5.8	5.5	100
Top	62.8	7.9	22.9	2.3	4.1	100

B. By Income Level

Percent

Income	Number of Returns (in thousands)	Wage and Salary	Small Business	Capitalist	Other	Total
$1–$30,000	71,586	72.7	3.5	6.7	17.1	100
$30,000–$50,000	21,051	80.3	3.0	4.9	11.7	100
$50,000–$100,000	17,963	80.0	4.2	5.7	10.0	100
$100,000–$200,000	3,403	67.3	11.2	12.0	9.5	100
$200,000–$500,000	880	56.0	18.2	19.8	6.0	100
$500,000–$1,000,000	148	47.7	20.9	27.2	4.1	100
Over $1,000,000	68	29.6	26.1	42.4	1.9	100
Total	115,099	72.3	6.8	9.5	11.3	100

NOTES: *Small business* = self-employment income, including business, profession, farm, partnerships, S-corporations. *Capitalist* = interest, dividends, capital gains, rent, royalty, estate, and trust. *Government transfers* = all cash benefits, including Social Security, Supplemental Security Income, public assistance, veterans benefits, and so on.

SOURCES: Part A computed from U.S. House 1992: 1399. Part B computed from Cruciano 1996 (part B).

source for high income households. (Note that this table adds percentages horizontally rather than vertically. Each row across refers to a different income level.) The table shows that government transfers such as Social Security and public assistance provide nearly 40 percent of income for the poorest fifth (quintile) of the population.[4] Wage and salary sources predominate at middle income levels, but their relative weight declines above $100,000. At the highest income levels, small business income sources and capitalist sources (such as interest and dividends) become increasingly important. These sources account for almost 70 percent of the income of people who make over $1 million annually.

---

[4] Middle and high income households also receive government transfers. In fact, the dollar amount received by the top quintile of the population is about the same as that received by the bottom quintile, though it accounts for a rather small proportion of income (U.S. House 1992: 1499).

FIGURE 4-3 Shares of Aggregate Income Received by Quintiles of Households, 1994 (pretax)

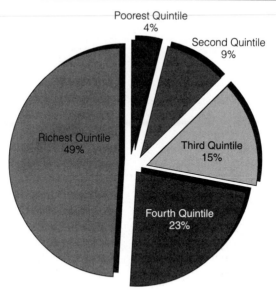

SOURCE: U.S. Census 1996a.

INCOME SHARES

Aside from parades, the distribution of income is typically analyzed in one of two standard formats: (1) *the distribution of households (or families)* across ranges of income. (2) *The distribution of income shares* among stratified segments of the population. Table 4-1 is a clear example of the first approach, which was the basic source of for the income parade. The income shares approach could be called a slices-of-pie distribution. It conceives the total income of all households as a national income pie, which has been sliced into pieces ranging from stingy to generous. This way of looking at income distribution measures the extent to which income is concentrated toward the top.

Figure 4-3 depicts the distribution of income shares among household income quintiles. The figure indicates the share of the total income pie that goes to each quintile. The first (poorest) quintile, for example, receives 3.6 percent of aggregate income. (Obviously, if the distribution of income were perfectly equal, each quintile would receive exactly 20 percent.)

Figure 4-3 reveals a remarkable concentration of income. The income share claimed by the richest quintile of households is 14 times that received by the poorest quintile. More detailed data indicate that the top 5 percent of households alone receive almost six times the income share of the bottom

20 percent (U.S. Census 1996a). Such comparisons belie the common notion that most income is received by the middle class. Indeed, a modern Robin Hood could transform the lives of the poor and near-poor by transferring income from the richest 5 percent of households to the poorest 20 percent until he had equalized the incomes of the two groups. In 1994, this tactic would have raised the average income of the bottom group from roughly $8,000 to $43,000.

TAXES AND TRANSFERS: THE GOVERNMENT AS ROBIN HOOD?

But isn't the government Robin Hood? Doesn't the government use progressive taxation to equalize incomes. The income distributions we have presented so far are all based on pretax data. What would they look like if we took taxes into consideration? The answer is: not very different.

We do have a federal personal income tax that is somewhat *progressive* in its overall effect. That is, people with higher incomes pay a greater *proportion* of their total income to the Internal Revenue Service than do those with lower incomes. However, other taxes operate in the opposite direction. Chief among these *regressive* taxes are the sales taxes, which are levied by states and localities. A sales tax imposes the same tax rate on a pair of children's shoes whether the purchasing parent is a welfare mother or a millionaire. But because a welfare mother spends a much higher proportion of her family's income on consumer items than does the millionaire (who is likely to reserve some income for investment), she loses a higher percentage of her income to the sales tax. Federal payroll taxes are also regressive in their impact. For example, a worker earning $30,000 in 1994 paid about $2,000 in payroll taxes. An executive earning $300,000 paid $8,000 in payroll taxes—obviously a much smaller portion of total earnings.

On the federal level, regressive taxes such as the payroll levies virtually cancel out the progressive effect of the personal income tax (U.S. House 1990: Tables 6 & 7). The Census Bureau factored in some state and local taxes to produce the before-and-after estimates illustrated in Figure 4-4. As the bar graph suggests, the combined impact of federal income and payroll taxes, plus state income taxes and residential property taxes was only modestly progressive: After these taxes, the income share of the top quintile was 4 percent smaller and the share of the bottom three quintiles was 4 percent higher. However, the Census Bureau estimate does not take into account the very regressive effect of the sales and excise taxes imposed by states and localities. A recent study concluded that, on average, such taxes absorb 7 percent of the income of the bottom quintile, 4 percent of the income of the middle quintile, but less than 2 percent of the income of the top 5 percent of households (Citizens for Tax Justice 1996: 1).

The government has another way of playing Robin Hood with income, and that is through transfer payments. Because transfer payments are counted as part of cash income, they are, unlike taxes, already reflected in

FIGURE 4-4 Income Shares Before and After Taxes, 1994

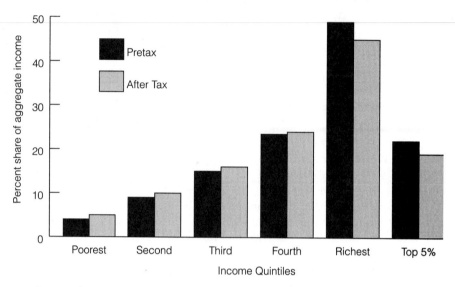

NOTE: Income after taxes excludes federal income and payroll taxes, state income tax, and property tax on home.

SOURCE: Unpublished Census Bureau CPS tabulation.

the income data we saw in the last section. (Noncash benefits, such as food stamps and Medicare, which are not counted as income, are treated in Chapter 10.) The influence of government transfers is generally progressive for the obvious reason that some are specifically designed to help the poor and for the less obvious reason that a large share of transfer income goes to the elderly who tend to be at the lower end of the pretransfer income distribution. In general, transfer payments raise the living standard of the poorest households but do not dramatically alter the overall distribution of income (See Figure 4-5).

HOW MANY POOR?

Our discussion of income distribution and redistribution has skirted an important issue: How many people have such low incomes that they can be considered poor? The easiest answer is based on official government statistics, which recorded 36 million poor Americans in 1994, about 14 percent of the population (U.S. Census 1996d). However, any count of the poor depends on the standard or definition of poverty used by the counters. Some researchers would adjust these figures upward or downward because they are skeptical of the standard employed by government statisticians. We will

FIGURE 4-5 Income Shares of Households Before and After Cash Transfers, 1994

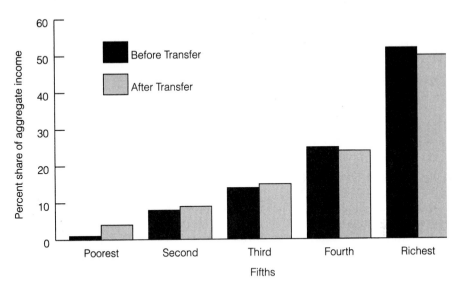

SOURCE: U.S. Census 1996a and unpublished census data.

take up the problem of defining poverty in Chapter 10, so that readers will be able to draw their own conclusions.

WOMEN AND THE DISTRIBUTION OF HOUSEHOLD INCOME

One lesson we drew from the income parade concerned the way women's situations changed with rising income. Women were (1) overrepresented at the beginning of the parade by elderly females living alone and female heads of households, (2) typically full-time housewives in lower-income, dual headed families, (3) usually working wives at higher income levels, (4) and a mix of working wives and housewives among the wealthy. In this section we will take a closer look at women's relationship to household income.

The elderly women we saw early in the parade were typically widows older than 75. Because women traditionally have had lower earnings and shorter, less continuous work histories, they have weaker claims on retirement income. Older women tend to be dependent on a husband's pension or Social Security check, and can lose all or part of that income in the event of their divorce or his death. Since women tend to outlive men, they are more likely to survive long enough to use up their savings. Although the more generous Social Security benefits of recent years have sharply reduced poverty rates among the elderly, older women living alone, especially if they are over 75, have relatively high poverty rates. Only 5 percent of wives over 65 are poor, according to the Census Bureau. But 25 percent of women over 65 and living alone are poor and most survive on less than $10,000 (U.S. Census 1996a).

Largely as a result of elevated divorce rates and the growing proportion of children born out of wedlock, nearly 20 percent of all U.S. families are now female-headed, compared with 10 percent in 1970 (U.S. Census 1995a: 57). The women who head these households, especially if they have children, face multiple disadvantages. Most single mothers and divorced or separated wives do not receive child support or alimony payments, and very few get substantial payments. The economic situation of men tends to improve after separation, while that of women deteriorates (Hoffman 1977; U.S. Census 1994:33). A 1992 Census Bureau child support survey revealed the following: (1) Mothers are 86 percent of custodial parents, (2) only 37 percent of custodial mothers received child-support payments in 1991, (3) for those who did, the average payment in 1991 was about $3,000, and (4) custodial mothers have an extraordinarily high poverty rate—35 percent (U.S. Census 1995b).

Women who go to work to support their families often find themselves in the lower-paying pink-collar jobs described in Chapter 3. Although women's earnings have advanced relative to men's in recent years, a significant gap remains: Women employed full-time, all year in the mid-1990s earned about two thirds of what similarly employed men earned. Of course, responsibilities at home, especially for mothers of young children, prevent many female heads of households from working full-time, year-round; only 46 percent of female heads of households under 65 managed to work the full-time all year in 1994 (U.S. 1996a).

Working wives face the same problems in the labor market as female heads of families and a similar clash between the roles of nurturer and bread winner. Nonetheless, the percentage of married women in the labor force has been rising continuously since the 1920s. By 1994, 74 percent of wives worked, 41 percent full-time year-round (Beeghley 1996: 231; U.S. Census 1996a).

Married women still earn less than their partners. The median earnings of employed wives is about half the median for employed husbands: $15,900 versus $31,400 in 1994. Fewer than one in four wives out earns her husband, but that gap has been closing.[5] But, in an era when men's earnings are declining, the rising earnings of women are becoming increasingly crucial to family incomes (U.S. House 1992: 1395).

Just how crucial is revealed in Table 4-4, which examines women's contribution to change in household earnings in the 1980s. Without the growth

---

[5] Unpublished Census Bureau tabulation from CPS survey.

TABLE 4-4. Women's Contribution to Change in Household Income

Income Quintile	Percent Change in Household Earnings, 1979–1989		
	Actual	Less Increase in Women's Earnings	Difference (Women's Contribution)
Lowest	1.0	-3.1	4.1
Second	-1.4	-6.8	5.4
Middle	3.9	-4.1	8.0
Fourth	7.7	-2.0	9.7
Highest	16.7	6.6	10.1
TOTAL	9.6	0.7	8.9

NOTE: Inflation adjusted.
SOURCE: U.S. House 1992: 1434-1435.

in women's earnings, total earnings would have fallen for the bottom 80 percent of households. Total income, (including transfers, interest, dividends, and so forth) would have fallen for the bottom 60 percent. Instead earnings and income more or less stagnated for the bottom 40 percent and rose at higher levels. (See Table 4-7). Note that the effect of women's expanded contribution to income was not evenly distributed: the higher the quintile, the greater the boost from growth in women's earnings. In other words, the expansion of women's earnings, because it is more rapid toward the top, is contributing to the growing inequality of household incomes.[6]

THE DISTRIBUTION OF WEALTH

We now turn from the distribution of income to the distribution of wealth. In Chapter 1, we distinguished between these two concepts. *Income* is the inflow of money over a *period* of time. *Wealth* is the value of assets held at a *point* in time. One year's wages, interest, and dividends, such as might be reported on a Federal income-tax return, are examples of income. The real estate, bank account, and stock shares someone owned on, say, December 31, are examples of wealth.

Wealth, we might say, is nothing more than accumulated income. True, but this underestimates the significance of wealth as a distinct dimension of class inequality. Wealth, unlike income, is not typically consumed in paying for daily necessities. It enhances what Max Weber called "life chances" in more basic ways. Wealth offers safety net protection against a sudden drop in living standard in the event of job loss or other emergency. Wealth can be

[6] This generalization holds for all households and all families, but probably does not hold for married-couple families considered separately. See Cancian 1993 and Ryscavage, et al. 1992.

converted into home ownership, business ownership, or college education. As we saw earlier in the chapter, people with very high incomes typically derive most of their income from wealth—stocks, bonds, commercial real estate, and other capitalist assets.

Wealth provides an important mechanism for the intergenerational transmission of inequality. As we will see in Chapter 8, about half of the wealthiest people in America inherited family fortunes. On a more modest level, the high school student who knows that there is money in the bank to pay for her college education and the young couple who purchase a house with help from their parents are the beneficiaries of the wealth accumulated by previous generations. But wealth, even more than income, is highly concentrated. Most Americans can expect, at best, a modest inheritance. From this perspective, it is hardly surprising that successful African-Americans— many of whom have achieved middle-class status in the last generation— are relatively close to the white average in income but lag far behind in wealth (Oliver and Shapiro 1995).

Wealth is measured in two ways: gross assets and net worth. The concept of *gross assets* refers to the total value of the assets someone owns. *Net worth* is the value of assets owned minus the amount of debt owed. A recent Census Bureau survey, summarized in Table 4-5, reveals that the net worth of most households is quite modest. According to the table, the majority of households are worth less than $50,000—a figure that includes home equity. The survey also showed that most households have little in the way of capitalist assets. They derive most of their net worth from three asset types: home equity, car equity, and bank deposits (U.S. Census 1995c; U.S. Census, n.d.)

Using data from the Census survey and a Federal Reserve survey conducted a few years earlier, we can distinguish three broad classes of wealth holders:[7]

1. *The Nearly Propertyless Class,* constituting 42 percent of the population and worth less than $30,000 in 1993. Many members of this class have little or no net worth; some actually owe more than they own. Others have built up precarious equities in homes and automobiles. Families worth $5,000 to $10,000 are better off than average in this class. Ninety percent of the families in this range own cars, with an average equity of $5,000. Only 36 percent own homes, with an average equity of just $5,500. Most have bank accounts with modest balances. As the low rate of home ownership suggests, younger households are overrepresented here. Seventy percent of African-Americans are in this wealth class.

---

[7] The framework for this section and the description of the first two classes draws on U.S. Census, n.d. The description of the "investor class" is based on the Federal study (Kennickell, McManus, and Woodburn 1996 and Wolff 1996).

TABLE 4-5 Distribution of Households by Net Worth, 1993

Mean Value	Percent of Households	Cumulative Percent of Households
Negative or Zero	11.5	11.5
$1 to $5,000	13.7	25.2
$5,000 to $10,000	6.3	31.5
$10,000 to $25,000	10.8	42.3
$25,000 to $50,000	12.2	54.5
$50,000 to $100,000	16.5	71.0
$100,000 to $250,000	18.5	89.5
$250,000 to $500,000	6.9	96.4
$500,000 and over	3.6	100.0
TOTAL	100	

SOURCE: U. S. Census, n.d.

2. *The "Nest-Egg" Class*—about 45 percent of the population, with net worths ranging from $30,000 to $300,000 in 1993. The families in this class might be described as savers rather than investors. For example, virtually all households in the $50,000 to $100,000 range have built up a cushion of safe, interest-earning assets, such as passbook savings accounts, CDs, and government bonds, worth, on average, more than $10,000. But less than a quarter hold corporate or mutual funds, and just 11 percent own a small business or professional practice. Homes and automobiles are their most substantial assets, accounting for about 70 percent of net worth.

3. *An investor class* comprising just 10 percent of families, worth more than $300,000. The households in this class own most of the privately held investment assets and typically control large, diversified portfolios. The average family's holdings of stocks, bonds, and bank accounts is worth about $385,000. Many members of this class have interests in small businesses, professional practices, and rental properties. On the other hand, equity in homes and autos contributes only 15 percent of average net worth. The investor class is not debt ridden. Although it controls nearly two thirds of gross assets, it is responsible for only one third of total liabilities.

As the privileged finances of our top class suggest, wealth is highly concentrated—much more concentrated than income. For example, in the early 1990s, the top 1 percent of income earners received about 16 percent of aggregate income, while the top 1 percent of wealth holders owned about 33 percent of net worth (Wolff 1996:38; Table 4-6). The latter figure is from the 1992 Federal Reserve survey referred to earlier. The Fed survey revealed that *the concentration of wealth has become so great that the net worth of the top 1 percent is roughly equal to that of the bottom 90 percent.*

TABLE 4-6   Concentration of Wealth, 1992

	Share Held by			
	Top 1%	*Next 9%*	*Bottom 90%*	*Total*
Widely Held Assets				
Principal Residence	9.0	27.1	63.9	100
Deposit Accounts	22.4	37.3	40.3	100
Automobiles	4.6	20.3	75.2	100
Pension Accounts	16.4	45.9	37.3	100
Concentrated Assets				
Stocks	49.6	36.7	13.6	100
Bonds	62.4	28.9	8.7	100
Trusts	52.9	35.1	12.0	100
Business Equity	61.6	29.6	8.9	100
Non-home Real Estate	45.9	37.1	17.0	100
Net Worth	**33.2**	**35.3**	**31.5**	**100**

NOTE: Asset distribution from Wolff 1996 except for Automobiles from Kennickell, McManus, and Woodburn 1996. Net worth includes all assets listed plus assets and liabilities not shown separately. Distribution of net worth is average of similar estimates by Wolff 1996 and Kennickell, McManus, and Woodburn 1996, from same Federal Reserve data.

Another basic conclusion that can be drawn from studies of wealth is that ownership of investment assets, such as stocks, bonds, and rental property, is more concentrated than ownership of consumption-oriented assets, such as automobiles and owner-occupied homes. Our sketch of wealth classes reflected this. According to the Fed survey (Table 4-6), the assets of the wealthiest 10 percent of the population include 36 percent of the value of all owner-occupied residences, but 83 percent of other (largely commercial) real estate and 86 percent of corporate stock. As the table indicates, an even smaller, wealthier minority, constituting 1 percent of households, controls almost half of corporate stock (including mutual funds) and close to two-thirds of small-business assets. The average net worth of these households exceeds $5.5 million (Kennickell, McManus, and Woodburn 1996). In Chapter 8, we will consider some political implications of this remarkable concentration of economic power.

TRENDS IN THE DISTRIBUTION OF WEALTH

Sometime in the early-1970s, a great shift began in the distributions of wealth and income, paralleling the growing disparities in job earnings we examined in the last chapter. We recognized this transformation in the distinction we made earlier between the Age of Shared Prosperity (from World War II to the early-1970s) to the Age of Growing Inequality (since the early 1970s).

FIGURE 4-6 Share of Net Worth Held by Top 1 Percent of Households

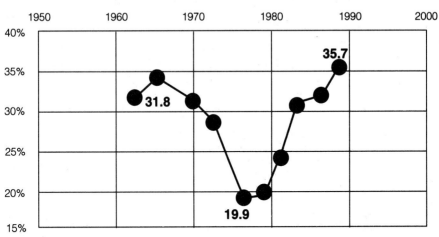

SOURCE: Wolff 1993.

The trend toward increasing inequality in the distribution of wealth was especially notable in the 1980s. During that decade, the combined net worth of the people on *Forbes* magazine's list of the 400 richest Americans more than doubled in real dollar value. By 1996, the poorest person on the *Forbes* list—who happened to be daytime TV star Oprah Winfrey—was worth $415 million (*Forbes* 1996).

On a broader scale, the wealth of the richest 1 percent of households rose spectacularly in the 1980s as did the concentration of wealth in this small stratum. Figure 4-6, which we previewed in Chapter 1, traces the proportion of aggregate net worth held by the top 1 percent since the 1960s. The curve assumes the familiar U-shape trajectory, bottoming out in the 1970s and then climbing steeply in the 1980s. This figure is based on an analysis of successive wealth studies by economist Edward Wolff, who concluded that the concentration of wealth in the top 1 percent was higher in 1990 than at any time since the 1930s (Wolff 1993).

What accounts for these spectacular increases in the concentration of wealth? Three factors seem especially important: (1) the rapid growth of incomes at the top of distribution, which has enabled those with high incomes to accumulate new funds for investment while incomes at lower levels have stagnated, (2) declining tax rates for the top income bracket, allowing high income households to retain more of what they make, and (3) the generally rising stock market since the early 1980s, which has swelled the value of securities held by the wealthy. We examine the changes in income distribution and tax rates in final sections of this chapter.

FIGURE 4-7 Income Ratio: Top 5 Percent versus Bottom 40 Percent

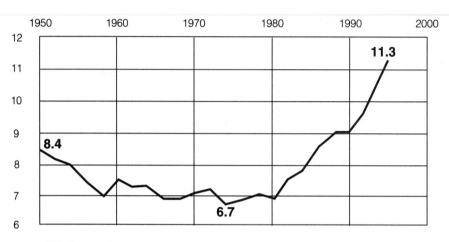

SOURCE: U.S. Census 1996e.

TRENDS IN THE DISTRIBUTION OF INCOME

During the Age of Shared Prosperity, family incomes at all levels were growing at a brisk pace and gradually becoming more equal. After the early seventies, income growth tapered off, and the fortunes of American families began to diverge.

The transition from the Age of Shared Prosperity to the contemporary Age of Growing Inequality is reflected in Figure 4-7, another of the curves we previewed in Chapter 1. This graph uses income ratios to measure the advantage of the richest 5 percent of families over the bottom 40 percent. For example, in 1950, the average income of the top 5 percent (in the valuable dollars of that era) was $13,200, which is more than 8 times the $1,600 average income of the bottom 40 percent in 1950. The graph shows that this ratio between the top 5 and the bottom 40 dipped in the 1950s and 1960s when low income families were gaining on the rich; it climbed after the mid-1970s as income growth accelerated at the top and low-income households were left behind.

Keep two things in mind as we discuss changes in family income: (1) Family income can come from varied sources, including jobs held by one or more family members, capitalist assets, and government transfers; change in the level or distribution of any of these income flows will affect the distribution of family income, and (2) income comparisons over long periods of time, like the earnings comparisons we made in Chapter 3, typically require adjustments for the distorting effects of inflation; for example, the incomes in Table 4-7 are expressed in "real" 1994 dollars.

TABLE 4-7   Mean Family Income, by Quintiles and Top 5 Percent, 1950–1994
(in 1994 dollars)

Year	Bottom	Second	Third	Fourth	Highest	Top 5%
1994	$10,387	$24,575	$38,808	$57,366	$115,608	$198,336
1990	$11,150	$26,006	$40,051	$57,598	$107,044	$167,957
1986	$11,139	$25,720	$39,846	$56,934	$102,547	$155,725
1982	$10,549	$23,935	$36,548	$51,847	$89,588	$128.177
1978	$11,913	$25,835	$38,793	$53,449	$90,696	$132,906
1974	$11,993	$25,317	$36,961	$50,639	$85,328	$124,459
1970	$10,995	$24,311	$35,152	$47,417	$81,539	$124,161
1966	$9,873	$21,918	$31,362	$41,984	$71,654	$110,026
1960	$6,878	$17,482	$25,507	$34,392	$59,182	$91,138
1950	$4,855	$12,948	$18,774	$25,248	$46,072	$74,665

Source: U.S. Census 1996c

Table 4-7 tells us what the 1970s income shift meant for people at different income levels. Imagine the fate of three hypothetical families—one in the bottom quintile, one in the middle quintile, and one in the highest quintile—over the period from 1950 to 1994 covered by the table. The low-income family's starting income of about $5,000 would double by 1966, and, then, more or less stagnate for the next three decades. Similarly, the middle income family's initial $18,800 income would double by 1974, but grow only slightly and erratically thereafter. In contrast, the family in the highest quintile would experience healthy income gains in the 1950s and 1960s, relative stagnation in the 1970s, but notable growth in the 1980s and 1990s. Its 1950 income of approximately $46,000 would climb to more than $115,000 in 1994.

Although the most rapid income gains came at the very top of the distribution, many households were attaining relative prosperity. As Figure 4-8 shows, the proportion of American families with incomes greater than $75,000 has grown considerably since the 1970s.

INCOME DYNAMICS

The three hypothetical families whose fortunes we traced from 1950 to 1994 had one thing in common: Their relative positions in the income distribution were stable, even when their incomes were changing. We assumed, for example, that the family that started in the bottom quintile was still there several decades later. We almost automatically make this kind of assumption when we talk about shifts in the distribution of income. But the government income surveys we have been analyzing in this chapter do not follow families over time. They are, in effect, periodic snapshots of the income

FIGURE 4-8 Households with Incomes over $75,000

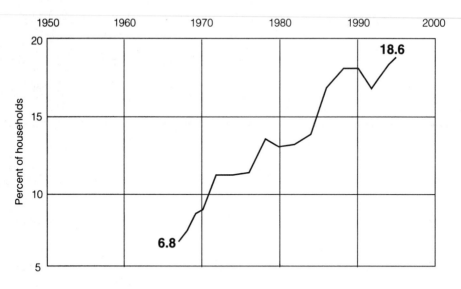

SOURCE: U.S. Census 1996e.

distribution, which tell us nothing about the degree to which particular families are moving up or down relative to one another.

Following the incomes of specific families over time is difficult and expensive. The few studies that have done so reveal a surprising amount of movement. From one year to the next, family incomes may change dramatically because someone lost a job or a spouse rejoined the labor force. Over longer periods, earnings tend to expand with experience and successive promotions. For example, during an academic career, the salary of a college professor might double in real dollar terms.

The best study of incomes over time is the University of Michigan's Panel Study of Income Dynamics (PSID), which has followed a national sample of some 5,000 families since 1967. Economist Stephen Rose recently analyzed the PSID data, comparing the patterns of growth of family income in the 1970s and the 1980s. For the two periods, he calculated the real gain or loss in family income for prime working age adults (men and women 22 to 58).[8]

---

[8] Some technical details: The sample for each decade consists of people 22 to 48 in the first year. Income was adjusted for family size to better represent family living standards. To smooth out short-term income changes, such as periods of unemployment, comparisons are between the average income of the first three years in each decade and the last three years. The comparisons points are business cycle peaks: 1967-1969, 1977-1979, and 1987-1989.

In both decades, Rose found, the majority of the people in the study experienced gains in family income as they aged and presumably advanced in careers. However, the gains were smaller in the 1980s than the 1970s, and the proportion of people whose incomes dropped from one decade to the next was larger. During the 1980s, one third of the sample actually lost ground. Losses were especially notable among families in the bottom 40 percent of households.

Rose's findings are further summarized in Figure 4-9. The bar graph shows big growth rate disparities among income quintiles in the 1970s and even bigger disparities in the 1980s. During the second decade, the bottom quintile actually lost ground and the middle quintiles experienced sharp declines in growth, but the top quintile sustained an income growth rate in excess of 60 percent. These trends inevitably led to growing income inequality. Contributing to the general pattern of change were the rising growth differentials shown between whites and blacks and between well educated and less educated adults.

In sum, Rose delivers good news and bad news. In the 1970s and 1980s, most working age people could still look forward to a rising family income. At the same time, the trend toward income inequality was accelerating— because of rapid gains at the top and relative stagnation toward the bottom. In the 1980s, the growth disparities widened. Lower-income households, African-Americans, and the less educated faced sharply diminished economic prospects. It is easy to see why many people felt that the American dream was slipping out of their hands.

NEWS FROM ANOTHER PLANET

The PSID and Census survey data give us broad sense of income trends in recent decades. Unfortunately, these sources are poor indicators of the change at the very top—where change has been most dramatic. The Congressional Budget Office (CBO) uses federal income tax data to supplement the Census survey estimates of income trends. According to the CBO, between 1977 and 1990, the average income for all households rose just $3,400 (in 1990 dollars). But the average income of the top 1 percent of households increased $254,000, climbing from $295,000 to $549,000 (U.S. House 1990: 28). This leap in income at the top is not just one more exotic fact about the rich. It tells us that most of the 1980s increase in personal income over that 13 year period was captured by a relatively small group of households.

To get a sense of what happened in the 1980s, we conjure up something even stranger than the income parade: a planet where annual income is retained for the year as body fat at the rate of a dollar per pound. One hundred people live on this planet, the economy is slowly growing, and, after a few years, the aggregate income of all the planet's inhabitants has increased by $100. This could mean that everybody gains one pound. Instead, the

FIGURE 4-9 Growth in Family Income

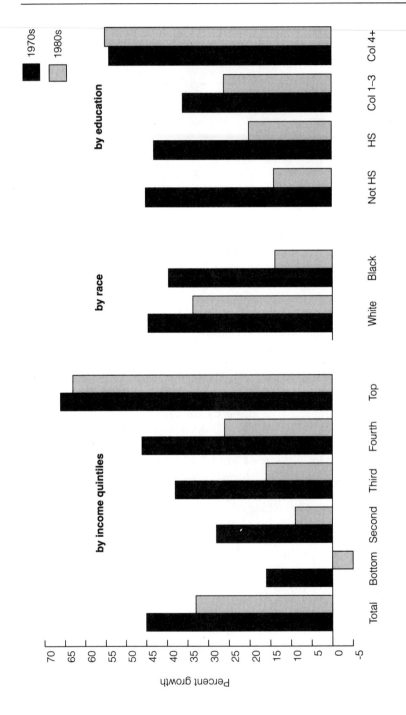

NOTE: Quintiles based on 10-year average of family income. Percent growth is aggregate for quintile.
SOURCE: Rose 1993.

TABLE 4-8   Federal Total Effective Tax Rates for Households, 1977-1996

	Percent of Income Paid in All Federal Taxes		
Income Quintile	1977	1989	1996
Lowest	9.3	9.3	5.0
Second	15.4	15.6	14.9
Middle	19.5	19.3	19.7
Fourth	21.9	22.0	22.6
Highest	27.3	25.6	28.1
TOP 1%	35.6	26.8	32.7

SOURCES: Michel, et al. 1997:110, based on Congressional Budget Office data.

richest person on the planet balloons up 75 pounds. The remaining 25 pounds are distributed unevenly among the other 99 inhabitants. Something like this happened on planet USA between 1977 and 1990. During this period, aggregate household income increased several hundred billion dollars. According to the CBO income data, 75 percent of this gain was concentrated in the top 1 percent of households. The rest was distributed unevenly among the remaining 99 percent of families.[9]

CHANGING FEDERAL TAX RATES

The trend toward greater income inequality in the 1970s and 1980s was magnified by regressive changes in the federal tax system. (Census Bureau income figures are, as noted earlier, pre-tax.) The top tax rate on personal income plunged from 77 to 28 percent. Corporate and inheritance taxes, whose main effects are felt by the wealthy, were also reduced. At the same time, payroll taxes, paid largely by lower and middle-income workers, were jacked up (U.S. House 1991a). The result of these policies can be seen in Table 4-8, which traces the evolution of *effective tax rates*—that is, the proportion of income lost to the *combined effect of all federal taxes*.[10]

The table misses the effects of significant tax reductions implemented before 1977, but allows us to draw some conclusions about subsequent trends: (1) The most dramatic changes were felt at the very top. Effective tax rates for the richest 1 percent of households dipped sharply in the 1980s, under conservative Republican administrations. For these wealthy taxpayers, the reduction in total taxes, from 1977 to 1989, was equivalent to a $88,000 tax break on on every $1 million of income. Their rates rose

---

[9] Authors' calculations based on U.S. House 1990: 28.

[10] Effective tax rates should not be confused with the marginal tax rates (or tax brackets) which are the basis of the personal income tax.

again in the first Clinton administration, but did not return to 1977 levels. (2) At the other end of the income parade, the already low effective tax rates for households in the poorest quintile were cut in half under Presidents Bush and Clinton. (3) For most households, the outcome of many changes in tax law from 1977 to 1996 was, surprisingly, a wash. Reductions in personal income tax rates were apparently offset by increases in payroll tax rates.

CONCLUSION

In this chapter, we added evidence of growing inequality in the distribution of income and wealth since the early-1970s to the evidence of growing inequality in earnings that we explored in the last chapter. The polarization of incomes is all the more remarkable because it reverses a well-documented trend toward greater income *equality* from the 1930s into the 1970s (Miller 1971; U.S. Census 1996a). How can we account for the shift? This question is an enlarged version of the one we asked at the end of Chapter 3. There we were interested in the increasing disparity in earnings. Here we are concerned with all sources of income and with whole households rather individual workers. In this section we will review what we have learned in this chapter, giving particular attention to developments that can help explain change.

We began the chapter with an imaginary income parade, a device to visualize the income distribution and the factors that shape it. The parade began with a long line of little people—not just the poor but also millions of families living marginally on the earnings of low-wage workers. At the end of the parade, we were struck by the abrupt increase in the size of the marchers, who grew in a matter of minutes to astronomical proportions.

The distribution of occupations in the parade was about what we expected from a postindustrial society. More surprising were the other factors that powerfully influenced where people marched in the ranks—in particular, the sources of household income and the number of workers in a family. Most households, of course, depend on job-earnings. We noticed that single worker families were common in the first half of the parade. Much later in the procession, among people with incomes around $80,000, we found very few families without two employed adults.

On the other hand, we discovered that jobs were less significant for those at the beginning and the very end of the procession. The first marchers were heavily dependent on government transfers, from Social Security to public assistance. The very last marchers—especially those with incomes above $1 million—depended less on jobs than investments for their incomes.

If those at the end of the parade are most dependent on financial wealth for their incomes, the rising concentration of wealth is certainly strengthening income inequality. Of course, the accumulation of wealth is also a result of income inequality—as well as the changes in effective tax rates that enabled those with the highest incomes to retain a higher proportion of their incomes.

The parade focused attention on the social factors leading to increased income inequality. We noted, for example, that families headed by females were crowded into the early part of the parade. The prevalence of such families is growing as a result of increased divorce and out-of-wedlock births, contributing inevitably to the growth in income inequality. Ironically, the rise in the labor force participation and earning-power of women has had a similar effect. The earnings of working wives are increasing fastest at the highest income levels, exacerbating existing inequalities among households.

These social trends are strengthening the economic pressures toward polarization that we discussed in Chapter 3. The result is what we have called an Age of Growing Inequality, which we will describe more broadly in the following chapters.

SUGGESTED READINGS

Hacker, Andrew 1997. *Money: Who Has How Much and Why?* New York: Scribners.
  An engaging book, loaded with intriguing facts.
Levy, Frank 1995. "Incomes and Income Inequality." In Reynolds Farley, ed. *State of the Union: America in the 1990s.* V. 1. New York: Russell Sage Foundation.
  Long term trends in household income for different social groups.
Oliver, Melvin and Thomas Shapiro 1995. *Black Wealth/White Wealth: A New Perspective on Racial Inequality.* New York: Routledge
  Systematic study of wealth as a critical dimension of racial inequality.
U.S. Bureau of the Census. *Money Income of Households, Families, and Persons in the United States.* Current Population Reports. Consumer Income. Series P-60.
  Published annually, the reports in this series are the basic source of detailed information on the distribution of income. (The title varies.)
U.S. House of Representatives, Committee on Ways and Means. *Overview of the Federal Tax System.*
  Published periodically. A clear, easy to use source of basic information on the federal tax system.

5

Socialization, Association, Lifestyles, and Values

Let me tell you about the rich. They are different from you and me.

F. Scott Fitzgerald (1926)

Yes, they have more money.

Ernest Hemingway (1936

Since Max Weber, many students of stratification have thought of prestige classes as "communities" characterized by more or less distinctive lifestyles and values. Sociological research has examined differences among classes in areas as diverse as consumption patterns, role relationships in marriage, sexual behavior, and language usage. Two of the variables outlined in Chapter 1 are critical for the emergence and maintenance of such differences: socialization and association. When people of similar class position associate more often with one another than with persons of lower or higher classes, they create identifiable class subcultures. Since people tend to inherit the class position of their parents, the socialization of successive generations of the community into the patterns of thought and behavior characteristic of their elders contributes to the crystallization of these subcultures.

Although this chapter will treat association and socialization as processes played out within social communities, we will find that both are shaped by occupational life and can in turn shape it. Patterns of association through friendship, marriage, residence, and membership in formal organizations are structured by occupation. "I'm a carpenter," comments one of Coleman and Rainwater's Kansas City informants, "and I won't fit with doctors and lawyers or in country club society" (1978: 81). The professionals and business executives who do fit may be able to enhance their careers through informal contacts made in country-club society. Furthermore, research demonstrates that occupational position influences the way that parents socialize their children and that childhood socialization, in turn, cultivates attitudes and abilities that affect adult occupational achievement.

CHILDREN'S CONCEPTION OF SOCIAL CLASS

Socialization is a social learning process that prepares new members of a society for adult life. The knowledge imparted ranges from how to hold a fork to conceptions of appropriate male or female behavior. The child also absorbs adult notions about class differences. For example, an early study of primary school students in a New England town of 15,000 showed that by the sixth grade, children had a fairly sophisticated awareness of the class significance of items such as an English riding habit, an elegantly furnished room, tattered clothing, and different occupational activities, all presented to them in pictures. And when they were asked to place their peers in one of three classes, the sixth graders agreed 70 percent of the time with adults who rated parents from the same households (Stendler 1949). Simmons and Rosenberg (1971) demonstrated that young children have a clear conception of occupational prestige differences. Even the third graders in their sample from the Baltimore city schools ordered fifteen occupations from the NORC list in a way that correlated almost perfectly with the rankings in a national survey of adults. (The children did tend to raise the relative position of their fathers' occupations, much as adults inflated the standing of their own occupations.)

Subsequent studies have concentrated on the development of conceptions of class distinctions, beginning in early childhood. Tutor (1991) showed first, fourth, and sixth graders photographs of upper, middle, and lower-class people. She asked the children to group the adults and children depicted into families and match them with the corresponding pictures of cars and houses. The first graders did substantially better than chance at this task; the sixth graders produced near-perfect scores; and the fourth graders were not far behind.

Leahy (1981; 1983) probed the thinking of children ages 7 to 17 about "poor people" and "rich people." He found that young children conceive of the rich and the poor in overt, physical terms, while older children think in terms of psychological characteristics of individuals and their positions in the society. Here are some of their observations:

Joe, age 6:

> [Poor people have] no food. They won't have no Thanksgiving. They don't have nothing... [R]ich people have crazy outfits and poor people have no outfits.

Mary, age 6:

> [People can become rich by going] to the store and they give you money... [Or] if your husband gives you money, and your grandmother or your grandfather.

Pete, age 10:

> [People are rich] because they save their money and they earn it. They work as hard as they can and don't just go around and buy whatever they want.

Dean, age 12:

> I think that [rich and poor people] should all be the same, each have the same amount of money because then the rich people won't think they are so big.

In general, these studies show that even preschool children are aware of class differences. They suggest that as children grow older their ideas about stratification become more consistent, abstract, and "accurate." By the time they reach 12 years of age, children are not very different from adults in their thinking about class.

KOHN: CLASS AND SOCIALIZATION

While studies such as those just reviewed approach socialization through the child's developing conception of the social world, a separate research tradition focuses on the childraising practices of parents. For decades, studies of the latter type have recorded class differences in the way people raise their children (Bronfenbrenner 1966; Gecas 1979). The most intriguing work along these lines is that of Melvin Kohn and his associates. What makes Kohn's work special is his effort to isolate the particular characteristics of the lives of adults that could explain the class variation in styles of parenting. In brief, Kohn was able to show that the values parents attempt to inculcate in their children reflect the character of the parents' occupational experience. His findings suggest, moreover, that the transmission of these values plays an important role in the perpetuation of the existing class order.

Kohn analyzed a series of studies of parental values, conducted from the mid-1950s to the mid-1970s, including a local survey of about 500 parents conducted in Washington D.C. and two much larger national surveys. (Kohn 1969, 1976, 1977; Kohn and Schooler 1983). In these studies, parents were asked to select from a list of characteristics those they considered most desirable for a child of the same age and sex as their own child. Here are a few examples (Kohn 1969:218):

> That he is a good student.
> That he is popular with other children.
> That he has good manners.
> That he is curious about things.
> That he is happy.

Although the studies found consensus across class levels about the importance of many values, there was less agreement about others. For example, parents at higher class levels were more likely to choose "curiosity," and those at lower class levels were more likely to select "honesty." The top panel (A) of Table 5-1 compares values characteristically cited by parents in the upper and lower halves of the class structure (for convenience labeled *working class* and *middle class*). Parental views were quite varied at every class level, but the values we are calling working-class become increasingly common at lower class levels and those we have labeled middle class become more common at successively higher class levels.

Kohn and his associates stratified families on the basis of father's occupation. The researchers found that the class patterning of parental values was even more salient when the occupations of employed mothers were taken into account. Among women married to working-class men, mothers with white-collar jobs were closer to middle-class mothers in their answers, and those with manual jobs presented the sharpest contrast with middle-class orientations.

Kohn's interpretation of the class patterning of parental values is based on the idea that the middle-class parents who stress the values of self-control, curiosity, and consideration are cultivating capacities for self-direction and empathetic understanding in their children, while working-class parents who focus on obedience, neatness, and good manners are instilling behavioral

TABLE 5-1   Typical Class Patterns in Parental Values and Occupational Experience

Middle-Class Pattern	Working-Class Pattern
A. Parents' Values for Children	
Consideration of Others	Manners
Self-Control	Obedience
Curiosity	"Good Student"
Happiness	Neatness, Cleanliness
Honesty	
B. Parents' Own Value Orientations	
Tolerance of Nonconformity	Strong Punishment of Deviant Behavior
Open to Innovation	Stock to Old Ways
People Basically Good	People not Trustworthy
Value Self-Direction	Believe in Strict Leadership
C. Job Characteristics	
Work Independently	Close Supervision
Varied Tasks	Repetitive Work
Work with People or Data	Work with Things

conformity. The middle-class pattern, particularly in the emphasis laid on happiness, curiosity, and consideration, is oriented toward the *internal* dynamics of the person—both the child and others. The working-class pattern, on the other hand, assumes fixed *external* standards of behavior. This general difference is neatly illustrated in the top panel of Table 5-1 by three pairs of contrasting values, beginning with "self-control" and "obedience." In each case, the first (middle class) choice favors internal development, and the second (working class) emphasizes conformity to rules.

An additional finding substantiates this interpretation. Parents were asked about the specific sorts of misbehavior for which they would discipline their children. Their responses revealed that middle-class parents were more likely to punish a child for the *intent* of his or her behavior, in contrast with working-class mothers, who were more likely to discipline for the *consequences* of behavior. For example, a middle-class mother might penalize her child for throwing a temper tantrum, while a working-class mother penalizes for boisterous play. The first suggests a loss of internal control, the second a violation of external standards.

Kohn noted that working-class and middle-class mothers also differ in their attitude toward sex-role socialization. The surveys indicate that working-class mothers are more likely to choose separate "masculine" values (dependability, ambition, school performance) for boys and "feminine" values (cleanliness, happiness, good manners) for girls. Middle-class mothers are less inclined to make such gender distinctions. Kohn argued that this difference is also related to the distinction between internal dynamics and external standards, because the working-class emphasis on traditional sex roles implies an external focus. However, this connection is certainly not as intuitively obvious as the one regarding occasions for punishment.

Kohn labeled the two underlying patterns *self-direction* and *conformity*. He demonstrated a consistent relationship between the class position of parents and the values they chose for their children. At successively higher class levels, parents value self-direction more and conformity to external standards less. But what are the roots of these differences? Kohn suggested that they reflect generalized value orientations that develop out of a specific aspect of social class: occupational experience. He reasoned, for example, that people who hold professional and managerial jobs, which are relatively unsupervised and require considerable exercise of individual judgment and initiative, are more likely to value self-direction than those who work at highly routinized blue-collar jobs. In brief, self-direction at work should produce self-direction in values.

Evidence from several studies, including a U.S. national survey and another done in Turin, Italy, supports this interpretation. The surveys showed that judgments about authority, deviance, and the goodness of human nature were related to social class. In particular, Kohn noted that "authoritarian attitudes" stressing "conformance to the dictates of authority and intolerance of nonconformity" become more frequent at lower class levels (Kohn 1969: 79). The surveys also confirmed that attitudes toward work are related to class. At lower class levels, respondents were more likely to rate jobs by their *extrinsic* qualities (for example, pay or hours). At higher class levels, respondents were more likely to emphasize *intrinsic* job qualities (how interesting the work is, the amount of freedom it has, the chance it affords to use abilities).

Finally, the surveys demonstrated that these general attitudes are systematically related to the character of respondents' occupational experience: Men whose work is (1) closely supervised, (2) repetitive, and (3) oriented toward things rather than people or data, are the most likely to subscribe to authoritarian values and to judge jobs on their extrinsic qualities. Of course, what are ordinarily considered working-class jobs are most likely to fit this occupational pattern—though some (plumber) do not fit it as well as certain office jobs (data entry operator). Kohn additionally notes that the wives of men who share this occupational experience tend to hold similar values.

The results and interpretative logic of Kohn's research are summarized in Table 5-1. Remember that the middle-class side of this chart represents patterns that are increasingly frequent at higher class levels and the working class side describes patterns that become more frequent at lower class levels. We are dealing with statistical tendencies here, not absolute contrasts between classes. Reading the table from the bottom up on the middle-class side, we find that parents at higher class levels are more likely to work at jobs requiring intellectual flexibility and independent judgment, more open to innovation and tolerant of nonconformity, and more likely to encourage self-direction in their children. The parallel finding on the working-class side is that parents at lower class levels are more subject to authority and routinization at work, more authoritarian in their judgments, and more likely to favor conformity in their children. These two contrasting patterns fit the causal chain that Kohn had anticipated to explain the relationship between social class and socialization patterns: Occupational experience gives rise to general value orientations, which in turn shape parental value preferences for children.

Further scrutiny of the data revealed that a second aspect of social class, level of education, also correlates with value orientations and parental value preferences and that this reinforces the class patterning of socialization. Kohn observed that education appears to "provide the intellectual flexibility and breadth of perspective that are essential for self-directed values …"(1969: 186). Education, of course, is closely related to occupation. It does not, however, "explain" the correlation between occupational experience and values. Kohn found that education and occupational conditions have independent impacts on parental values, although the effect of occupational conditions is substantially stronger.

Kohn's research on class differences in the socialization of children has important implications for our understanding of the class system as a whole. Since Marx, sociologists have been aware that life experience, especially occupational experience, shapes social values. Kohn observed that "the essence of higher class position is the expectation that one's decisions and actions can be consequential; the essence of lower class position is the belief that one is at the mercy of forces and people beyond one's control, often, beyond one's understanding" (1969: 189). If this is true, we should expect people in top positions to learn to value self-direction and those at the bottom to learn to value conformity to authority. We might also anticipate that they will teach these values to their children. At this point the larger significance of Kohn's work becomes clear. When parents inculcate values that reflect their experience of the class system, they are preparing their children to assume a class position similar to their own and, by so doing, are contributing to the long-term maintenance of the class system. Kohn explicitly rejects the notion that these outcomes reflect the conscious intentions of parents. Working class parents may, in fact, believe that the values they are encouraging in their children will help them to ascend the class hierarchy.

SCHOOL AND MARRIAGE

The class differences in values inculcated by parents are reinforced by the tendency of children and adolescents to associate with others of like class background as they are growing up. Because neighborhoods tend to group people of similar economic means, the kids on the block are likely to be of the same social class. Local schools reproduce the class patterns of the neighborhoods they serve. Many upper- and upper-middle-class families make sure that their children's classmates will be from similar households by sending them to private schools.

Although large public high schools in some communities bring together students of diverse backgrounds, they do not necessarily mix freely. Often class differentiation within the school is institutionalized through "tracking" systems that separate students on the basis of their ability or postgraduation aspirations. Because there is a substantial correlation between social class and academic track, the system segregates students by class (Colclough and Beck 1986: 489; Oakes 1985).

Students' own preferences contribute to the class patterning of association. A series of somewhat dated studies of adolescent cliques and friendships show that students are inclined, but by no means certain, to choose friends who share their own class backgrounds (Hollingshead 1949; Cohen 1979; Duncan, Haller, and Portes 1968). From 1988 to 1994, one of the authors of this book surveyed groups of students at a selective college regarding their associations in high school. Approximately 100 students were asked the occupations of the parents of their three best friends and most significant romantic interest in high school. Students also provided information on the occupation and income of their own parents. Families were stratified according to the Gilbert-Kahl model introduced in Chapter 1. The results are summarized in Table 5-2, which shows that the high school associations of this relatively privileged group of students were largely restricted to people with class backgrounds similar to their own but quite different from the class distribution of Americans generally . The vast majority of the students and most of their friends and dates were from families in the top 15 percent of the class structure. They had few social ties to the bottom 55 percent. The strong pattern of class segregation among these affluent teenagers is, as we will see later in the chapter, consistent with the highly restricted association patterns of upper-middle class adults.

Like friendship, mate selection is influenced by social class.[1] Sociologist Martin K. Whyte (1990) confirmed this conclusion from earlier studies in a survey of women in the Detroit metropolitan area. Whyte asked the respondents about the class positions of their parents and in-laws, at the time the women and their husbands were in high school. Given a choice of five classes, fifty-eight percent of the respondents placed their parents and in-laws in the same class. Occupational comparisons between fathers and fathers-in-law produced similar results.

In an earlier study of a Massachusetts community, Edward Laumann (1966) found that the inclination toward class endogamy (in-group marriage) was especially notable in the upper-middle class (Laumann's "top professional and business" grouping). Although this category constituted a relatively small proportion (24 percent) of the population surveyed, creating limited opportunities for endogamous marriages, 60 percent of the sons of top professionals and businessmen married women who shared their class

---

[1] See Hollingshead 1950, Dinitz, Banks, and Pasamanick 1960, Laumann 1966, Rubin 1968, and Whyte 1990.

TABLE 5-2 Association Patterns of High School Students Bound for Selective College

		Class Distribution (in percent)		
	All Americans	Students Surveyed	Student's 3 Best Friends	Student's Best Date
Capitalist/Upper Middle	15	69	52	52
Middle	30	23	35	32
Working or below	55	8	13	16
Total	100	100	100	100

SOURCE: Cumulative surveys of approximately 100 college students enrolled in social stratification course, 1989 to 1994.

background (Laumann 1966: 74–81). A couple of quotations from Laumann's informants neatly reveal the attitudes underlying this phenomenon (1966: 29):

> What sort of husband would a carpenter be? What sort of education would he have? My viewpoint would not jibe with a carpenter. Marriage is based on equals. I would want my daughter to marry in her own class. She would go to college and would want her husband to be educated. I would want to be able to mix with in-laws and converse with them.

> I was born into a family of great privilege—but I think I have a responsibility. Family responsibilities will be better held if the background of the members are similar. My own interests are not manual. Our family relations are very close. It would be very demanding to have an unskilled or factory worker in the family. The tradition in our family is the professions or land-owning. I would not turn out a daughter who married a machinist but—a machine operator is bound to have a modest education—probably not interested in intellectual things. Even with the best will in the world, a family relationship would be very difficult to accomplish.

The pattern of class endogamy observed by Laumann and others has an important implication for class differences in childhood socialization: If parents are from the same social class, they are likely to transmit a consistent set of class influences to their children.

MARRIAGE STYLES

Social class not only channels mate selection, but also shapes the character of marital relationships. In general, sociological studies of husbands and wives have found clearer sex role distinctions at lower class levels and greater intimacy, equality and companionship at higher class levels (Langman 1987: 222–234). A classic study conducted by Lee Rainwater (1965) supports this view.

Rainwater analyzed the marital role relationships of several hundred couples. Following Elizabeth Bott's (1964) earlier work on English couples, Rainwater distinguished three types of relationships, on a continuum: joint, intermediate, and segregated. Joint relationships focus on companionship and de-emphasize the sexual division of labor. Husbands and wives with joint role relationships share the planning of family affairs, carry out many household duties interchangeably, and value common leisure activities. Even when responsibilities are parceled out by gender (wife-homemaker, husband-breadwinner), each partner is expected to take a sympathetic interest in the concerns of the other. In segregated relationships, there is clear differentiation of concerns and responsibilities, which minimizes the husband's involvement with household matters and the wife's with the world of (the husband's) work. Husband and wife are likely to have distinct leisure pursuits and separate sets of friends. Intermediate relationships fall between these two poles.

Based on answers to questions about family decision making, duties of husbands and wives, interests and activities of the partners, and the general character of the relationship, Rainwater placed each couple in one of the three categories. Of course, all couples were in some sense "intermediate." So the ratings were based on relative differences. When Rainwater compared the distributions of these three types of role relationships at four class levels, unmistakable differences emerged (Table 5-3). Joint relationships predominate in the upper-middle class and segregated relationships in the lower-lower class, while the classes between them exhibit a neat gradient.

The relationship between social class and marital role types is more than a matter of academic curiosity. Reported marital happiness increases with class level, especially for women (Bradburn 1969: 156), and this phenomenon is tied to the character of the organization of marital roles. In Rainwater's study, middle-class couples reported greater sexual satisfaction than lower-class couples, but the difference was largely a function of the level of role segregation. For example, most lower-class wives in segregated relationships evaluated their sexual experience in marriage negatively, but lower class wives in intermediate relationships were generally positive (Rainwater 1965: 28). In national surveys, companionship in marriage (which would appear to be similar to joint organization) is positively correlated with social class and with marital happiness (Bradburn 1969: 163).

Let's take a closer look at conjugal role types by examining how they function in upper-middle-class and working-class families (Sussman and Steinmetz 1987: 226-231; Rubin 1976). In important ways, the very character of upper-middle-class life lends itself to the joint role relationship. College life, generally a prologue to upper-middle class careers, delays marriage and encourages informal, relatively egalitarian association between men and women. High rates of social and geographic mobility are typical of this class. Husbands and wives are isolated from kin and removed from successive sets of friends as they move from community to community and up the career ladder. They must look to each other for support and companionship.

TABLE 5-3 Social Class and Conjugal Role Relationships

Class	Number	Role Relationships (percent)			
		Joint	Intermediate	Segregated	Total
Upper-middle class	(32)	88	12	—	100
Lower-middle class	(31)	42	58	—	100
Upper-lower class	(26)	19	58	23	100
Lower-lower class	(25)	4	24	72	100

SOURCE: Rainwater 1965: 32.

Spouses are typically drawn into the career-oriented social life that is one of the keys to success for ambitious executives and professionals.

Upper-middle class wives are expected to be "gracious, charming hostesses and social creatures, supporting their husbands' careers and motivating their achievements" (Kanter 1977:108). The traditional result has been the "two-person career" that links a husband's advancement to his wife's unpaid efforts. A more recent phenomenon, typical of younger couples, is the "dual-career" family, in which both spouses pursue demanding professional or managerial careers. A survey of 1,000 working-age women in Chicago (Lopata, et al. 1980) found that dual-career couples are as likely as single-career couples to mix social and professional life. About 60 percent of wives employed as managers or professionals reported that their husbands helped them with career-related entertaining at home. Husbands with professional or managerial jobs were somewhat more likely to receive such help from their wives. (It made little difference whether the wife was employed). On the other hand, women employed in blue-collar jobs or married to blue-collar men reported little job-related entertaining.

As the Chicago study suggests, the career-oriented social life that becomes a shared endeavor for upper-middle class couples has no working class equivalent. Working-class men and women do develop social ties on the job, but these tend to segregate rather than join husbands and wives. For example, many of the workers in a New Jersey chemical plant studied by David Halle (1984), drank together after work and joined coworkers on fishing trips and at sports events. Working-class occupations are less likely to require geographic mobility. Remarkably, most of Halle's chemical workers were born within two miles of the plant where they worked (1984:303). Under such circumstances, it is easier for spouses to maintain ties with kin and friends from adolescence and early adult years. Dependency on the couple's parents is intensified by the economic insecurity that is especially typical of young working-class families (Rubin 1976: 69–92). These social ties tend to draw husband and wife to separate sources of support and companionship outside the marriage.

Bott's (1964) work in England showed that couples who come to a marriage with tight-knit networks of friends and kin (that is, the people each spouse knows tend to know one another) and maintain these ties are the most likely to develop segregated marital relationships. Her data suggest that social networks of that sort are least typical of professionals and most typical of manual workers. A variant on this theme can be seen in the work of Carol Stack (1974), who found that lower-class black women in the United States develop extensive networks of kin and friends as a shield against economic vicissitudes. Such networks provide security that many erratically employed black men of this class cannot give. But networks also maintain an insistent claim on the loyalties of their members, which undermines strong relationships between women and their lovers or husbands.

We have dealt with the origins of joint and segregated role relationships in experiences typical of top and the bottom of the class order. What can we say about the mix of marriage types Rainwater found in the middle of the class structure (Table 5-3)? Two social factors seem relevant. One is social mobility: People moving up or down in the class structure may carry with them lifestyles acquired in their class of origin. Thus, the upper-middle-class origins of many lower-middle-class couples (especially younger couples) can help explain the predominance of joint relationships among them. An analogous argument can be made for the spread of segregated relationships upward. The second factor is the tendency of upper-middle-class lifestyles to become generally fashionable models and filter downward. Through these processes, couples are exposed to conflicting influences, which may be reflected in intermediate role relationships.

Sex-role socialization is another source of class differences in marital-role organization. As we noted in our discussion of Kohn's work, working-class parents are more likely than middle-class parents to hold separate sets of expectations for boys and girls. A study of college-age women (Vanfossen 1977) found that college-age daughters of working-class fathers are more likely than their middle-class peers to subscribe to traditional sex-role values as expressed in questionnaire items such as "A woman should not expect to go to exactly the same places or have the same freedom as a man." Such women are the most likely to find the segregated marital role acceptable. On the other hand, boys of all classes are taught to be more controlled, more instrumental, less emotional, and less empathetic than their sisters, but the distinction is made much more emphatically in blue-collar families. Lillian Rubin, a sociologist and psychotherapist who conducted lengthy interviews with working-class and upper-middle-class couples, noted big differences in the behavior of their sons:

> Not once in a professional middle-class home did I see a young boy shake his father's hand in a well-taught "manly" gesture as he bid him good night. Not once did I hear a middle-class parent scornfully—or even sympathetically—call a crying boy a sissy or in any way reprimand him for his tears. Yet, these were not uncommon observations in the working-class homes I visited. Indeed, I was impressed with the fact that, even as

young as six or seven, the working-class boys seemed more emotionally controlled—more like miniature men—than those in the middle-class families (1976: 126).

Boys who are taught to be "manly" in this way are less likely as adults to feel comfortable with joint role relationships in marriage.

BLUE-COLLAR MARRIAGES AND MIDDLE-CLASS MODELS

Two intimate studies of working-class life, Rubin's book and another by E.E. LeMasters (1975), published about the same time, caste new light on the differences between working and middle-class marriages. Rubin, who interviewed young parents in their Northern California homes, and LeMasters, who spent five years getting to know the somewhat older patrons of the Oasis, a "family-type" working-class tavern in Wisconsin, reached surprisingly similar conclusions. Both found that marital norms filtering down from the upper-middle class were creating enormous strains in blue-collar marriages.

One of LeMaster's informants, a woman married for thirty years, bitterly described the traditional pattern of segregated blue-collar marriages:

> The men go to work while the wife stays home with the kids—it's a long day with no other adult to talk to. That's what drives mothers to the soap operas—stupid as they are.
>
> Then the husband stops at some tavern to have a few with his buddies from the job—not having seen them since they left to drive home ten minutes ago. The poor guy is lonely and thirsty and needs to relax before the rigors of another evening before the television set. Meanwhile the little woman has supper ready and is trying to hold the kids off "until Daddy gets home so we can all eat together." After a while, she gives up this little dream and eats with the kids while the food is still eatable. About seven o'clock, Daddy rolls in, feeling no pain, eats a few bites of the overcooked food, sits down in front on the TV set, and falls asleep.
>
> This little drama is repeated several thousand times until they have their twenty-fifth wedding anniversary and then everybody tells them how happy they have been. And you know what? By now they are both so damn punch drunk neither one of them knows whether their marriage has been a success or not (LeMasters 1975: 42).

Such dissatisfaction was probably nothing new, but as the tone of her comment suggests, expectations were changing. By the 1970s, the traditional pattern was being challenged by notions of intimacy, companionship, sharing, and equality received from above. The problem was and still is that these ideals do not appeal equally to wives and husbands. Women are prepared for them by their socialization and in many cases by contact with a middle-class world through white-collar employment. They are exposed to the newer conceptions of marriage through women's magazines, television soap operas, and daytime talk shows. Men, on the other hand, are likely to be satisfied with established role relationships, which they have regarded as part of the natural order of things. The traditional women's role is, according to one of LeMaster's informants "natural for them so they don't mind it" (Rubin 1976: 105). Men sense, of course, that many women do "mind it," but they are inclined to think that women's complaints are groundless. At the Oasis, a construction worker asks LeMasters,

> What the hell are they complaining about? My wife has an automatic washer in the kitchen, a dryer, a dishwasher, a garbage disposal, a car of her own—hell, I even bought her a portable TV so she can watch the goddamn soap operas right in the kitchen. What more can she want? (LeMasters 1975: 85).

But behind the bluff there is fear. From the less "macho" setting of his living room, one of Rubin's informants phrased the problem differently:

> I swear I don't know what she wants. She keeps saying that we have to talk, and when we do, it always turns out I'm saying the wrong thing. I get scared sometimes. I always thought I had to think things to myself; you know, not tell her about it. Now she says that's not good. But it's hard. You know, I think it comes down to that I like things the way they are, and I'm afraid I'll say or do something that'll really shake things up. So I get worried about it, and I don't say anything (Rubin 1976: 121).

For their part, working-class women in these studies are very dissatisfied but also frightened and confused and occasionally given to wondering whether asking a man to be more than a conscientious provider is indeed asking too much.

> I'm not sure what I want. I keep talking to him about communication, and he says, "Okay, so we're talking, now what do you want?" And I don't know what to say then, but I know it's not what I mean. I sometimes get worried because I think maybe I want too much. He's a good husband; he works hard; he takes care of me and the kids. He could go out and find another woman who would be very happy to have a man like that and who wouldn't be all the time complaining at him because he doesn't feel things and get close (Rubin 1976: 120).

A second aspect of blue-collar marriage was under strain in the 1970s: sexual adjustment. Problems in this area were not new. For instance, husbands and wives had long clashed over the desirable frequency of sexual intercourse. LeMasters heard this complaint among patrons of the Oasis (1975: 101). However, difficulties of more recent origin, deriving from the sexual revolution of the 1960s and 1970s, were evident in the comments of the younger couples interviewed by Rubin.

In the 1970s, working-class sexual behavior was moving closer to middle-class norms. For example, working-class couples had nearly caught up with middle-class couples in their willingness to engage in once-exotic sexual variants such as cunnilingus and fellatio; working-class men had become similar to middle-class men in their concern for their wives' sexual satisfaction (Rubin 1976: 134–135, 137–148). But change has psychological costs. Again, differential receptivity to new standards was creating stress for working-class marriages. In this case men were more open to change. The blue-collar workers whom Rubin interviewed wanted freer, more expressive, more mutually satisfying sexual relationships with their wives, as their remarks show:

> I think sex should be that you enjoy each other's bodies. Judy doesn't care for touching and feeling each other, though.

> She thinks there's just one right position and one right way—in the dark with her eyes closed tight. Anything that varies from that makes her upset.

> It's just not enjoyable if she doesn't have a climax, too. She says she doesn't mind, but I do (Rubin 1976: 136).

But their wives responded warily. This comment from a woman married twelve years is typical:

> He says I'm old fashioned about sex and maybe I am. But I was brought up that there's just one way you're supposed to do it. I still believe that way, even though he keeps trying to convince me of his way. How can I change when I wasn't brought up that way; [*with a painful sigh*] I wish I could make him understand (Rubin 1976: 138).

The emphasis here is on "how I was brought up." Working-class women were traditionally socialized to fear or deny their own sexuality and to expect men to divide women into "bad girls" they use and "good girls" they marry. Even after marriage, such attitudes persisted. Rubin reports a different situation among her upper-middle-class couples. While the working-class women had generally remained at home until they were married, the middle-class women had the more open atmosphere of their college experience to free themselves psychologically from traditional sexual constraints. Nearly all the couples in both groups had premarital intercourse, but the middle-class wives were less likely to express guilt about having done so. In fact, the middle-class women were more likely to feel guilty about their sexual inhibitions than about their sexual behavior (1976: 60, 62).

Rubin and Masters portrayed the powerful influence of upper-middle class models on working-class behavior, even in intimate realms of experience like marital gender roles and sexuality. But has the behavior of working-class men and women continued to change along the path they described in the 1970s? Unfortunately, we do not have truly comparable studies from the 1980s and 1990s, but two books from this period contain some clues: Halle's (1984) study of chemical plant workers referred to earlier and Rubin's recent book on working-class life in an age of economic insecurity (1994).

Halle observes that most of the men he studied endorse the modern marital ideals of companionship and mutually fulfilling sexuality. But, by and large, these ideals were not reflected in their own conflict-ridden marriages. Halle stresses that husbands and wives typically do not share leisure interests, and the men are drawn to a male leisure culture that excludes women and absorbs much of men's time. The following complaints, the first from a wife and the second from one of the chemical workers, are typical:

> I never see Jimmy. He's out a lot, and when he's home he likes to watch sports on TV. And I hate sports, especially football.

> Susan keeps saying, "Why don't we go away together? Just you and I." But that's boring. I like to go fishing and drink a few beers when I'm on vacation, but . . . she doesn't like to fish, and she thinks I drink too much (Halle 1984: 55-6).

Such comments suggest differences more deep-seated than incongruent leisure interests. In fact, only one third of Halle's respondents claim to be happy with their marriages. This minority does stress the importance of companionship. One worker tells the author,

> I don't care what they [the other workers] say, marriage is a beautiful thing. But you have to work at it. You can't just leave your wife alone like a lot of these guys do. . . I'll tell you what makes a good marriage. It's when you do things together (Halle 1994: 56).

Rubin's recent book (1994), published two decades after her original study, portrays a very different world. The lives of working-class women have been transformed, reflecting changed conceptions of womanhood in the broader culture and the new economic role of women. In the 1990s, as we noted in the last chapter, wives and mothers are more likely to hold jobs and to work full time. As the real wages of working-class men have declined, their families have become increasingly dependent on the earnings of women. People have abandoned the notion—widely held in the 1970s—that men could and should support families by themselves or that women's earnings were merely supplementary to the household budget.

Rubin reports that the expanded economic role of women is altering expectations about the domestic roles of husbands and wives. Working-class women are now more likely to demand equality and reciprocity from their husbands. Even the relatively traditional young secretary who rejects "feminism" and says a women should make her man "feel like a king," tells Rubin that she will not "bow down" to a man or be like her mother who "waits on my father all the time." "Why should I," she asks. "I'm as equal as him" (Rubin 1994:74). The attitudes of men are also changing. In the 1970s, Rubin interviewed few working-class husbands "who would even give lip service to the notion of gender equality" (Rubin 1994:77). By the 1990s, such views were common, especially among younger men. But Rubin, like Halle, found a wide gap between professed ideals and everyday behavior. In particular, she notes that the household division of labor has been slow to change and has become the subject of growing contention between husbands and their hard-pressed working wives.

Rubin also interviewed the teenage children of the working-class families she studied in the 1990s. Among the daughters especially, a transformation is evident. Their mothers dreamed of little but marriage. They married young because they were pregnant or anxious to escape the smothering authority of traditional families. Now the daughters are more independent and have wider horizons. They also have fewer illusions about what young men, in the economy of 1990s, can provide. They expect to marry "some day," but first they want to work, to live on their own, to travel and experience the world. One daughter, single and pregnant like her mother before her, gets an abortion rather than follow her mother's path into a bad marriage. When they do marry, these young women will likely be very different from their mothers

The studies we have examined in this section suggest that working-class marriage has been changing over the last generation, under economic and cultural pressure. At least at the level of values, working-class couples are moving toward upper-middle class models. But without new studies that make systematic class comparisons, like Rainwater's survey (1965) or Rubin's original book (1976), we cannot be sure of the current extent of class differences. Both classes are affected by the changing roles of women and the transformation of the economy, but they may be reacting to these changes in different ways.

INFORMAL ASSOCIATION AMONG ADULTS

Warner, whose classic Yankee City study we examined in Chapter 2, considered patterns of association so critical to understanding the class system that he sometimes appeared to define class in terms of association. A social class, he suggested, is a group of people who belong to the same social cliques, intermarry, dine in each other's homes, and belong to the same organizations. Warner defined a social clique as "an intimate nonkin group," with no more than 30 members. Warner's research team collected elaborate data on the clique membership of families in Yankee City. They found that most cliques joined people of the same or adjacent classes (Warner and Lunt 1941: 110–111, 350–355).

The idea that social class is, in some sense, about "who you hang out with" is widely shared. Asked to discuss the basis of social class differences, almost half of the skilled blue-collar workers and two-thirds of the white-collar workers in a Providence, Rhode Island, study referred to patterns of association. Their comments suggest that people belong to the same class if they "run around together," intermarry, "belong to the same churches, clubs, organizations," "live in the same neighborhoods." or send their kids to the same schools (Mackenzie 1973: 148).

Numerous studies suggest that patterns of association are shaped by so-cial class—though association is also influenced by factors that cut across class lines, including gender, race, age, religion, and shared interests (Allan 1989; Smith and Macauley 1980; Argyle 1994:66–92). Earlier in this chapter, we saw that adolescent friendships and marital mate selection reflect class backgrounds. Adult friendships are also patterned by class, according to surveys done in Providence, the Boston area, and metropolitan Detroit (Mackenzie 1973; Laumann 1966, 1973).

The findings of Laumann's Boston area survey are typical. Laumann asked 422 white male residents of Cambridge and Belmont the occupations of their three closest friends.[2] Table 5-4, drawn from the survey, compares the occupations of respondents and their friends. The figures in each col-umn show the distribution of friends for respondents from a single occupa-tional class. For example, 50 percent of the friends of semi/unskilled workers are drawn from the same class; an additional 23 percent are skilled workers. The italicized figures on the diagonal (from upper left to lower right) are the percentages of friends drawn from the respondents' own oc-cupational class. Three general conclusions can be drawn from the table:

1. *Self-selection* As the figures on the diagonal indicate, there is a high level of class self-selection in friendships. This tendency is strongest at the extremes and weakest in the middle. The top grouping—roughly upper-middle class—is especially exclusive in its friend-ships.
2. *Class patterning* Although many friendships do cross occupational class lines, men are quite likely to choose close friends from their own or an adjacent occupational class; 80 percent of friendships reflected in the table meet this criterion.
3. *Blue collar/white collar* The traditional gap between manual and non-manual workers is evident in friendship patterns. Only 25 percent of all close friendships are between blue-collar and white-collar work-ers. On the other hand, skilled workers and men in the clerical/small business class have many friendships across the divide. Thus, we cannot draw a sharp distinction between a white-collar middle class and a blue-collar working class based on friendship patterns.[3]

Again, it is worth quoting Laumann's informants for the attitudes be-hind associational choices. The following remarks are in response to ques-tions eliciting subjective reactions to association with people in occupations ranging from janitor to physician (Laumann 1966: 28–29). Unfortunately, Laumann does not indicate the speaker's own occupation, though it can usually be inferred.

---

[2] Laumann's study (which we used earlier in our discussion of marriage patterns) exclud-ed areas of Cambridge near academic institutions to avoid overrepresentation of students.

[3] Mackenzie 1973: 173 also supports this conclusion.

TABLE 5-4 Cambridge-Belmont Study: "Best Friends'" Occupations by Respondent's Occupation

Friends' Occupations	Respondent's Occupation					
	Top Professional, Business	Semi-professional, Middle Business	Clerical, Small Business	Skilled	Semiskilled, Unskilled	Total
	Percent Distribution					
Top prof., business	74.2	22.9	19.0	9.0	3.9	24.4
Semiprof., mid. business	17.1	47.9	26.3	16.2	11.1	22.4
Clerical, small business	2.5	13.1	19.6	13.1	12.2	11.8
Skilled	3.8	8.9	21.8	34.8	22.8	18.3
Semiskilled, unskilled	2.5	7.2	13.4	26.7	50.0	23.1
Total	100.0	100.0	100.0	100.0	100.0	100.0

SOURCE: Laumann 1966: 65.

> I am not a snob, but it is a fact of life that in most of these occupations we would have nothing in common. There must be an intellectual ground for my associations.

> I have had experiences with factory workers as a class of people and they're rough. They tend to go on the rough side of living—"hurray for me and the hell with the other guy" is their attitude . . . Top executives tend to, when that high up, to get standoffish and independent, and I can't afford to have independent friends.

> The doctor has his trade; the machine operator also has his trade. As long as a man puts his mind into his trade, it is OK. A tradesman is not much of a big man compared to a top executive. The top executive would try to be uppity—would try to lord it over me. I have a little bit of this in the family already. This guy (my relative) is too big for his own family. His wife is not good enough for him any more. Just because you are a top executive does not mean that you can act uppity.

Not all informants accepted class segregation in friendships as desirable. One man commented:

> I just don't believe it is possible to know anything about a man just by knowing his occupation. If he is a good guy, I don't care what he does . . . The job isn't important—it's what he's like himself.

Social class, then, channels friendship choices. Research shows that it also influences the extent and character of informal association. The literature suggests that people at higher class levels: (1) have more friends and more active social lives; (2) are less likely to preserve friendships from their youth; (3) spend proportionately less time with relatives; (4) are more likely to entertain friends at home and, in particular, to host dinner parties; (5) are more inclined toward couple-oriented social activities; (6) are more likely to develop (nonromantic) cross-sex friendships; and (7) are more likely to mix career and social life (Allan 1989; Argyle 1994:66–92; Dotson 1950; Curtis and Jackson 1977:169; Kahl 1957:138; Kanter 1977; Rubin 1976, 1994; Shostak and Gomberg 1964; Whyte 1952).

How can we explain the class patterning of informal association? Why do people tend to marry and maintain friendships with class peers? What accounts for the class differences in the character of social life? Two obvious but powerful factors are money and propinquity (physical or social proximity). Dinner parties can be costly affairs and guests are expected to reciprocate in kind. Skiing and sailing are more expensive than bowling. These price-of-admission differences segregate leisure activities and the people who engage in them by ability to pay. In everyday life, people tend to encounter others who are close to them in status. They live in neighborhoods and send their children to neighborhood schools that are relatively homogeneous in household income. Their coworkers have similar jobs—except for bosses, whose authority places them at a social distance. In short, daily life is structured in ways that limit the opportunities to develop social ties across class boundaries.

Beyond money and propinquity are a series of more subtle factors—matters of prestige, style, interests, values, and comfort level. Their influence on patterns of informal association is suggested by the comments of Laumann's respondents, quoted earlier. One man indicates that he has "nothing in common" with people in lower occupations, another characterizes factory workers as "rough," and a third complains about the "uppity" attitudes of a relative who is a successful executive. These men seem uncomfortable with disparities in social prestige. But their attitudes also reflect objective differences in areas including education (which is strongly correlated with class) and occupational experience. Education produces contrasts in language usage, attitudes, and personal interests. Adults with limited education are likely to be uneasy in the presence of the well educated. Occupational experience, as Kohn's studies of socialization demonstrate, contributes to class differences in values. Halle (1984) emphasizes that working-class people typically have dull "jobs," while upper-middle class people have engaging "careers." The former are inevitably less interested in conversations that revolve around work and generally less inclined to mix work and leisure.

FORMAL ASSOCIATIONS

Like informal ties, participation in formal associations is patterned by social class. Formal associations are large groups or organizations with explicit purposes and rules of membership, including the YMCA, the Elks Club, the Teamsters union, the Burning Tree Country Club and the Boy Scouts (Hodges 1964: 105–115; Mackenzie 1973: 81–84; Smith 1980; Warner, et al 1949a).

From its beginnings as a nation, the United States has been characterized as a nation of joiners. This generalization turns out to be somewhat less than half true. Most working- and lower-class Americans have little or no participation in formal associations. Even the participation of the lower-middle class is modest. The true joiners are members of the upper-middle and upper classes, who are especially likely to participate in civic and charity organizations.

Members of these top classes are not just the most likely joiners. They are also the most active participants in organizations and, even when organizational membership cuts across classes, the most likely to serve in leadership positions. The reasons for this phenomenon are not hard to imagine. These managers and professionals enjoy the prestige attached to high class position. They have more education. At work, they develop organizational skills and confidence as leaders. Finally, many see active participation in community organizations as a way to bolster their careers.

Associations often draw their membership from a limited range in the class structure. Country clubs and exclusive men's clubs such as New York's Links or Boston's Sommerset draw from the upper and upper-middle classes. Service organizations such as Rotary or Lions, fraternal orders like the Elks Club, and patriotic organizations like the Veterans of Foreign Wars recruit members from successively lower class levels.

Even churches—institutions supposedly rejoicing in the common brotherhood of a common good—are class typed. According to the existing research, people of higher status are likely to attend those Protestant denominations that feature services of quiet dignity and restrained emotion, such as the Episcopal or Unitarian groups. Middle status people are more often seen at the Methodist and Baptist churches, where the services are more vigorous, or in Catholic churches (reflecting their origins as part of the "new" immigration from southern and eastern Europe). Lower-status individuals who go to church at all are most likely to join revivalistic and fundamentalist sects (Demerath 1965; Laumann 1966: 55; Smith and Macauley 1980: 514).

SEPARATE LIVES

Americans are increasingly segregated by social class. No one has made this point better than Tom Wolf in his novel *Bonfire of the Vanities* (1987). The novel's protagonist, Sherman McCoy, is a smug, young Wall Street trader with a $3 million Park Avenue apartment and few redeeming qualities. Driving into the city in his Mercedes one evening, accompanied by his mistress, Sherman blunders into an impoverished ghetto neighborhood, where he is involved in a fatal hit and run accident. Subsequently McCoy finds himself locked up in a court house holding cell in the unwanted company of dozens of tough young men—poor and dark-skinned like the victim of his Mercedes. These events initiate a downward spiral in McCoy's life. By the end of the novel, a year after the accident, McCoy is separated from his wife, his mistress, his money, and his lawyer, who has resigned from the case because McCoy is broke.

The action of *Bonfire of the Vanities* is driven by its satisfying but improbable premise: Sherman McCoy has smashed through the wall that normally separates privileged people like the McCoys from people like the accident victim and Sherman's cellmates—or for that matter, from the $36,000-a-year assistant D.A. who prosecutes the case. "If you want to live in New York," a friend advises McCoy, "you've got to insulate, insulate, insulate" (Wolf 1987:55). Before the accident, McCoy used his money to do just that. His world was as insulated from the grimy reality of the city as the posh cabin of his Mercedes from the asphalt below.

Wolf's novel reflects the growing disparities of the current era. It portrays a society, whose members, divided by class and race, live in increasing isolation from one another; they no longer share (despite McCoy's strange fate) a common destiny. Journalist Mickey Kaus develops this theme in *The End of Equality* (1992), arguing that rising economic inequality has been accompanied by rising *social* inequality. Money, fear of the poor, and an inflated sense of their own superiority are motivating prosperous Americans to develop separate lives. "An especially precious type of equality—equality not of money but in the way we treat each other and live our lives—seems to be disappearing" (Kaus 1992:5).

Kaus looks back at the post-World War II era as (with the "evil" exception of race) "a golden age of social equality." The war, perhaps more than any event in our history, provided Americans with a common experience and a sense of shared destiny. Wealthy 26-year-old John F. Kennedy served on a small PT boat in the South Pacific with men who had been machinists, factory workers, truck drivers, and night school students. Seventy percent of able-bodied young men, most of them drafted, served in the military (Kaus 1992:50). Some, like Kennedy's brother Joe, did not survive. Those who returned brought with them a network of friendships, forged under the threat of death, with little regard for class differences.

After the war, the GI Bill offered veterans of diverse backgrounds access to college and the opportunity to buy their own homes. In the midst of the shared prosperity of the 1950s, there was a sense that the social distance between Americans of different classes was shrinking. Today, it appears to Kaus that the just the opposite is happening. Against a backdrop of growing economic inequality, Americans worry about the emergence of what they take to be a permanent underclass. The opulent lives and social pretenses of the rich—objects of ridicule in a more egalitarian age—inspire fawning articles in glossy magazines aimed at upper-middle class readers in search of role models. Professionals with merely comfortable incomes see themselves as "not just richer, but more civilized, better educated, wittier, smarter, cleaner, prettier" than the average American (Kaus 1992: 27).

"Who killed social equality?" asks Kaus. Oddly, he rejects the most obvious suspect, rising economic inequality, and insists that the guilty party is "the decline of the public sphere." What he has in mind is the reduction of the social realm where Americans of different classes meet on more or less equal terms. He points to the end of the draft and the replacement of a broad-based citizen military with a volunteer force, which recruits few soldiers from the upper end of the class structure. But most of his examples revolve around residential segregation by class. As the rich and relatively rich retreat to exclusive suburbs (sometimes even to "gated private communities") they separate themselves from the less privileged. Here public spaces—the mall, the supermarket, the drugstore, the coffee shop—are largely inhabited by other members of the privileged classes. There is no "Cheers," a democratic place where the postman and the psychiatrist can chat (or argue) over a beer. Above all, children attend school with others of the same class, even if they do not enroll in private academies. And, not surprisingly, their parents, who vote for budget slashing state and national politicians, do not hesitate to support local school bond issues. They know their own kids will benefit.

Journalist Kaus is examining the same fractured social reality as novelist Wolf. Has he proven anything? Is the social distance between Americans of different classes actually increasing, and is it doing so for the reasons that Kaus suggests? *The End of Equality* book is, unfortunately, full of vivid examples but short on systematic evidence.

This chapter has examined a series of studies demonstrating that Americans are, in fact, separated by class in their everyday lives. But the attentive reader will have noted that many of the studies are rather dated. There are few recent studies of class patterns of association and hardly any that would allow dependable comparisons with the past. (Of course, we need such comparisons to think seriously about the issues of change Kaus raises). There is, however, an important category of exceptions: the studies of changing residential patterns based on the decennial U.S. Census. By comparing the geographic distribution of households at various class levels from one census to the next, we can determine whether class segregation is actually increasing.

Figure 5-1, based on analyses of census data by Douglas Massey (1996) and others, traces changes in the residential segregation of neighborhoods in metropolitan areas from 1970 to 1990. The upward trajectories of the three lines indicates that both class and racial segregation have been increasing. In the graph, "Poor" refers to members of households with incomes lower than $13,400 in 1990 dollars, and "Affluent" refers to households with incomes at least four times the poverty figure or above $54,000 in 1990.[4] All three lines in the graph are based on an isolation index, measuring the extent to which people who are poor or affluent or black are concentrated in particular

---

[4] The cutoff figures for each year are, respectively, the Federal poverty line for a family of four and four times that amount. Income figures were not adjusted for family size, as they would be for Federal poverty statistics. For a discussion of the Federal poverty standard, see Chapter 10.

FIGURE 5-1 Residential Isolation Index

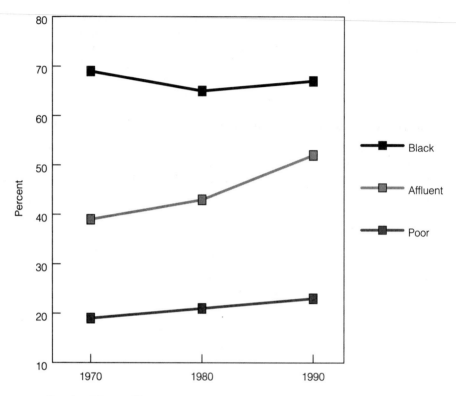

SOURCE: Based on Massey 1996.

neighborhoods. For example, the 1990 index of 52 for affluent indicates that the average member of an affluent household lived in a neighborhood whose population was 52 percent affluent.

The residential isolation of the affluent is high and has been rising quite steeply since 1970—a trend that would not surprise journalist Kaus. The poor are less segregated than the rich—though the relatively low segregation index for the poor (22 percent) may be misleading because of meager income standard on which it is based. A neighborhood that is 22 percent poor by this measure is likely to have many more low-income families that would be considered poor by the average observer. Like the residential concentration of the rich, the concentration of the poor has been rising.

These data point to rising class segregation in the 1970s and 1980s, in contrast with an earlier study showing that class segregation was relatively stable in the 1950s and declined in the 1960s at the lower end of the class structure (Simkus 1978). The contrast between the two periods is consistent with our claim in earlier chapters that sometime around 1970, the United States left an Age of Shared Prosperity for an Age of Growing Inequality.

Figure 5-1 also reveals quite clearly that Americans are much more segregated by race than by class and that the country was becoming more racially segregated in the 1980s. But despite the continuing separation of blacks and whites, the residential patterns of blacks have undergone a real transformation, according to a recent study that considered race and class simultaneously. Paul Jargowsky (1996) separately tracked class segregation for whites, blacks and Hispanics in several hundred metropolitan areas. His measure, based on income, showed that from 1970 to 1990, class segregation increased for all three groups. The increase was especially great for blacks. This finding should not be surprising. It reflects the process of class differentiation among African-Americans that we alluded to in Chapter 3. More concretely, it means that middle-class families are leaving urban ghettoes for prosperous black communities in the suburbs.

In sum, Americans are segregated by class—and to an even greater extent, by race. Segregation on both dimensions is increasing. These findings provide at least some support for Kaus' argument about growing social inequality, which emphasizes residential separation. It appears that even Americans of relatively modest affluence are heeding the advice of Sherman McCoy's friend: "Insulate, insulate. . . ."

CONCLUSION

This chapter has explored the social implications of class structure, emphasizing socialization and association. The life cycle has served as a guiding thread. We learned that children are precociously aware of class distinctions and that they are socialized according to class patterns that reflect the occupational experience of their parents. Adolescents tend to form friendships and romantic ties with others who share their class background. Young adults typically marry class equals or near equals. Adult friendships, marital styles, residential choices, organizational activities, and even church membership are all patterned by social class.

These observations bring us back to Max Weber's idea, invoked at the beginning of the chapter, that prestige classes (or "status groups," as he called them) are social "communities," characterized by distinctive lifestyles and values. To a remarkable extent our social lives and outlooks are molded by class position. Consider the typical differences we have found between the upper-middle and working classes. Members of the upper-middle class generally share the life-shaping experiences of college, late marriage, geographic mobility, and a career-oriented social life. They are drawn to joint models of marriage and values of self-direction and tolerance, which they stress for their children. Their exclusive choices of friends and mates suggest that this class is relatively isolated from the rest of the population. In contrast, members of the working class tend to get married younger, retain stronger ties with kin and friends from adolescence and young adulthood, and maintain stronger separation between social and economic life. They are likely to develop segregated marriages and to inculcate values of discipline and conformity in their children.

Are these social differences between classes becoming more or less pronounced? We might expect social differences to widen as economic differences grow, and we know that residential segregation by class has intensified since 1970. But we have little other evidence about recent trends.

Whatever the direction of change, the American class system is far from becoming a series of discrete class cultures. The differences we have described in this chapter are loose statistical tendencies, not absolute contrasts. Americans of all classes are influenced by a national culture and share many key values. They are exposed to the same ideas and lifestyles in pervasive mass media. At the same the time, the diversity of American society and (as we will see in the next chapter) relatively high rates of social mobility guarantee that the membership of any social class will be quite varied.

SUGGESTED READINGS

Argyle, Michael 1994. *The Psychology of Social Class*. New York: Routledge.

Summary of social-psychological literature on class. Areas covered include social relationships, lifestyles, and personality differences. Extensive bibliography.

Cookson, Peter, and Caroline Hodges Persell. 1985. *Preparing for Power: America's Elite Boarding Schools*. New York: Basic Books.

How the upper class is educated.

Fussell, Paul. 1983. *Class: A Guide Through the American Status System*. New York: Ballantine.

One man's view of the class system: idiosyncratic, witty, perceptive, arrogant. Emphasizes class differences in lifestyle, covering such areas as dress, speech, recreation, and home interiors.

Hollingshead, August B. 1949. *Elmtown's Youth*. New York: Wiley.

This classic work from the Warner era includes a comprehensive study of patterns of association among high school students. Demonstrates strong correlation between social class and social cliques.

Kohn, Melvin, and Carmi Schooler. 1983. *Work and Personality: An Inquiry into the Impact of Social Stratification*. Norwood, N.J.: Ablex.

Synthesis of their research.

Lamont, Michele. 1992. *Money, Morals and Manners: The Culture of the French and the American Upper-Middle Class* Chicago: University of Chicago.

How successful professionals and managers in the U.S. and France define what it means to be a "worthy person" and distinguish themselves from others.

Langman, Lauren. 1987. "Social Stratification." In Marvin Sussman and Suzanne Steinmetz, eds. 1987. *Handbook of Marriage and the Family*. New York: Plenum.

Wide-ranging essay summarizing literature on class differences in family life.

Rubin, Lillian Breslow. 1976. *Worlds of Pain: Life in the Working-Class Family*. New York: Basic Books.

_____ 1994. *Families on the Faultline: America's Working Class Speaks About the Family, the Economy, Race, and Ethnicity*. New York: HarperCollins.

Sensitive portrayals of working-class life. First volume makes revealing comparisons with upper-middle class.

6

Social Mobility: The Structural Context

Some people's money is merited
And other people's is inherited.

Ogden Nash, The Terrible People

With regard to social stratification, no society is completely open or completely closed. A completely open society would permit the members of each new generation to achieve a position based wholly on their merits and motivation, without reference to the status of their parents. An absolutely closed society would fix the future position of each child at birth. Societies do, however, differ in the degree of openness they allow. They change over time as economic shifts or political transformations enlarge or reduce opportunities. India today is a more open society than it was when the caste system radically constricted occupational and marital choices. The United States may be a less open society than it was when a young industrial economy was creating new opportunities for people at all levels of the class system.

Social mobility, our topic in this chapter and the next, is a measure of openness. In Chapter 1, we defined social mobility as the extent to which people move up or down in the class system. We will approach the topic from two distinct perspectives, structural and individual. The structural perspective focuses on shape of the class structure and the trends that influence it. Here we are concerned with the relative proportions of positions at high, middle, and low levels of the system, and with the tendencies that are changing those proportions. For instance, an aristocratic society that is made up of a mass of peasants dominated by a handful of large landowners and their administrators is structurally like a very steep pyramid: broad at the base, narrow in the middle, and pointed at the top. When such a society industrializes, the landed aristocrats are replaced by industrial and commercial capitalists at the top, probably in greater numbers than the landlords they push aside. Over time, the middle sectors grow, with new engineers, teachers, and skilled workers, who are needed in profusion to make the system function with efficiency. The pyramid becomes more squat in shape.

A hierarchy with few positions at the top allows only a few people to move up, regardless of how fair the system is, but a hierarchy with many positions at the top is likely to generate more movement. Particularly during the period of transition from one type of society to the other, the very expansion of new positions, which are disproportionately placed in the middle and upper levels, creates mobility. With the industrialization of a traditional agricultural society, many sons and daughters of peasants have a chance to climb into newly created jobs and thus raise themselves above their parents. In such a situation, it is possible to have a lot of upward mobility without much downward slippage. Thus, changes in the structure can increase opportunities for everybody. Children of people near the top can stay there, and at the same time, many children from below get a chance to improve their position, because the upper strata are expanding.

The second perspective on mobility, which we will take up in the next chapter, concerns the particular factors that influence the careers of individuals. Within a given structure of opportunities, why do some individuals move ahead, while others fall behind? How much difference does education make? How much influence does family background have on a person's career? How important are gender and race in shaping individual opportunities?

In this chapter, we begin with a discussion of the gross amount of mobility in the United States and of the structural facts and trends that explain it. We will look at studies comparing fathers and sons. Because women have entered the labor force in massive proportions only in recent years, mobility research has focused on men. We will, however, examine some evidence on the mobility of women.

Ideally, we would like to have detailed data gathered from both fathers and sons. Unfortunately, that is impossible, if only because many of the adult sons we might interview do not have living fathers. We are reduced to studying a sample of sons and asking them about themselves and their fathers. We cannot expect adult sons to recall too much detail about the careers of their fathers. Sons are not likely, for example, to recall how much their fathers earned in the past. But they will recall their fathers' occupations—a key piece of information, which is the starting point for most mobility studies.

HOW MUCH MOBILITY?

How much social mobility is there? Unfortunately, no one has conducted a full-scale national mobility survey since 1973. But we persuaded Professor Robert Hauser, one of the top authorities in the field, to estimate mobility in the 1970s and 1980s, using data from the General Social Survey (GSS). The GSS, a national survey conducted annually and covering many topics, asks respondents their occupations and the occupations of their fathers (or other family head). Using this information, Hauser could answer questions like these: What chance does the son of an unskilled worker have of attaining a professional or managerial position? What are the social origins of the people in high-status occupations?

To obtain statistically trustworthy answers to such questions, Hauser needed very large samples. His analysis aggregates the data on working-age men in successive GSS surveys and uses a simplified five-level occupational hierarchy. The data cover all men, aged 25 to 64, who are employed or who have work experience and are looking for a job.

Table 6-1 is an "outflow" table that starts with fathers and shows what happened to their sons. It is drawn from Hauser's analysis of GSS polls from 1982 through 1990. To answer the question about the sons of unskilled workers, find "lower manual" under "father's occupation" on the left side of the table and read across. The percentage breakdown shows that many of these sons (27 percent) attained "upper white-collar" (professional or managerial) positions, but the largest group (37 percent) followed their fathers into unskilled "lower manual" jobs. (See the notes at the bottom of the table for a fuller description of the five occupational categories.)

TABLE 6-1 Outflow from Father's Occupation to Son's Occupation, 1982–1990

	Son's Occupation (in percent)					
Father's Occupation	Upper White-Collar	Lower White-Collar	Upper Manual	Lower Manual	Farm	Total
Upper white-collar	60	14	11	14	1	100
Lower white-collar	39	18	19	23	1	100
Upper manual	32	10	33	24	2	100
Lower manual	27	11	25	37	1	100
Farm	19	10	22	33	16	100
Total ($N = 4,632$)	36	12	23	27	3	100
	Up	**Stable**	**Down**	**Total**		
	44	36	20	100		

NOTE: Occupational categories are upper white-collar—professionals, managers, officials, non-retail sales workers; lower white-collar—proprietors, clerical workers, and retail sales workers; upper manual—craftsmen and foremen; lower manual—service workers, operatives, and non-farm laborers; farm—farmers and farm workers.

SOURCE: GSS data from Davis and Smith 1990; analyzed for this volume by Robert Hauser.

Several general conclusions can be drawn from these data. (1) There is a high level of occupational succession, sons following fathers, in American society. This is especially true at the top: Well over half of the sons of managers and professionals hold similar positions. (2) But there is also considerable movement up and down the occupational ladder from one generation to the next. The figures at the bottom of table sum up the results. Forty-four percent of sons moved up, about half as many slipped downward, and the rest remained at the same level as their fathers. (3) There is a barrier, albeit a fairly porous one, between manual and nonmanual jobs (that is, between the bottom three categories and the top two). Roughly two thirds of those born on either side of this divide will remain there. The other third moved up or down across the divide (calculated from the original data).

If Hauser had been able to use a more detailed occupational hierarchy, he would have uncovered a large volume of short-distance occupational mobility: the son of an unskilled construction laborer who becomes a semi-skilled factory worker, or the son of a salesman who becomes a social worker. Both moves would go undetected by the five-step occupational hierarchy we are using here. Large-scale studies in 1962 and 1973 suggest that mobility often takes this form—a few steps up or down the ladder (Blau and Duncan 1967; Featherman and Hauser 1978).

Table 6-2 presents the same data in an altered form that allows us to answer a different set of questions. It is an "inflow" table, which groups the survey respondents according to their *own jobs* and asks *where they came from*. (The previous "outflow" table grouped people according to their

TABLE 6-2   Inflow to Son's Occupation from Father's Occupation, 1982–1990

Father's Occupation	Upper White-Collar	Lower White-Collar	Upper Manual	Lower Manual	Farm	Total
Upper white-collar	36	25	11	11	4	21
Lower white-collar	14	19	10	11	5	12
Upper manual	22	20	35	22	11	24
Lower manual	20	24	29	37	5	27
Farm	8	12	15	19	75	15
Total (N= 4,632)	100	100	100	100	100	100

NOTE: See Table 6-1 for source and definitions.

fathers' jobs and asked *where they ended up*). The difference is evident in the way the tables are percentaged. The "inflow" table calculates percentages down the columns, instead of across the rows.

For example, if we read down the "farm" column on the right side of Table 6-2, we find, to no one's surprise, that most farmers are the sons of farmers. It is similarly unremarkable that most manual workers are the sons of manual workers. But reading down the upper-white-collar column, we discover that the people in these better occupations (who constitute about one third of all sons, according to Table 6-1) are of very diverse social origins. Nearly half are from farm or manual homes. We get this result, despite the high level of father-to-son succession at the top, because the relative number of professional and managerial jobs has been growing, opening opportunities for advancement from below.

The mixed social origins of people at the upper end of the occupational structure affects a question we have raised at various points in this book: the degree of consistency or clarity in the class system. Social mobility appears to reinforce consistency toward the bottom, while undermining it at the top. In practice, this might limit the homogeneity of values or political outlook among the people in the upper occupations.

SOCIAL MOBILITY OF WOMEN

Studies of social mobility have focused almost entirely on men. Until recently, this emphasis seemed justified because most women did not hold jobs. But it also reflected the dubious notion, which sociologists until recently shared with much of the public, that women's work did not matter because so many women held part-time or sometime jobs just to supplement family income.

Studying women's mobility presents a special set of problems. For one, with whom do we want to compare women workers—their mothers or

TABLE 6-3   Outflow from Father's Occupation to Daughter's Occupation, 1976–1985

| | Daughter's Occupation (in percent) | | | | | |
Father's Occupation	Upper White-Collar	Lower White-Collar	Upper Manual	Lower Manual	Farm	Total
Upper white-collar	51	36	2	11	0	100
Lower white-collar	39	43	3	15	0	100
Upper manual	27	45	3	25	0	100
Lower manual	24	38	3	35	0	100
Farm	21	32	3	44	1	100
Total (N = 2,462)	32	39	3	27	0	100

NOTE: See Table 6-1 for definitions.

SOURCE: Data from Hout 1988: Appendix.

their fathers? Many mothers in previous generations did not work or worked only part-time or intermittently; they were not the economic mainstay or foundation of social status for their households. If the comparison is with fathers, we face the problem that women workers are distributed across occupations in a very different fashion than men. Thus, if we see that many daughters of lower-manual men have moved "up" to white-collar work, are we looking at a difference between generations or between genders? If we find that the daughters of upper-white-collar men are concentrated in upper-white-collar positions, can we assume that they hold the same sort of jobs as their fathers? Probably not, given the segregation of women workers into a relatively small number of occupations. We know that the women in this category are much more likely than men to work as teachers, nurses, and social workers.

Table 6-3 does not resolve these problems (a task beyond the scope of this book), but it does give us a preliminary picture of women's mobility. The data employed were aggregated from successive GSS surveys in the same fashion as the data on men we just examined. The one conclusion that we can confidently draw from this table is that *women's occupational achievement, like men's, is powerfully influenced by class origins.* For example, more than half of the daughters of upper-white-collar men, but only 24 percent of the daughters of lower-manual workers, hold upper-white-collar positions.[1] Few daughters of white-collar workers hold blue-collar jobs, but nearly 40 percent of the daughters of lower-manual men hold such positions.

---

[1] By referring to daughters and fathers in this discussion, we are, of course, using the same shorthand we used in our exposition of men's mobility. In the GSS surveys, occupational origin refers to father or, in a minority of cases, other family head—such as mother or grandfather.

CHANGING CAUSES OF MOBILITY

We have been looking at the pattern of intergenerational movement up and down the occupational hierarchy. But what are the causes of this mobility? To simplify the problem, imagine a completely closed society of male workers in which all sons replicate the positions of their fathers. Consider the factors that could open it up. There are four basic factors:

1. *Circulation.* Some sons slip down the scale and thereby make room for others to climb up. In this instance, mobility is a two-way street: There must be some who lose for others to gain.
2. *Occupational change.* If technological and organizational changes are occurring in an asymmetric way that creates jobs at a faster rate in the middle and upper levels of the hierarchy than in the lower levels, then some sons will have the chance to climb into the new positions without displacing anybody. In this instance, mobility is a one-way street, and everybody gains. (Of course, occupational change could also reduce opportunities and force some people down.)
3. *Differential reproduction.* If men in upper levels of the system have fewer children than those in lower levels, then the system will reproduce itself in an asymmetric way that allows some sons from the bottom to climb into higher positions without forcing anybody to decline. This is another one-way street.
4. *Immigration.* If many immigrants enter the system at the lower levels, they make it possible for the system as a whole to grow in ways that give native-born sons more chance to move into higher positions than they would otherwise have.

Complexities of data and measurement make it impossible to give precise answers on how much mobility at any given moment comes from each of the four causes, but some partial estimates can be made and some rough trend lines drawn (Duncan 1966). One of the most obstinate complexities could be called the *multiplier effect:* If a new job is created near the top of the hierarchy, it is likely to be filled by someone from the middle, and that person's job in turn is taken by someone from below. Thus, one new opening can create two or more moves in a step-by-step progression. Another problem is that many men change jobs during their careers; intergenerational and intragenerational mobility are happening simultaneously, adding to the puzzle.

Of course, a man who has advanced in the world as compared with his father has no idea which of the causes made it possible for him to get ahead, and he probably would not care if he were told. Thus, analyses of these separate factors cannot be used in a direct way to explain class consciousness or other subjective phenomena. But a student who looks at the system as a whole to measure the amount of mobility that exists wants to understand the main causes of that mobility and current tendencies among them.

TABLE 6-4   Male Occupational Distributions, 1940 and 1990

	In Millions		
	1940	1990	Percent Change
Total	39.2	64.4	+64
Professional/technical	2.3	9.7	+322
Managers/proprietors	3.4	8.9	+162
Sales/clerical	4.8	11.0	+129
Craftsmen/foremen	6.1	12.5	+105
Operatives	7.1	9.2	+30
Service workers	2.4	6.3	+163
Laborers, except farm	4.7	4.0	−15
Farm occupations	8.5	2.1	−75

NOTE: Because of modifications in Census Bureau occupational categories and the shift from 14 to 16 as the minimum age for which occupational data are typically tabulated, these 1940 and 1990 distributions are not strictly compatible (see Table 3-4, especially 1972 tabulations). But they are an appropriate basis for the rough comparisons made here.

SOURCES: U.S. Census 1975; U.S. Labor 1991.

Let us begin with occupational change. We have already discussed the trends in Chapter 3, where it was shown that some occupational categories have been expanding faster than others. For example, in the decades from 1940 to 1990, professional and technical jobs for men increased more than 300 percent, whereas farm jobs decreased by 75 percent. The average growth for the whole male labor force was 64 percent, as shown in Table 6-4. (Remember that in earlier discussions, we dealt with the full labor force, including women, but at the moment, we are focusing on fathers and sons.)

Concentrating on the nonmanual jobs, most of which put men into the middle and upper classes of society, we note that the categories of professional and technical, managers and proprietors, and clerical and sales expanded in absolute numbers from about 10.5 to 29.6 million positions in three decades, or a growth of 182 percent. If those positions had expanded at the same rate as the average for the labor force as a whole, there would have been only 17.1 million men in those jobs in 1990. The difference of the actual over the "expected" growth was 12.5 million positions: that many men had the chance to move up in the system to fill the newly created jobs. Those men constituted 19 percent of the male labor force in 1990. And this estimate is an absolute minimum because we have no way of adding those men who moved up as a result of the multiplier effect described earlier. Also, we are not counting here the sizable expansion in jobs at the craftsman and foreman level, which is often called the "aristocracy" of manual labor.

Similar calculations for the three decades preceding 1940 show that the average growth for the male labor force was 31 percent, but the growth of the nonmanual jobs was 73 percent. The relatively faster growth of the better jobs allowed at least 9 percent of the labor force to experience upward mobility, a somewhat smaller percentage than occurred in more recent decades. Consequently, the rate of mobility caused by occupational or structural change appears to have sped up during the middle years of the century.

We have only limited information about the relative reproduction rates of men at different levels of the occupational hierarchy. But we do know that before World War II, the families of white-collar workers were smaller than those of blue-collar workers, especially farmers. To take a specific example to illustrate the difference, it was estimated that 1,000 professional men had 870 sons, whereas 1,000 farmers had 1,520 sons (Kahl 1957: 257). Thus, the sons of professionals were too few in number to take over their fathers' occupational positions, to say nothing of filling the expanding number of professional jobs in the system. It was calculated that if fathers at all status levels had generated families of equal size, the mobility of blue-collar sons would have been reduced by some 7 percent in the decades before the war. Since then the differences in reproduction rates among the strata have declined markedly, thus reducing the mobility opportunities arising from this particular cause.

Until recent decades, the natural growth of the American population produced by an excess of births over deaths was not sufficient to meet the needs for labor in an expanding economy. In the eighteenth and nineteenth centuries, we had a wilderness to conquer, and millions of farmers were enticed or forced to come from Europe and Africa to clear and cultivate the land. Around the beginning of the twentieth century, the farm population began to stabilize, and the new demand for workers was in the mines and urban factories. "In 1907, the peak year of the pre–World War I immigrant inflow, one-third of the nearly one million immigrants who reported an occupation on arrival were classified as farm laborers and nearly three-fourths of the total were classified as either farm or nonfarm laborers, as domestics, or as other service workers" (Wool 1976: 63). But most of those who arrived as farmers took jobs in the cities. In other words, the expanding needs for unskilled and semiskilled labor were still being partly met by immigrants. Their presence was essential for economic growth and permitted native-born citizens (including the sons of foreign-born parents) to climb into the ranks of skilled manual and nonmanual workers.

The First World War was a major turning point. Immigration was sharply curtailed by the conflict and by subsequent restrictive legislation. But internal migrants kept the stream of unskilled labor flowing to the cities. As machinery replaced farm labor, the sons of farmers moved to the city seeking work. In 1910, there were 10.4 million male farm owners and laborers; in 1940, there were 8.5 million, and by 1970, only 2.2 million. These migrants were better prepared for city life than the average immigrant: They spoke English, most were literate, and some were well educated. But the majority of them moved (at least at first) into the lower levels of skill in the city labor force. This was particularly true of the blacks who left the farms of the South (where 90 percent had lived) and moved to the cities of both North and South.

Today there is little migration from the farm to the city because there are so few farmers. Immigration, especially from Latin America and Asia, has expanded in recent years. But even counting illegal immigrants, the flow is modest relative to the national population and certainly not comparable in its impact to the massive immigration in the years before the First World War. It is doubtful that the new immigrants are significantly affecting the national pattern of social mobility, though their concentrated presence may undermine the job prospects of low-skilled workers in certain areas of the country (Higham 1963; U.S. Labor 1978; Bagby 1997: 36–37).

TRENDS IN SOCIAL MOBILITY

In earlier chapters, we examined trends in income, wealth, and occupational structure. What can we say about trends in social mobility? Is there more or less upward mobility than there was in the past? How has the pattern of mobility changed?

We have two landmark studies of mobility: *The American Occupational Structure* (1967) by Peter Blau and Otis Dudley Duncan, and *Opportunity and Change* (1978) by David Featherman and Robert Hauser. These books present sophisticated analyses of social mobility, based on massive national surveys conducted in 1962 and 1973. In the next chapter, we will look at their conclusions about the determinants of individual mobility, but here we will focus on their measurement of the overall pattern of mobility. Table 6-5 presents the gross mobility results from these two surveys, together with Hauser's estimates for the 1970s and the 1980s compiled for this book. Unfortunately, data from the first two are not perfectly compatible with the data from the GSS polls used by Hauser. One problem is that Hauser's estimates refer to men aged 25 to 64 at the time of the survey, whereas the earlier surveys cover men 21 to 64. We will have to be cautious about any direct comparisons between the first two and the second two. Luckily the two series overlap, so we can get some notion of general trends

For each period, the bottom row of numbers *(Total)* gives us the occupational distribution of all sons. These figures, followed from period to period, provide a way of tracing structural mobility. They indicate—as we might expect—that more sons are finding upper-white-collar jobs, and fewer are working on the farm. This shift in occupational structure creates what we might describe as a mobility updraft, drawing people into the higher positions. On the other hand, this phenomenon is playing itself out as the proportion of farm fathers declines and the proportion of fathers in the top stratum increases. By the 1980s, more respondents' fathers were upper-white collar workers than farmers (The rightmost column gives the distribution of respondents' fathers).

TABLE 6-5 Outflow from Father's Occupation to Son's Occupation, 1962–1990

	Son's Occupation (in percent)						
Father's Occupation	Upper White-Collar	Lower White-Collar	Upper Manual	Lower Manual	Farm	Total	Row %
1962							
Upper white-collar	54	17	13	15	1	100	17
Lower white-collar	46	20	14	18	2	100	8
Upper manual	28	13	28	30	1	100	19
Lower manual	20	12	22	44	2	100	27
Farm	16	7	19	36	22	100	29
Total	28	12	20	32	8	100	100
1973							
Upper white-collar	52	16	14	17	1	100	18
Lower white-collar	42	20	15	22	1	100	9
Upper manual	30	13	27	29	1	100	21
Lower manual	23	12	24	40	1	100	29
Farm	18	8	23	36	15	100	23
Total	30	13	22	31	4	100	100
1972–1980							
Upper white-collar	65	11	11	12	0	100	14
Lower white-collar	45	22	15	18	0	100	12
Upper manual	29	11	33	26	1	100	24
Lower manual	25	11	23	40	1	100	28
Farm	16	9	24	34	18	100	23
Total	32	12	23	27	5	100	100
1982–1990							
Upper white-collar	60	14	11	14	1	100	21
Lower white-collar	39	18	19	23	1	100	12
Upper manual	32	10	33	24	2	100	24
Lower manual	27	11	25	37	1	100	27
Farm	19	10	22	33	17	100	15
Total	36	12	23	27	3	100	100

	Up	Stable	Down	Total
1962	49	34	17	100
1973	49	32	19	100
1972–1980	50	35	16	100
1982–1990	44	36	20	100

	Blue-Collar to White-Collar (as % of Sons with Farm or Manual Fathers)	White-Collar to Blue-Collar (as % of Sons with White-Collar Fathers)
1962	31	31
1973	35	34
1972–1980	34	28
1982–1990	37	32

NOTE: See Table 6-1 for definitions of occupational categories.

SOURCES: Featherman and Hauser 1978 and GSS (Davis and Smith 1990).

Judging from the summary statistics at the bottom of the table, the overall pattern of mobility has not changed dramatically over the last three decades. The differences between 1962 and 1973 are minor. The most notable change seems to be between the 1970s and 1980s, when upward mobility dropped off, and downward mobility correspondingly increased. Despite the decline in upward mobility, sons were still much more likely to move up than down relative to their fathers.

Since the biggest differences are the most recent, we should take a closer look at the last two periods. One way to do this is to concentrate on workers in the first half of their careers. These younger workers are the most likely to feel the effects of change. Focusing on them reduces the birth-cohort overlap that dilutes the contrast between successive surveys. (For example, the cohort of men who were 45 to 54 in 1980 would be 55 to 64 in 1990 and therefore be represented in surveys both years.)

Using the same GSS data he employed for the general mobility analysis, Hauser looked at the experience of men, 25 to 44, in the 1970s and 1980s. The results are reproduced in Table 6-6. Two differences are immediately apparent. First, in the 1980s, upward mobility was less common for young men, and downward mobility was more common. Second, young men found it harder to move into the top occupational category. Compare the upper-white-collar columns for the two periods. In virtually every case, a smaller proportion of sons reached this level. Even the sons of upper-white-collar fathers had a tougher time retaining their fathers' occupational status. The sons of men in both white-collar categories were more likely to be bounced down into lower positions than they were in the 1970s.

What about the mobility of younger workers across the blue-collar/white-collar line? In the 1980s, the proportion of white-collar jobs increased slightly. Nonetheless, upward mobility across the line *decreased*, and downward mobility *increased*.

The 1980s drop in overall mobility was, in good measure, the result of the increasing status of fathers. From the 1970s to the 1980s, the proportion of fathers in the top two categories rose, and the proportion in the bottom three fell—thereby reducing the possibilities for upward movement by sons.[2] This brings us back to the question of structural mobility. How has the changing occupational structure affected the overall level of intergenerational mobility?

In the last section, we looked at this question using national occupational statistics going back to the early twentieth century. In Table 6-7, we chart

---

[2] If, hypothetically, the sons of the 1980s had the fathers of the 1970s (that is, had the same social origins as workers in the 1970s), overall mobility in the second decade would have looked like this: 45.0% up; 34.2% stable; and 20.7% down. These figures were calculated by taking the numbers of men in each category of occupational origin in the 1970s and assigning them destinations in the proportions recorded in the 1980s.

TABLE 6-6 Mobility of Younger Workers (25–44): Outflow, 1972–1990

| Father's Occupation | Son's Occupation (in percent) | | | | | | |
	Upper White-Collar	Lower White-Collar	Upper Manual	Lower Manual	Farm	Total	Row %
1972–1980							
Upper white-collar	61	10	15	14	0	100	24
Lower white-collar	47	26	12	15	1	100	7
Upper manual	32	10	32	26	1	100	25
Lower manual	28	9	23	39	1	100	28
Farm	15	8	27	33	17	100	16
Total (N = 1,780)	36	11	23	27	3	100	100
1982–1990							
Upper white-collar	54	11	15	19	1	100	31
Lower white-collar	40	25	13	21	1	100	9
Upper manual	30	9	34	26	2	100	24
Lower manual	26	11	26	37	1	100	26
Farm	16	9	22	33	20	100	11
Total (N = 2,282)	36	12	23	27	3	100	100

	Up	Stable	Down	Total
1972–1980	47	36	18	100
1982–1990	38	37	24	100

	Blue-Collar to White-Collar (As % of Sons with Farm or Manual Fathers)	White-Collar to Blue-Collar (As % of Sons with White-Collar Fathers)
1972–1980	45	19
1982–1990	40	29

NOTE: See Table 6-1 for source and definitions.

an index of structural mobility based on the difference between the occupational distributions of sons and fathers. The index is the minimum percentage of sons who have moved because of changes in occupational structure between generations.[3] The fall of the index from 1962 to the 1980s suggests that structural mobility has been declining, especially in recent years. The trend is clearest among younger workers. We can conclude that shrinking structural opportunities for mobility were contributing to the decline in overall mobility during the 1980s.

[3] The index is calculated by subtracting the percentage of fathers from the percentage of sons for each occupational category in which the latter exceeds the former. The sum of these differences is the value of the index.

TABLE 6-7   Index of Structural Mobility, 1962–1990 (in percent)

	All Sons	Sons 25–44
1962	22	—
1973	19	—
1972–1980	19	18
1982–1990	14	10

SOURCE: See Table 6-5.

TABLE 6-8   Mobility of African-American Men: Outflow, 1962 and 1973

	Son's Occupation (in percent)						
Father's Occupation	Upper White-Collar	Lower White-Collar	Upper Manual	Lower Manual	Farm	Total	Row %
1962							
Upper white-collar	13	10	14	63	0	100	4
Lower white-collar	8	14	14	64	0	100	3
Upper manual	8	11	11	67	3	100	9
Lower manual	7	9	11	71	2	100	37
Farm	1	5	7	66	20	100	48
Total (N = 1,122)	5	8	9	68	11	100	100
1973							
Upper white-collar	44	12	8	36	0	100	4
Lower white-collar	20	21	13	46	1	100	4
Upper manual	16	14	16	54	0	100	10
Lower manual	12	12	14	61	1	100	46
Farm	5	7	17	63	8	100	36
Total (N = 3,493)	12	11	15	59	4	100	100

SOURCE: Featherman and Hauser 1978: 326.

TRENDS IN BLACK MOBILITY

The 1962 and 1973 mobility surveys we have been examining were large enough to include a significant sample of African-American men. The general results reveal a remarkable change over a relatively short period (Table 6-8). Upward mobility increased, especially mobility from blue-collar to white-collar positions. Slippage from white-collar to blue-collar became much less common. Data broken down by age show that even larger gains were made by men 45 or younger (Featherman and Hauser 1978: 326).

In the 1973 survey, fathers in higher occupational categories were much more likely to pass their advantage on to their sons. In 1962, for example, only 13 percent of the sons of upper-white-collar men had similar occupations. By 1973, 43 percent did. Nonetheless, the mobility chances of black men still lagged behind those of whites, even when they started at similar levels. For example, in 1973, at all five levels of occupational origin, white men had a better chance than African-American men of attaining upper-white-collar status (compare the right-hand columns for 1973 in Tables 6-5 and 6-8).

Unfortunately, no mobility data are available for African-Americans in the 1980s. It would be especially interesting to see how blacks who entered the labor force after the civil rights revolution of the 1960s have fared. However, the occupational data that we examined in Chapter 3 suggest continuing improvement in mobility prospects for African-Americans.

CONCLUSION

Four basic factors described in this chapter influence the amount of social mobility:

1. *Circulation.* Some sons slip, creating opportunities for others to rise.
2. *Occupational change.* As the economy evolves, the occupational structure is altered, opening and closing opportunities.
3. *Differential reproduction.* Class differentials in fertility can create room at the top.
4. *Immigration.* New immigrants take bottom positions as the native born move up.

Mobility data from the 1980s reveal a high level of father-to-son succession. Nonetheless, about two-thirds of sons have moved up or down from their fathers' occupational level, and upward mobility is far more common than downward mobility. Daughters' mobility cannot be described in quite the same terms as sons', but similar data show that father's occupation is a strong influence on daughter's occupational destination. We do not have recent data on the mobility of African-American men. But we do know that in the early 1970s, they still had slimmer mobility chances than white men from similar occupational backgrounds. Nonetheless, their situation had improved from what it was in the 1960s.

Over the last generation, the overall pattern of mobility has been relatively stable. Upward mobility, for example, has consistently outweighed downward mobility. Succession to father's status is little different now than it was in 1962. Upward mobility has declined, but only slightly. Movement either way across the blue-collar/white-collar divide has remained about one-third of the workers in the surveys.

But when we sharpen our focus and look at the experience of younger men in recent years, we find evidence of stagnating opportunities. Significantly more workers are downwardly mobile, fewer are moving from blue-collar backgrounds to white-collar careers, and it is becoming harder to move into the upper-white-collar stratum. Furthermore, it is evident that the structural change that has fueled mobility in the past (#2 in our list) is waning. These findings regarding the experience of younger workers are our best window on the future. What they suggest is not encouraging.

These trends could have powerful political implications. Already many young workers fear that they will not have the same opportunities for advancement that their parents enjoyed and that they might not even be able to maintain the living standard that they grew up with. Especially in periods of economic stagnation, such feelings could turn them against politicians they identify with the status quo.

SUGGESTED READINGS

Erikson, Robert, and John Goldthorpe 1992. *The Constant Flux: A study of Class Mobility in Industrial Societies.* New York: Oxford University Press.

> *Comparative mobility study covering the era of prosperity following World War II. American pattern seems quite similar to European.*

Featherman, David, and Robert Hauser. 1978. *Opportunity and Change.* New York: Academic.

> *Detailed report on the last major national mobility survey (1973). Comparisons with the 1962.*

Hout, Michael. 1988. "More Universalism, Less Structural Mobility: The American Occupational Structure in the 1980s." *American Journal of Sociology* 93: 1358–1400.

> *Trends in mobility of men and women from 1972 to 1985. Hout is writing an updated version of this article.*

Newman, Katherine S. 1988. *Falling From Grace: The Experience of Downward Mobility in the American Middle Class.* New York: Free Press.

> *How ex-managers, laid-off skilled workers, unemployed professionals, divorced women, and their families respond to their fall from economic grace.*

7

Family, Education, and Career

The transmission of property from generation to generation, in the same name, raised up a distinct set of families, who, being privileged by law in the perpetuation of their wealth, were thus formed into a Patrician order, distinguishable by the splendor and luxury of their establishments. From this order, too, the king habitually selected his counselors of State; the hope of which distinction devoted the whole corps to the interests and will of the crown. To annul this privilege, and instead of an aristocracy of wealth, *of more harm and danger than benefit to society, to make an opening for the* aristocracy of virtue and talent, *which nature has wisely provided for the direction of the interests of society, and scattered with equal hand through all its conditions, was deemed essential to a well-ordered republic.*

Thomas Jefferson (1821: 38)

In the quotation that begins this chapter, Thomas Jefferson hopes to replace the corrupt old "aristocracy of wealth" with a new "aristocracy of
virtue and talent." The aristocracy of wealth, Jefferson thought, could be
eliminated by changing the laws of inheritance so that men would divide
their landed estates equally among all their children and thus eventually arrive at small holdings. The aristocracy of virtue and talent—characteristics
Jefferson assumed were "scattered with equal hand" among all classes—
could be cultivated by creating a public school system that would provide
primary education for all citizens and then select the best students for further training in high schools and universities. In his later years, Jefferson
founded the University of Virginia as the capstone to this system and was
so proud of it that he ordered that his tombstone should record his two
greatest accomplishments: the writing of the Declaration of Independence
and the founding of the University of Virginia.

From Thomas Jefferson forward, American political leaders have en-
dorsed high rates of social mobility. In doing so, they have reflected the general values held by most Americans, which accept some inequality in
society, believing that people should get different rewards for different
kinds of work, but also believing that each generation should start fresh and
compete in a "fair" way for those rewards. In other words, we believe that
young people ought to have careers based on their own talents and desires
rather than have their lives determined by the class positions of their
parents; they should have "equality of opportunity" but not "equality of
result." Accepting the fact that some aspects of the talent and desires of the
children are inherited from or shaped by the parents, we nevertheless believe that a high degree of equality of opportunity can be achieved through
the school system. If all children have access to good schools at all levels, *regardless* of the financial resources of their families, then the graduates
should be able to compete on a reasonably fair basis.

The practical means for achieving this end began in our early history
with free elementary schools in every village and town, supported at first
by churches (more interested in equal access to the Bible than to economic
success) but later by local tax money. Later, free high schools were added,
and, in most small towns, they are now the most prominent buildings to be
seen. In the mid-nineteenth century, free or inexpensive state universities
were founded with federal aid (Jefferson was a generation ahead of his
time). In the late twentieth century, educational opportunity was expanded
through programs ranging from Head Start classes for poor youngsters before first grade to college scholarships and other forms of affirmative action
for bright but economically disadvantaged students.

Although public expenditures on education have grown steadily over
the years, Americans have begun to lose the faith they inherited from
Jefferson in the socially redemptive power of education. In particular, they
are increasingly skeptical of the proposition that schooling can overcome the disadvantages (or advantages) of family background in the lives
of individuals.

The connections among family background, education, and career success, often the focus of heated debate, will be our main concern in this chapter. Here we shift away from the systemic or structural perspective of the last chapter to highlight the experience of individuals. Instead of asking how much mobility exists in the system as a whole, we will ask why some individuals are successful and others are not. Obviously these are closely related topics. An individual's chances for success will depend on both the available opportunities and his or her personal characteristics, including family background and educational achievement.

In what follows, we will often be interested in establishing the strength of the connections between variables. We will use two yardsticks for this purpose: *simple correlation* and *variance explained*. Both tell us how accurately we can predict the value of one variable by knowing the values of preceding variables. Correlation is measured by a coefficient that ranges from 0.0 (there is no relationship between two variables) to 1.00 (we can predict the second variable by knowing the first with perfect accuracy). In the social sciences most correlation coefficients are intermediate. For example, the correlation between father's education and son's education is around 0.45. Variance explained, which is closely related to correlation, is typically invoked when we are trying to measure the influence of a series of prior variables on a single dependent variable. It is expressed as a percentage of total variance in the dependent variable, with higher percentages expressing stronger relationships. We might for example, determine that father's occupation and education together "explain" 30 percent of the variance in son's occupational success—meaning that the differences in occupation and education recorded among fathers statistically account for 30 percent of the differences in occupational level observed among sons.

BLAU AND DUNCAN: ANALYZING MOBILITY MODELS

What makes the study of the social mobility of individuals complicated and debatable is this: Everything is related to everything else. We know, for example, that a son's occupational attainment is significantly correlated with his father's occupational status and his own education (both approximately 0.50). These relationships are depicted in the first diagram in Figure 7-1 But, as the second diagram illustrates, the father's occupational status is also correlated with the son's education. How are we to interpret this? One possibility, portrayed in the third diagram, is a simple causal chain. Concretely, sons with high status fathers have the opportunity to get a good education and, as a result, get good jobs; sons with low status fathers cannot get a good education and end up with crummy jobs. The effect of father's socioeconomic status on the son's career is mediated through education.

But things might be more complicated—they often are. Imagine that a son with a high status father completes his education and then gets an additional career boost— the old man hires his son or finds him a good job

FIGURE 7-1 Causal Relationships in Occupational Attainment

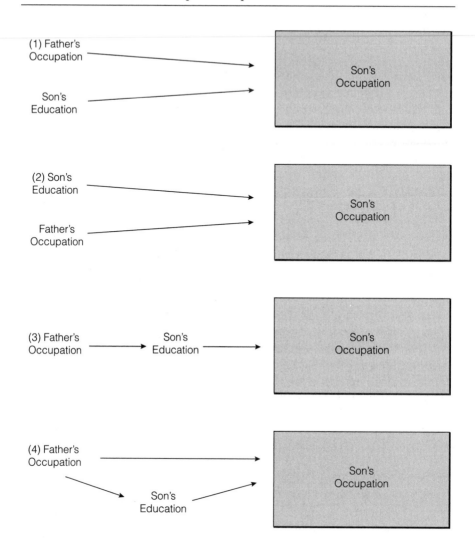

through a family connection. The fourth diagram adds another arrow to represent this possibility. But the additional arrow opens a new mystery. Which is more important, education or easy access to the job? Perhaps education did not really count for much and was just the inevitable result of a privileged background like, say, driving an expensive car; in the end, what really mattered was the job opportunity. Of course, the opposite is also possible or some unequal combination of the two influences.

Things would get even more complicated if we considered additional factors, for example, race and parents' education. A black child might get less encouragement in school than white schoolmates, end up quitting school early, only to face discrimination in the job market. This outcome would be less likely if the child's parents were well educated themselves. Of course, well-educated parents would be likely to have high status jobs, an additional career advantage for the son or daughter. If we added new arrows to the diagram to represent these and other interrelated influences, we would soon have a tangle of causal pathways resembling a New York subway map—a complicated image for the complex realities of social mobility.

In 1967, Peter Blau and Otis Dudley Duncan introduced an innovative methodology for analyzing the complexities of career success and failure. Their book, *The American Occupational Structure*, which analyzed the results of a large-scale national mobility survey, inspired much of the subsequent research in the field. Blau and Duncan depicted career development in formal "path diagrams"—more elaborate versions of our drawings—that became the basis for their statistical analysis of the survey data. Their goal was to sort out the influence of each variable *independent* of the others. Path analysis, as this method is called, helped Blau and Duncan conceptualize two key aspects of the problem: *chains of causation* (father's occupation influences son's education, which in turn, shapes his career prospects) and *multiple causal pathways* (father's occupation influences son's education and later directly influences his job search).[1] Note that path analysis emphasizes the temporal element in social mobility. Blau and Duncan's analysis dealt sequentially with family background, education, son's first job, and job at the time of the survey, examining the influences operating at each stage.

Figure 7-2 summarizes a later national mobility study that used the methods introduced by Blau and Duncan (Featherman 1979; Featherman and Hauser 1978). The study is based on representative sample of adult men, who answered questions about their family background, education, and career experience. Each respondent's occupation at the time of the survey was rated on a scale of *socioeconomic status* (SES) developed by Duncan and akin to the occupational rankings we discussed in Chapter 2. The analysis sorts out the determinants of occupational attainment as measured by SES.

Our very simplified path diagram and the related pie graph emphasize the study's the main conclusions. The diagram indicates that both family background and education, as expected, contribute to son's socioeconomic status. It shows *direct* and *indirect* (through education) connections between family background and son's SES. The pie graph parcels out responsibility

---

[1] All our examples here are sons because the mobility surveys we are discussing did not cover women.

FIGURE 7-2 Factors in Social Mobility

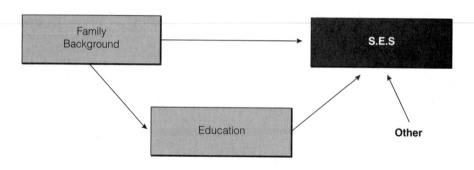

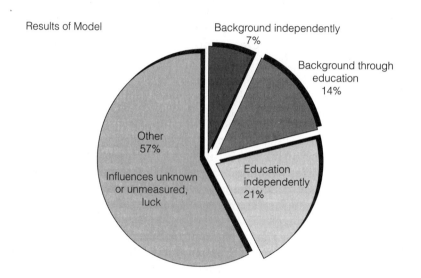

NOTE: Family background includes father's S.E.S., father's education, broken home, and race.

for the final result (son's SES) among the factors involved. Specifically, it reports the proportion of total variance explained by each factor—that is, the degree to which the differences in SES scores among the sons in the study are statistically attributable to the various causal elements in the path model.

What do we learn here? (1) The main effect of family background is its influence on education, which, in turn, contributes to career success (14 percent of total variance). In other words, rich kids generally get good educations, poor kids normally do not. (2) Family background also has a smaller, independent effect on SES (7 percent). This refers, for example, to the father who hires his own child or gets him into the union, as well as the negative effects of racial discrimination. (3) Education has a substantial independent effect (21 percent). This tells us that educational achievement does more than just reflect the privileges or disadvantages of family background. Some rich kids flunk out. Some not-so-rich kids graduate from college. In both cases, they defy the odds.

The model suggests that family background and education (independently) are equal in their influence on career success. Together they account for 42 percent of total variance. This leaves 58 percent of the variance, corresponding to "other" in the diagram, "unexplained." As we will see later in this chapter, there are more elaborate models that reduce the amount of unexplained variance, but significant unexplained variance always remains. There are two basic reasons for unexplained variance. One is that we cannot always measure our variables as precisely as we would like to. For example, the father's or son's occupation may be better or worse than it sounds from the rough occupational titles we use. The other is that there are many influences that we do not, often cannot, include in the model. For example, some individuals are more ambitious and able than their academic accomplishments suggest. Personal charm often contributes to career success, but it is not easy to measure.[2]

Some of the unexplained variance in mobility models is certainly the result of nothing more than dumb luck. Someone starts a career at just the right moment and gets ahead fast. Someone else experiences a traumatic event and never quite gets over it. This is all very frustrating for sociology, but from a more romantic viewpoint, its nice to know that life is more than a little unpredictable.

JENCKS ON EQUALITY

Blau and Duncan inspired a large new literature on "status attainment," including two important books by Christopher Jencks and a research team at Harvard University. In the first book, Jencks attempted to integrate most of what was known into one grand model that would include more variables than Blau and Duncan had studied. The book was titled *Inequality: A Reassessment of the Effect of Family and Schooling in America.* (1972). It created quite a stir.

Jencks did not collect original data but extended Blau and Duncan's results. He interpolated a lot of material from other studies and organized it into a series of discrete stages corresponding to points in the life cycle, instead

---

[2] Unfortunately, many of the variables we might add turn out to be largely "redundant" because they overlap with variables already included. For example, this model does not include mother's education, but mother's education is likely to be highly correlated with father's education and occupation, which are included. Therefore, this additional variable would not contribute much to explained variance.

of relying on a single path diagram that summed up the whole process. Jencks also added income as the final dependent variable in the causal chain, after SES. He found that there was only a modest correlation between occupation (SES) and income. This may seem surprising, but as we noted earlier, the occupational titles we use are fairly crude measures of what people do at work. Moreover, pay scales vary around the country, workers generally get higher pay as they accumulate seniority on the job, and unionization or the economic strength of a firm can have big effects on earnings.

As the Harvard researchers worked backwards in the causal chain they found ever weaker connections with income. A son's income correlated 0.44 with his SES, 0.35 with his education, and 0.29 with his father's SES. Every time we add another variable to a chain of causes and effects, we weaken the connection between the earliest predictors and the final outcome, because at each step in the process, other factors are at work (or "luck" keeps reappearing). Consequently, this new model, which uses income as the final dependent variable instead of occupational prestige, reduced the measured impact of education on life chances

The Harvard Team also wanted to measure somewhat more sharply the "pure" effect of years of schooling, independent of the qualities that students bring with them to the school building. To do so, they added another variable to the model: the son's IQ at age 11, on the assumption that it is the best available measure of the talent or "cognitive ability" of the child that reacts with the stimuli of the school.

In organizing the vast statistical material from many different studies that is integrated in *Inequality*, Jencks tended to stress a polemic point. Economic (occupational and income) inequality in the United States is very large, and simply improving the quality of bad schools and reducing the differences among individuals in the number of years they attend school will not go very far in eliminating the economic differences. The reason is that education is not a strong determinant of income. In fact, Jencks estimated that barely a quarter of the variance in the incomes of adult men could be predicted by combining all the usual predictors: family background, IQ score, years of education, and even job title. In other words,

> Economic success seems to depend on varieties of luck and on-the-job competence that are only moderately related to family background, schooling, or scores on standardized tests . . . The fact that we cannot equalize luck or competence does *not* mean that economic inequality is inevitable. Still less does it imply that we cannot eliminate what has traditionally been defined as poverty. It only implies that we must tackle these problems in a different way. Instead of trying to reduce people's capacity to gain a competitive advantage on one another, we would have to change the rules of the game so as to reduce the rewards of competitive success and the costs of failure (Jencks et al. 1972: 8).

In other words, if we really propose to equalize incomes, we must do something that *directly* alters the operation of the labor market to reduce the range of incomes: Corporation presidents should receive less, and people who fry hamburgers should receive more. The amount of government intervention required to produce such a result is often call socialism, and Jencks was not afraid of adopting the word. He implied that our national preoccupation with changing the schools was a distraction from the real issue.

The reaction to Jencks' conclusions was strong; he was not faulted on technical grounds, but many critics were unhappy about the implications of the words *luck* and *socialism*. Jencks and colleagues set out on a second round of research, including the integration of several new surveys of careers (including those of brothers) that had since become available. The second book, *Who Gets Ahead?*, (1979), was much less controversial, for at least two reasons. First, Jencks gave up the noble attempt of the earlier volume to communicate with the general public and instead offered dry and statistical prose in a highly technical style. Second, he avoided the discussion of policy implications for the most part. Jencks no longer emphasized the difficulty of creating more equality of result; he simply reported on the amount of variance in schooling, job, and income that could be explained by different background characteristics.

These policy weaknesses of the book were offset by a prodigious amount of new work, using many new sources of data, that certainly increased the accuracy of the calculations, although it did not dramatically alter the conclusions. Let us report a few of the more significant analyses.

Who Gets Ahead? made a number of changes in procedure, but the most interesting for our purposes comes from a recognition that the standard measures of family background, such as parents' education and occupation, are crude estimates of the total influence of social origins. Although Jencks gave those measures, he also added, where possible, a measure based on the similarity of the careers of brothers. Jencks assembled some new data of his own as well as material available in other studies in order to increase knowledge about what happens to boys reared in the same home, although the difficulty of tracing both brothers keeps the samples much smaller than one would like. The resemblance between brothers is a total measure of family influence that is useful but not very specific: It covers shared genes, shared environment in the home and the neighborhood, and the possible influence of one brother on the other. It predicts career outcome somewhat better than conventional SES measures of family status, thus reducing the size of that bothersome residual variance.

Jencks and his colleagues also rounded up data on personal characteristics that went beyond the usual IQ scores, including measures of personality. They found that no one of the additional indicators was particularly powerful, but a combination of them predicted career outcome about as well as the IQ score by itself and seemed to be adding something new to the mix.

Let us start with the overall conclusion and then break it down into a series of sequences that follow one another. *The resemblance between brothers, or total family background, explained almost half the variance in the occupational statuses of men and a little less than a third of the variance in their incomes.* (If we could measure the lifetime earnings of men instead of using only one year as the dependent variable, the predictability would be even higher because temporary ups and downs would be averaged out.) Obviously, this result is much more powerful than Blau and Duncan's original model and suggests more inheritance of position than was previously believed.

The most important single indicator of total family background was the father's occupational status, but that accounted for only a third of the resemblance between brothers. Adding in a string of other demographic variables (education of both father and mother, income of both, family size, and race) adds another third. The remaining third of the variance "is presumably due to unmeasured social, psychological, or genetic factors that vary within demographic groups" (Jencks 1979: 214).

If the schools are to be equalizers, then talented children of poor families must stay in school as long as talented children from rich families. Otherwise, the schools are just mediators or the transmission belts used by privileged families to obtain privileged careers for their children. If we measure talent at a relatively early age and then follow the subsequent paths taken by the children, we can estimate how much of school attainment is a consequence of talent (as measured by the tests), regardless of the socioeconomic background of the students' families. Jencks and his colleagues reported that early cognitive test scores explain about 40 percent of the variance in ultimate educational attainment; about two-thirds of that is *independent* of the connection between the IQ scores themselves and family background. This implies that a lot of variation in talent is picked up on the tests, even among families of similar social status (indeed, even among brothers), and the variation in talent has a significant amount of influence on educational attainment. However, test scores have far less effect on educational attainment than a completely meritocratic system would require, and they have even less effect, further out the causal chain, on earnings, predicting only about 5 to 10 percent of their variance (all else being equal). The schools are already more "fair" than the labor market, according to these particular measures.

In a general sense, everybody knows that staying in school pays off— that is, people with more education have higher-status jobs and earn more money. But *how much* difference does it make? And is the difference linear, so that an extra year of high school is worth about the same as an extra year in college? Indeed, is the last year in college, which provides the coveted bachelor's degree, worth something extra?

Studies done after Jencks's first book was written indicate a slightly higher correlation between education and earnings than he first reported, probably more because of methodological improvements than to actual trends in society. The various surveys indicated an average correlation of about 0.40 between years of education and earnings in dollars, without controlling for other influences. The second book contained a chapter on education written by Michael Olneck (Jencks et al. 1979: chapter 6), which shows that the relationship between years and dollars is not linear: Years in college are worth more than years in high school, particularly if one stays long enough to pick up a degree. Four years of high school bring an increase in dollar earnings over elementary school graduates of about 40 percent; four years of college bring an increase of almost 50 percent over high school graduates. Thus, teachers appear to be correct when they tell students to stay in school because doing so will improve their careers.

But are they really so accurate? We know that family background affects both talent and the length of schooling; maybe the family influence is what counts, rather than the impact of the schooling by itself. Olneck estimated that if we control for both family background and IQ scores, the "pure" effect of a high school diploma is reduced by half, to about a 20 percent increase in earnings, and the effect of a college degree is reduced to an additional 35 percent in earnings. Notice that the reduction for the controls is greater for the high school years than for the college years.

Olneck commented,

> Our findings place a number of widespread presumptions in doubt. The most significant of these is that high school dropouts are economically disadvantaged because they fail to finish school. Our results suggest that the apparent advantages enjoyed by high school graduates derive to a significant extent from their prior characteristics, not from their schooling. Unless high school attendance is followed by a college education, its economic value appears quite modest . . . (Jencks, et al, 1979:189-190).

There appears to be a gradual reduction in the impact of background variables as one gets older. Thus, *once young men get into college,* they are on their own to a greater extent, and the payoff of higher education (especially for those who stay long enough to get the degree) is almost as great for those from poorer families and for those with less measured talent as it is for the others. In the earlier years, education is as much a reflector of family background as an equalizer that offsets family background, but in the later years, education has more independent influence.

These measured effects of education fit with our discussion in Chapter 3, which emphasized a rather sharp break in the occupational system, dividing men with college degrees from all others. Employers screen applicants through the simple device of academic credentials, and jobs at the managerial level, even for young people just starting to work who are being selected and trained for such jobs, are given to those with college degrees in hand. Over past decades, we have dramatically increased the level of education for all segments of the population: Almost everybody now finishes elementary school, and more than 80 percent finish high school. Among those who go no further, there is a large pool of workers, and the jobs they get and the earnings they receive depend more on family background, individual talent, and the oddities of work experience than on a few years more or less of schooling. But at the upper levels of the occupational system, the good jobs go primarily to those who have received a complete college education—about a quarter of young men and women. Within that select group, the further differential impact of family background on jobs attained or dollars earned is rather small: It is the degree that counts. Of course, background has a lot to do with the chance of getting the degree in the first place, so we reach a double conclusion: *College degrees both protect the privileges of people born into upper-status families and permit many from lower-status families to climb into the elite.*

WHO GOES TO COLLEGE?

Not all people with college degrees have outstanding careers, but few people (other than athletes and entertainers) now achieve important positions without them. The income gap between high school and college-educated workers is even larger than it was when Jencks was writing about education and men's careers. Thus, the question posed is obviously an important one: Who, exactly, goes to college?

William Sewell and his associates tackled this question in a massive study that followed the careers of 9,000 Wisconsin students who graduated from high school in 1957 (Sewell and Hauser 1975; Sewell and Shah 1977). They found that IQ, SES, and gender strongly and independently affected the graduates' chances of attending college. For example, a boy with below average intelligence might go to college if he came from a high status family. A boy from a low status family was unlikely to continue his education unless he had a very high IQ. Girls, however smart they were, had slim prospects of attending college.

Some twenty-five years later, an even larger, national study produced remarkably similar results. The only big change was the virtual disappearance of the gender gap in college enrollment. The new study, entitled "High School and Beyond" (HSB) was conducted in the 1980s by the U.S. Department of Education and the National Opinion Research Center at the University of Chicago. We analyzed part of the HSB data for this book. Because gender was no longer important, we focused on the joint influence of social class and mental ability.

The HSB researchers divided their national sample of high school seniors into four cognitive ability quartiles and four socioeconomic status quartiles. We have cross tabulated the two variables in Table 7-1, which records the proportion of students in each IQ-SES subgroup who were attending college two years after graduation from high school. The results at the extremes are predictable: Nearly all the high-ability graduates from high-status families, but very few low-ability graduates from low-status families went to college (83 versus 13 percent). But what of the privileged kids with dull minds or the very smart kids from low-status families? Fifty-seven percent of the high SES grads with lower than average ability (second quartile) went to college, compared with 51 percent of the high-ability, low-SES students. Thus, the HSB data confirm the Wisconsin study findings that mental ability and social class are strong, independent determinants of educational attainment.

TABLE 7-1   College Attendance by Social Class and Cognitive Ability

| | Percent in College | | | |
| | Family Socioeconomic Status Quartile | | | |
Cognitive Ability Quartile	Lowest	2nd	3rd	Top
Lowest	13	13	20	35
Second	23	24	40	57
Third	33	42	51	69
Top	51	63	74	83
Total (N = 11,995)	24	33	48	69

NOTE: Proportion of 1980 HS seniors attending college in 1982. Cognitive ability quartiles based on average of standardized scores on reading, math, and vocabulary tests.

SOURCE: Author's analysis of HSB data.

Class disparities in education have proven quite resilient, despite efforts dating back to Jefferson's time to eliminate them. Figure 7-3 tracks college attendance rates of high school graduates younger than 25 by income quartile. The chart reveals that the disparities shrank in the 1970s but grew again in the 1980s. As a result, the class gap in college attendance was actually wider at the end of the 1980s than it had been in 1970 (Mortenson and Wu 1990).

Despite persistent class differences, all the data we have examined in this section show that many young people from lower class backgrounds, especially those with high mental ability, do go to college. In the HSB study, a high-ability, low-SES graduate had a fifty-fifty chance of attending college. Can we go beyond the statistics and explain why some bright but under-privileged youths went to college, and others did not? What additional factors of motivation can we understand that shape individual choices in a group that seems to have so much leeway for personal decision?

The Wisconsin survey and other studies confirm that *within* any given status level, some families are more ambitious for their children than others (or, to be more precise, that the children perceive such differences and can report them on a questionnaire). The ambition is transmitted to the child through encouragement to succeed in school as a means of getting a better job later on. Children with such encouragement are likely to do a little better than others, particularly if they are talented; the approval and encouragement they receive from teachers and the higher grades they get in school are rewarding and begin to shape self-perceptions as persons who are "college material." That process in turn helps the motivated students to get into college preparatory courses, and such participation in turn increases the chance of making friends with other college-bound students. These encouraging factors of parental ambition, school performance, and influence of peers and teachers all lead to the formulation of college aspirations and

FIGURE 7-3 College Participation Rates of Young Adults by Income Quartiles, 1970–1989

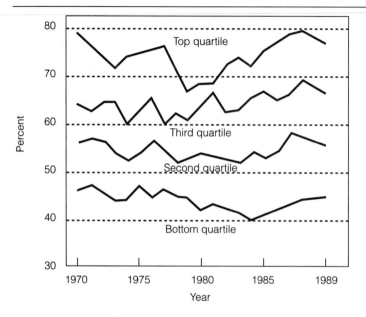

NOTE: Refers to unmarried high school graduates, ages 18–24, currently attending college or previously enrolled for at least one year.

SOURCE: Mortenson and Wu 1990: vii. Based on U.S. Census, *Current Population Survey,* Series P-20.

plans. For upper-status students, these pressures toward college are almost automatic, and only those with severe handicaps in intelligence or personality fail to succumb to them. Thus, for upper-status students, the factors described are the "mediating" variables that routinely translate the abstraction of socioeconomic status into particular realities that stimulate and guide behavior. But within lower-status groups, there is enough openness of choice to allow some students, particularly those of high talent with ambitious parents, to aim toward college (Kahl 1953; Karabel and Halsey 1977; Sewell and Hauser 1975; Sewell and Shah 1977).

The two basic factors of class and IQ not only influence the chance of continuing education after high school; they also influence the type of school chosen and the chance of graduating. A substantial proportion of college students start at two-year community colleges. These schools' curricula usually permit some students to transfer later to four-year colleges, but many community college students will drop out of school before they complete two years and others concentrate on two-year technical courses that prepare them for positions that are somewhere in the middle of the occupational status range: dental assistants, automobile mechanics, computer programmers, and the like. Indeed, the whole system of higher education is stratified according to the quality of the education provided and the particular career preparation emphasized, and that hierarchy is paralleled by the stratification of students' families. Two-year community colleges tend to draw their students from the lower half of the income distribution. Private colleges and universities recruit disproportionate numbers of students from high-income families. The admissions profile of public four-year colleges and universities places them somewhere in between (Berg 1970; Karabel in Karabel and Halsey 1977).

These studies of college attendance and graduation can be used to give partial support to both of two opposing ideological positions: (1) The American system of higher education is so big and so open that it provides major opportunities for talented youths from poor families to prepare themselves for successful careers that raise them above the level of their parents, and (2) the American system of higher education is sufficiently stratified that its main function is to reproduce for each generation of children the status positions held by their parents. Does the American system of higher education promote mobility or reproduction? The answer appears to be both.

COLLEGE AND THE CAREERS OF WOMEN AND MINORITIES

In recent decades, ethnic and gender disparities in college attendance have narrowed more than class disparities. Figure 7-4 is what Thomas Mortenson (1991), a researcher with the American College Testing Program (ACT), calls an "equity scorecard." He compares people aged 25 to 29 in each group with a reference population of the same age: females with males; minority members with whites; and the bottom income quartile with the top quartile. Rather than focusing on college matriculation, as the HSB and Wisconsin researchers did, Mortenson tracks differences in college graduation rates. The "equity scores" he charts from 1940 to 1990 are the ratios between the degree attainment rates of each group and the corresponding reference population. A score of 100 means parity with the reference group in the percentage of people who have received college degrees. So, for example, in the early 1950s, young women were about half as likely as young men to hold college degrees; by the late 1980s, they had attained near parity. The bottom quartile was far behind the top quartile in 1970 and even further behind twenty years later.

The abrupt gyrations of the black and female trend lines for the 1940s correspond to the reduction of the male college population during World War II and the generous college financial aid provided to returning soldiers under the G.I. bill. Black veterans were quick to respond to this opportunity, which was not available to the next generation of black high school graduates.

In the 1960s and early 1970s, the black-to-white ratio oscillated around 40 percent. It then jumped to roughly 50 percent, where it has remained.

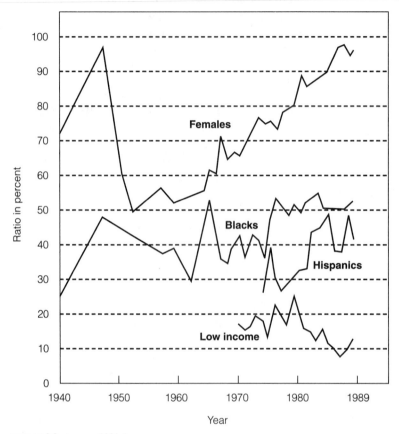

FIGURE 7-4 Equity Scorecard on Attainment of College Degrees: Disadvantaged
Populations Compared with Reference Groups, 1940–1989

SOURCE: Mortenson 1991: ix.

The data for Hispanics, though less complete, suggest a similar pattern,
starting 10 points lower. Given the lag time required to push up the
numbers of 25- to 29-year-old college graduates, these quantum leaps must
represent the college freshmen of the 1960s, when federal aid to higher education was increasing.

Even as valuable a credential as a college degree is not automatically
converted into career success. In general, education yields greater economic
returns to white males than to females or African-Americans. (See
Coverman 1988; Jaynes and Williams 1989: 299–301; Schiller 1989: 120–126;
Treiman and Roos 1983.) Both blacks and women face discrimination in
hiring and promotion. Women at all levels of education tend to be segregated into comparatively low-paying occupations. Women's earnings are

TABLE 7-2 Earnings Equity Ratios, 1995 (Full-Time, Year-Round Workers, by Education and Age, in Percent)

	Black/White		Female/Male
	Males	*Females*	*All*
All educational levels	74	89	72
High school	75	88	69
High school, 25–34	80	91	74
College	79	88	72
College, 25–34	83	90	77

NOTE: Based on median annual earnings of workers over 25, including wage, salary, and self-employment income.

SOURCE: Unpublished U.S. Census data.

further undercut by their tendency to move in and out of the labor force for family reasons. However, as we saw in Chapter 3, such differences have been narrowing.

By 1990, the average yearly earnings of college-educated African-American men were 73 percent of the earnings of their white counterparts (U.S. Census 1991a: Table 29). This was far from parity, but an improvement over 1949, when the ratio was just 52 percent (Jaynes and Williams 1989). One way to evaluate the effect of education of the earning power of disadvantaged groups is to make such ratio comparisons between populations with carefully matched characteristics. For example, we can compare black and white men with the same age, education, and work pattern, as in Table 7-2.

The table makes race and gender comparisons of median income for full-time, year-round workers, matched for age and years of education. People with exactly four years of high school or exactly four years of college are singled out, as are those twenty-five to thirty-four years old. We focus on these younger workers because their careers should most reflect the changes promoted by the civil rights revolution of the 1960s and the women's movement that began in the 1970s. Here is an example of how to read the table: The first figure in the bottom row tells us that a young (25–34), black man with a college degree earned 83 percent of what a similar white man made.

Not evident from the table, which is based on relative comparisons, is the fact that education pays an absolute dividend for all categories of workers. For most demographic groups (females, white females, young black men, and so on), a college degree is worth roughly $11,000 more in annual earnings than a high school diploma. Two encouraging findings can be seen in the table. (1) A college degree generally moves black earnings closer to white earnings for corresponding groups, and likewise brings women's earnings closer to men's. (2) Among younger workers (25–34) with college degrees, race and gender disparities are generally smaller than they are among their older counterparts. Young, college-educated blacks have closed much of the gap between themselves and whites. These generalizations re-inforce the idea that a college education pays off for disadvantaged groups and suggest that it will do so more in the future than it has in the past.

The gender gap in earnings, according to these data, is larger than the race gap. This fact of social life undercuts the encouraging black/white ra-tios among females. If black women earn close to what white women earn, that still leaves them well behind men. For example, the average young man with four years of college, fully employed in 1995, earned $36,500 if he was white and $30,300 if he was black. A similar female graduate earned $28,000 if white and $25,300 if black. A fully employed young, white male high-school graduate was not far behind at $24,600. But his demographic twin sister earned just $17,900 (U.S. Census, unpublished data).

We can conclude that education, especially college education, is an equalizer—but with two caveats: It is less a gender equalizer than a race equalizer, and it is only an equalizer if people have good access to educa-tion. We have seen that access for women is virtually equal to that of men, but access for blacks and Hispanics is well behind that of whites, and low-income young people have suffered a deterioration in their chances of get-ting a college degree.

CONCLUSION

The previous chapter focused on measuring the gross amount of intergener-ational mobility and explaining the broad social factors that create mobility opportunities. In this chapter, we shifted perspective and concentrated on the determinants of individual success within the framework of gross mo-bility. Two factors—family background and education—are strongly corre-lated with career success. But measuring their separate influence is complicated because the two are correlated with one another. In general, people from privileged backgrounds get a double boost for their careers: a direct advantage (for example, father's connections help son get job) and an indirect advantage (son gets more education). But education also exercises a strong influence, independent of background. In effect, this means that the son of a blue-collar family who manages to get a college education is likely to do well.

Jencks used comparisons between brothers to measure the effect of fam-ily background. Building on Blau and Duncan's work, he concluded that background variables (including father's occupation, parents' education, in-come, and race) account for nearly half the variance in occupational attain-ment. (This figure includes family influence on years of schooling.) If Jencks' estimate is correct, we can make a good guess at a boy's odds of oc-cupational success on the day he is born. A good guess, but not a sure thing: There is another 50 percent to be determined by factors ranging from indi-vidual initiative to pure luck.

The literature on status attainment suggests that the influence of education on occupational success is of similar magnitude to the influence of family background. Jencks' research improved on earlier studies that had simply measured the cumulative effect of years spent in school. He showed that the value of time spent in the classroom, measured by gains in lifetime income, varies considerably. A year of college is worth more than a year of high school. The final year of college, if it results in a degree, is worth much more than any of the preceding years. The power of the degree is such that among those who manage to graduate, the influence of family background is greatly reduced.

But if college graduates are equalized in this sense, access to college remains very unequal. There is, for example, an enormous gap in the college participation rates of young adults from high- and low-income families. The gap has actually increased in recent years. This finding is consistent with our conclusion in Chapter 6 that mobility among younger workers seems to be slowing.

What would happen if recent trends were reversed—if there were a significant increase in the proportion of high school graduates who complete college? Would this reduce the influence of family background on career chances? It might. But it could also contribute to "degree inflation" similar to what happened with a high school diploma—once a valuable credential, now almost universal and therefore devalued. If a high percentage of the population earned bachelor's degrees, the best jobs would probably be reserved (as they increasingly are) for people with postgraduate degrees, who would likely come disproportionately from privileged class backgrounds. The outcome, of course, would not simply depend on the supply of workers with education credentials, but also on the demand created by economic growth and the changing shape of the occupational structure. As we have seen, individual ambitions, abilities, and family advantages are only half of the mobility picture.

SUGGESTED READINGS

Corcoran, M. 1995. "Rags to Rags: Poverty and Mobility in the United States." *Annual Review of Sociology* 21: 237–267.

> *How being raised poor, on welfare, or in a poverty neighborhood affects adults attainments. Research review.*

Haveman, Robert and Barbara Wolf 1994. *Succeeding Generations: On the Effects of Investments in Children*. New York: Russell Sage Foundation.

> *Economists' study of effects of family background on fortunes of young adults — especially educational attainment, teen pregnancy, welfare dependency, labor force participation.*

Herrstein, Richard and Charles Murray 1994. *The Bell Curve: Intelligence and Class Structure in American Life*. New York: Free Press. Fisher, Claude, et al, 1996.

Inequality by Design: Cracking the Bell Curve Myth. Princeton, N.J.: Princeton University Press.

Psychologist Herrstein and conservative ideologue Murray contend that genetically determined IQ differences are responsible for class and perhaps racial inequalities. Fisher and his colleagues reanalyze their data and consider factors Herrstein and Murray chose to ignore, concluding that inequality is largely the result of social structure.

Jaynes, Gerald David, and Robin M. Williams, Jr., eds. 1989. *A Common Destiny: Blacks and American Society*. Washington, D.C.: National Academy Press.

See chapter 7, "The Schooling of Black Americans."

McLeod, Jay. 1987. *Ain't No Makin' It: Leveled Aspirations in a Low-Income Neighborhood*. Boulder, Colo.: Westview.

Engaging portrait of black and white teens in a public housing project. Shows how their occupational aspirations are shaped by peer group, family, and school.

8

Elites, the Capitalist Class, and Political Power

Those who hold and those who are without property have ever formed distinct interests in society.

James Madison (1787: 79)

Although we are all of us within history we do not all possess equal powers to make history.

C. Wright Mills (1956: 22)

A lexander Hamilton and James Madison were political opponents. But these two signers of the Constitution agreed on this much: Politics and social class are unavoidably linked. Why? Because there is an inevitable conflict between "the few and the many" (Hamilton), the propertied and the propertyless (Madison), "the rich and the mass of the people" (Hamilton).[1]

Chapters 8 and 9 examine the connection between politics and social class. In this chapter we deal with power, focusing on "the few," those at the top of the class structure and the apex of the political order. In chapter 9, we take up class consciousness, concentrating on "the many, the mass of the people." Our emphasis on the few is typical in studies of power and might appear to be guided by the inherent logic of the subject. After all, those at the top have the best opportunity to exercise power. But "the many" are far from powerless, especially—and here is where class consciousness comes in—when they are united by a sense of common identity and shared interests. Thus, we see these two chapters as complementary and hope that readers will do the same.

THREE PERSPECTIVES ON POWER

In Chapter 1 we defined *power as the potential of individuals or groups to carry out their will even over the opposition of others* and indicated that we would restrict our use of the term to broad political and economic contexts. Here we will emphasize community and national power structures, examining studies developed from three competing theoretical perspectives: elite, class, and pluralist. The *elite perspective* makes a sharp distinction between an organized minority (the elite) that rules and an unorganized majority that is ruled. The *class perspective*, which has its origins in Marxist theory, also focuses on a ruling minority, but class theory is more specific about the identity of the rulers and the structure that creates them: They are the owners of productive wealth, or the capitalist class. The *pluralist perspective* denies that power is concentrated in one group. It maintains that in democratic societies, there are multiple bases of power representing the interests of competing groups, and no minority can easily impose its will. Obviously, the first two approaches have much more in common with one another than with the third.

A final prefatory note. To avoid improper usage and conceptual confusion, we consistently employ the term *elite* as a collective noun like class or jury. Such terms refer to groups rather than individuals. In this chapter, *elite* (singular) alludes to some notable group (business executives) and *elites* (plural) connotes two or more such groups (executives, military officers,

---

[1] See the epigraphs to Chapter 1 and the current chapter quoting Hamilton and Madison.

and public officials). These terms *should not* be used to refer to individuals, as in, "three elites went to a movie."

HUNTER: POWER IN ATLANTA

Two provocative books published in the 1950s ignited a major controversy among social scientists concerning the degree of concentration of political power in the United States. C. Wright Mills' *The Power Elite* (1956), which we will discuss later in this chapter, concluded that national decision making was being concentrated in the hands of a small, unrepresentative elite of corporate, political, and military leaders. Floyd Hunter's *Community Power Structure* (1953) suggested that a parallel tendency was affecting local communities. Those who defended the Mills-Hunter position came to be known as *elite theorists,* in contradistinction to the opposing pluralists, who contended that power was so diffused among competing groups that no single group could dominate the others.

Hunter's book is a case study of the structure of power in Atlanta, Georgia (which he called Regional City). Using a new methodology, which we will describe shortly, Hunter concluded that a small circle of men, most of them executives in leading banks and corporations or lawyers with corporate practices, made the key decisions in the community. This elite exercised more power over the city's affairs than the mayor and other high ranking public officials, whom Hunter relegated to the second level in a four-tier system of power. As Hunter described the decision-making process in Regional City, important public initiatives originated out of public view, in the offices and private clubs of the city's corporate elite. From there, they filtered down to the second and third levels of the power system. By the time they became public knowledge, the key decisions had been made.

How did Atlanta's power structure relate to its class structure? Hunter did not use *social class* or related terms, but he was obviously suggesting that the city's upper class (represented by wealthy businessmen) stood at the top of the power structure. The role of other classes is left ambiguous. Hunter did attribute some influence over the city's affairs to certain labor and black community leaders. But, in general, he appeared to be suggesting that middle- and lower-class people in Atlanta were so isolated from the exercise of power that their influence was of minor significance.

A contemporary portrait of the structure of power in Atlanta would look very different from Hunter's. The city's economy has grown and diversified—perhaps beyond the point where it can be dominated by the small circle of business leaders Hunter described in the 1950s. A new black elite, which emerged in the wake of the civil rights movement of the 1960s, plays an important role in the city's affairs. But *Community Power Structure* and the responses to it are still worth reading because they raise important issues about local systems of power and the problems inherent in any study of power.

THE REPUTATIONAL METHOD AND ITS CRITICS

Hunter's book was influential in two ways, both of which gained it friends and enemies. One was its critical attitude toward American institutions. If Hunter was right, political democracy is weak in the face of economic inequality; middle- and lower-class citizens cannot hope to match the power of the wealthy. The other source of influence was the book's innovative technique for identifying community leaders.

The new procedure, which came to be known as the "reputational method," has been widely imitated and just as frequently attacked. Taking a clue from community prestige studies, Hunter systematically recorded the opinion of well-informed members of the community about power standings. To determine who were the most influential people in Atlanta, Hunter asked 40 of the city's ranking public officials, civic leaders, businessmen, and professionals whom they would choose "if a project were before the community that required *decision* by a group of leaders—leaders that nearly everyone would accept" (Hunter 1953: 62). This poll yielded a select group of approximately twenty leaders who were regarded as influential by the leaders themselves.

The reputational method provided a quick way to identify community leaders, which contributed to its popularity. However, pluralist critics have raised compelling objections to the procedure. They argue that Hunter and his followers:

1. *Assume that which needs to be proven.* By asking their informants *who* the decision makers are, researchers are implicitly assuming the existence of a small, decision-making elite. According to political scientist Nelson Polsby, the proper question is not Who runs the community? but Does *anyone* at all run the community? (Polsby 1970: 297).

2. *Equate a reputation for power with power itself.* The information that researchers using Hunter's technique receive from informants could well be inaccurate because informants misperceive the power structure.

3. *Assume that a single elite deals with a broad range of issues.* Decision makers might be specialized in their interests and powers, thus forming a system of multiple elites, perhaps representing different segments of the community.

4. *Assume that leaders form a cohesive elite.* The identification of leaders does not tell us whether leaders agree on issues and exert their power in a concerted fashion.

5. *Downplay the role of formal political institutions* by largely ignoring the role of city government, political parties, and elections.

6. *Base research on an asymmetrical (one-sided) conception of power,* which is virtually implicit in a narrow focus on the identity of leaders. Policy makers may, in fact, find themselves compelled to consider the

probable reactions of other groups as they make decisions. If this is the case, power is in some sense reciprocal.

Hunter's critics were formally less concerned with his conclusions than with how he had reached them. Yet many of their criticisms implied that community power is dispersed rather than concentrated and that the reputational approach is defective because it is incapable of detecting such dispersion. Pluralist researchers proposed a methodological alternative designed to meet these objections.

Instead of focusing on the identification of putative *decision makers*, pluralists insist that students of community power should study *decision making*, by examining how representative community issues are resolved. If it can be shown that a well-defined group consistently triumphs over opposition in these matters, the existence of a dominant elite has been demonstrated. Otherwise, the pluralists contend, it must be assumed that power is divided among competing groups and shifts from one to another as the issue that is the focus of debate changes. (For criticism of the reputational approach see Aiken and Mott 1970; Dahl 1967; Polsby 1970.)

DAHL: POWER IN NEW HAVEN

The best-known piece of research employing a decisional methodology is Robert Dahl's study of New Haven, Connecticut, *Who Governs?* (1961), which was presented as a pluralist refutation of Hunter's work and Mills's examination of the national power structure. The core of Dahl's book is an examination of community decision making in three issue areas: (1) party nominations for political office, (2) urban renewal planning, and (3) public education. Dahl was interested in establishing who the decision makers were in each of these areas and, more specifically, who had successfully introduced or vetoed key policy initiatives.

Dahl's most general conclusion was that New Haven did not have the pyramidal power structure that Hunter described for Atlanta; rather, power in New Haven was diffused among a variety of groups and individuals. Leaders in any one of the three issue areas were unlikely to be influential in either of the other two. Among fifty major leaders, only three exerted influence in more than one area. All were public officials: the mayor, his predecessor, and the chief of the redevelopment agency (Dahl 1961: 181). Leaders, in other words, are specialized. Moreover, the sources of leadership are diverse. In none of the issue areas were leaders predominantly drawn from a "single homogeneous stratum of the community" (Dahl 1961: 182).

According to Dahl, things had not always been this way in New Haven. *Who Governs?* presents an intriguing account of the evolution of municipal power in this New England town. In its early years, the city was ruled by professional men drawn from a small circle of Yankee patrician families. But by the twentieth century, Dahl finds, the patricians were largely irrelevant.

They were replaced by a diversified cast of political characters, including popular politicians who often had roots in New Haven's ethnic communities, and professional public administrators. Relevant political resources, once concentrated in the hands of the patricians, were now dispersed. No one group was able to monopolize wealth, prestige, knowledge, electoral attractiveness, and control of mass media.

Two "strata" were the focus of particular attention in Dahl's study. Shying away from social class or terms suggestive of class, he labeled them "economic notables" and "social notables." The notables were New Haven's top wealth and prestige classes. By Dahl's account, the two were discrete groupings with little overlap: only 5 percent of the individuals on his combined list of notables fell into both categories (Dahl 1961: 68). (This conclusion, to which we will return later, is clearly at odds with the findings of the community prestige studies we examined in Chapter 2.)

Dahl was interested in the notables because of the significance that had been attributed to them by the elite theorists. One test of their importance for New Haven's power structure was representation among the fifty leaders influential in the three issue areas. Dahl counted only seven economic and social notables on this list. To these few men might be added an additional four "leaders," which were actually corporations that had influenced specific decisions.

Dahl's research convinced him that the social notables had largely withdrawn from the public arena that had in earlier epochs been dominated by their patrician ancestors. The economic notables appeared to be more active, but their participation was largely limited to the one issue area that engaged their interest: urban renewal. Even in this context, their influence did not seem overwhelming. A careful examination of key urban renewal decisions led Dahl to conclude that the economic notables were simply one of several groups attempting, with varying degrees of success, to influence the city's policies.

Thus, Dahl did not find a concentration of community power in the upper class, but neither did he see an even distribution of influence throughout the class structure. He attributed little or no direct influence over important government decisions to those below the line "dividing white-collar from blue-collar occupations" (Dahl 1961: 230). None of the fifty key leaders, for example, had a blue-collar occupation. Influential individuals, Dahl reported, were typically drawn from "middling social levels" (Dahl 1961: 229). But from Dahl's point of view, the class origins of the powerful are less important than the fact that power is diffused, specialized, and based on diverse political resources, so that no social group can regularly impose its will on the community.

THE DECISIONAL METHOD AND ITS CRITICS

Dahl's book especially appealed to those who saw in it a sophisticated vindication of American democracy. If New Haven, as he described it, did not quite fit the idyllic world of the traditional high school civics text, it did maintain that power was not concentrated in a small, upper-class elite and that most citizens of whatever class had access to some significant political resource.

Who Governs?, like Hunter's *Community Power Structure*, popularized a method of inquiry. The decisional method was widely imitated, but like the reputational technique, it did not escape criticism. Among the principal objections raised were the following:

1. *The conclusions of a decisional community power study inescapably depend on the specific decisions selected for investigation.* Yet there are no generally accepted criteria for specifying a list of community issues that is in some sense "representative" of local power arrangements. Dahl, for instance, has been criticized for selecting education as an issue area because, as he concedes, the city school system is of little interest to the notables, who generally live in the suburbs. They are therefore prohibited from holding school board positions and send their children to private or suburban public schools (Dahl 1961: 70). The party nomination process is likewise a questionable test of community power because it involves a matter of institutional procedure rather than conflict over public policy alternatives.

2. *The decisional method is biased toward attributing power to the government officials and civic leaders who are most directly and publicly associated with each issue area.* It is least likely to uncover a small elite of the sort described by Hunter, which is quietly involved in a broad array of issues, setting the basic goals that a larger group of more specialized leaders publicly pursue.

3. *By focusing specifically on decisions, Dahl's method misses the significance of latent concerns that are not allowed to become public issues requiring decisions.* For example, when Hunter studied community power in Atlanta, most questions involving equity for black citizens were in this category of "nondecisions." By manipulating public opinion or institutional procedures (such as the operation of legislative committees), the powerful can frequently suppress consideration of matters they prefer to ignore. (For criticism of decisional method, see Aiken and Mott 1970; Bachrach and Baratz 1974; Bonjean and Grimes 1974; Domhoff 1978).

DOMHOFF: NEW HAVEN RESTUDIED

Despite such criticism, *Who Governs?* was so influential that two decades later one of the critics, G. William Domhoff, felt compelled to reexamine New Haven politics of the 1950s. Domhoff is a class theorist. His book, *Who Really Rules?* (1978), challenged several of Dahl's principal claims, including the sharp distinction between economic and social notables and the idea that the economic notables had relatively little influence over New Haven's urban renewal program.

Domhoff challenged the criteria that Dahl used to define his social and economic notables. Domhoff's top prestige class, delineated by the membership of city's prestigious private clubs, was broader than Dahl's social notables, though still limited to approximately 1 percent of the population. His top economic class is defined by a network of interconnected business enterprises, banks, and corporate law firms at the center of New Haven's private sector. The leaders of these organizations were Domhoff's economic notables. Defined this way, Domhoff's top economic class was larger than Dahl's, but included fewer small-business owners and fewer managers of big corporations headquartered outside New Haven. It consisted essentially of the leaders of the largest local enterprises—about 300 people.

By Domhoff's criteria, 60 percent of New Haven's economic notables were also social notables (Domhoff 1978: 30). Dahl, of course, had contended that the two groups were almost mutually exclusive—which he took as evidence that New Haven was not dominated by a single elite. By demonstrating an overlap between the social and economic elites, Domhoff had dislodged part of the support for Dahl's basic argument.

Domhoff also chose to reexamine Dahl's conclusions about decision making in urban renewal because he regarded it as the most significant of the three issue areas considered by Dahl. Although Domhoff did not point this out, urban renewal is also the issue most likely to affect the pocketbooks of the rich and thus the issue most likely to draw them into the decision-making process. (See also a book on the same subject by one of Dahl's associates: Wolfinger 1973.)

Dahl portrayed the urban renewal program as the work of a newly elected Democratic Mayor, Richard C. Lee, and his redevelopment chief. The role of the economic notables, by this account, was essentially passive. With the help of documentary material unavailable to Dahl, Domhoff told a very different story. He showed that the urban renewal effort in New Haven was actually shaped by the prior initiatives of certain upper-class interests, both national and local. Nationally, Domhoff traced the story back to the late 1930s. At that time, a national debate over urban renewal policy set those who would emphasize building adequate housing for the poor against those who were concerned with preserving downtown real estate values and anxious to replace central city slums with commercial and other nonresidential structures. The first alternative was backed by labor and liberal groups, the second by major business interests. By the early 1950s, it was clear that the second, conservative alternative had emerged victorious.

Domhoff stressed the influence of several national, business-dominated organizations, such as the Urban Land Institute and the U.S. Chamber of Commerce, which played key roles in defining and promoting the conservative urban renewal alternative. The Institute carried on research and advised cities. The U.S. Chamber lobbied on a national level and showed its local affiliates how to design renewal programs beneficial to downtown business interests.

Locally, Domhoff emphasized activities of the New Haven Chamber of Commerce, which, he concluded, was dominated by members of his network of economic notables. In the early 1940s, the chamber persuaded a reluctant mayor to appoint a special commission to develop a comprehensive renewal plan. The commission was headed by a business executive active in the chamber; it produced a plan that, without essential modification, became the basis for the renewal program of the 1950s. After the Second World War, an independent redevelopment agency was set up by the city. The agency's board was dominated by business figures. Its minutes suggest that the Chamber of Commerce was the only group that took an active interest in its proceedings. Immediately after Lee's election in 1953, the chamber began to press the mayor-elect for action on the urban renewal program. Lee was already familiar with the chamber's views on the matter because he had been employed by the organization in the early 1940s when its interest in urban renewal first emerged. Domhoff did not deny Mayor Lee's leadership role in the 1950s, but he showed that there was much more to the story.

In sum, Dahl's *Who Governs?* substantially underestimated the influence of business interests on urban renewal in New Haven. Dahl was apparently misled by the unjustifiably restrictive way he defined his economic notables, by his lack of access to relevant documentary evidence, and by his failure to consider national developments affecting local events. The last concern points to a significant problem of theoretical perspective. Dahl appears to have proceeded on the assumption that the power structure of New Haven was self-contained. Domhoff reminded us that communities are integrated into a larger society and therefore potentially subject to the decision making of national elites.

COMPARATIVE COMMUNITY POWER STUDIES

Ultimately, the power structures of Atlanta and New Haven are of minimal importance to people outside Atlanta and New Haven. They are of wider interest only if studying them can tell us something about community power generally. For this reason, scholarly interest in this field turned to comparative analyses of the dozens of individual community power studies that have been done since the publication of Hunter's book (Bonjean and Grimes 1974; Clark 1971; Walton 1970). The object of such studies is to search out systematic patterns in the variations among communities. Researchers have been especially interested in determining the community characteristics that might make it lean toward a pluralistic or pyramidal (elite-dominated) power structure.

One of the best comparative studies is John Walton's (1970) analysis of research covering sixty-one separate communities. Walton tested a long list of hypotheses about community power that had been suggested by earlier work. Among his principal conclusions was a rather discouraging one that confirmed the suspicions of many familiar with the literature: Investigators who employed Hunter's reputational technique (typically sociologists) were most likely to uncover pyramidal power structures, and those who used Dahl's decisional method (generally political scientists) were most likely to find pluralistic configurations. It would seem that many investigators were predisposed toward one or another image of the power structure before research had even begun and, perhaps unwittingly, chose the method that tended to verify their assumptions. Because of this tendency, Walton looked for generalizations that held true in both reputational and decisional studies.

One important finding relates directly to social class. Walton found a general agreement in the studies he analyzed that decision makers are "overwhelmingly" recruited from the upper-middle class and include a particularly high proportion of business leaders. Another key finding concerns the influence of local affiliates of large corporations. Walton determined that communities in which a high proportion of business firms are absentee-owned are the most likely to have pluralistic power structures. Local managers of national firms tend to stay out of town politics because their future careers are within the corporation not the community. On the other hand, local business leaders, like Atlanta elite studied by Hunter, have deep roots in the community, along with knowledge and connections that make them effective participants in politics. They are likely to have inherited a family tradition of involvement in community affairs.

The perspective of the affiliate manager is reinforced by the parent corporation's own attitude toward local politics. Within the community, the corporation's concerns can include taxation, water supply, labor relations, and police protection. But these matters are typically settled before the local plant is established, and the corporation can simply shift its operations elsewhere if it becomes dissatisfied with conditions. Firms engaged in extractive operations (for example, mining) are an important category of exceptions to this generalization. Such firms cannot easily pull up stakes and leave, and they are often deeply involved in local politics (Goldstein 1962).

Absentee firms appear to encourage pluralistic power arrangements by reducing community dependence on local business interests, without becoming directly involved in local politics. However, Walton's finding should not obscure the extent to which community affairs are indirectly but powerfully shaped by national corporate interests. A corporation's implicit threat to rechannel investments away from a community is a potent influence on local decision making, even when corporate mangers are not among community decision makers. Moreover, corporate interests can collectively influence local communities by shaping national policies that affect them. Domhoff's portrayal of the development of urban renewal policy is a relevant example. Thus, a community may be inclined toward pluralistic decision making, yet have few critical decisions to make. We must shift our focus to a national level of power if we are to understand fully the forces that shape local communities.

MILLS: THE NATIONAL POWER ELITE

C. Wright Mills' book, *The Power Elite* (1956), was published shortly after Hunter's community study and provoked a parallel controversy on national power arrangements. Mills, whose work on the middle class we referred to in Chapter 3, was a major critic of American civilization during the 1950s, a self-satisfied era in American life national life characterized by Mills as "a material boom, a nationalist celebration, a political vacuum" (Mills 1956: 326). Mills' conception of the national power structure centered on the growing significance of three major interlocking institutions: the modern corporation, the executive branch of the federal government, and the military establishment. He saw each of these institutions becoming enlarged and centralized. A few hundred major corporations were taking the place of thousands of smaller competing firms, which once typified the economy. The federal executive had gathered enormous powers and resources previously nonexistent or scattered among other units of government. The military, once small and decentralized, had developed into a colossal bureaucracy, commanding a war machine of unprecedented scale and destructive power. (For evidence supporting Mills with regard to these trends, see Dye 1995.)

As the corporations, the federal government, and the military grew, they eclipsed and subordinated other institutions (Mills 1956: 6):

> No family is as directly powerful in national affairs as any major corporation; no church is as directly powerful in the external biographies of young men in America today as the military establishments; no college is as powerful in the shaping of momentous events as the National Security Council. Religious, educational, and family institutions are not autonomous centers of national power; on the contrary, these decentralized areas are increasingly shaped by the big three in which developments of decisive and immediate consequence now occur.

The implication Mills drew from these trends was that the basis of national power had been reduced to control over the three key institutions. Those who sit at the "commanding heights" of the corporate, political, and military hierarchies make the critical national decisions. Who are they? In the corporations, an amalgam of very rich families with corporate-based fortunes, plus the ranking executives of the top national firms, whom Mills collectively labeled the "corporate rich"; in the federal government, the president, vice president, cabinet, heads of major agencies, and members of the White House staff; among the military, the generals and admirals. These three institutional elites together constitute Mills' *power elite*.

Mills argued that the emergence of a national elite undercut important traditional bases of power. Community elites decline in importance as the power of national institutions grows. Investment decisions that are crucial for a community may be made in a distant corporate boardroom. Even the significance of personal wealth as a power resource is reduced. Very sizable individual fortunes seem lightweight relative to the massive assets of any major national corporation. However, the largest family fortunes, such as those of the Rockefellers, Mellons, or Fords, are typically invested in national corporations, and in this form, personal wealth can retain its significance as a basis of power.

Mills conceived the national structure of power as consisting of three tiers (see Figure 8-1). The top tier is, of course, the power elite. The bottom tier, which Mills labels "mass society," encompasses the great majority of the population. Subject to large-scale national institutions beyond their control or comprehension and misinformed by media that are dominated by national elites, the members of the mass are passive participants in the political system. Between the mass society and the power elite are "the middle levels of power," comprising a multitude of competing interest groups from labor unions to the gun lobby, whose typical arena of conflict is the Congress. The middle levels of power are the source of most political news, but they are not, according to Mills, the locus of the most important political decisions. In the welter of competing interests, none can impose itself. This "semi-organized stalemate," as Mills characterized it, only reinforces the dominance of the power elite.

The key decisions are, of course, reserved for the power elite. But which are the key decisions? Mills is unambiguous on this point. Two issue areas are of sweeping importance: national economic policy and military affairs–foreign policy. These matters unmistakably, often brutally, intrude into the lives of ordinary men and women as boom or bust in the economy and peace or war abroad. Mills saw decision making in both areas shifting from the Congress to the executive branch of government, where the viewpoints of corporate and military elites are well represented. He was keenly aware of the terrible fact that a small circle of men had assumed the right to launch a cataclysmic nuclear war with little or no public debate on the matter. And, although most economic decisions under capitalism had always been private decisions, Mills felt that the leaders of major national corporations were much freer to determine their own course than were their smaller predecessors, subject to the restraints of a competitive market.

Mills' conclusions were clearly influenced by the period during which he was writing. The country had in recent memory suffered through World War II, followed by the Korean War, and was entering a period of protracted cold war. General Eisenhower, a military man, was in the White House, and other officer-heroes were serving in civilian posts. These transitory developments probably led Mills to overestimate the significance of the military sector of his power elite. In the postwar world, the country faced a series of key decisions in foreign policy and international economic policy, which

FIGURE 8-1 Mills' Conception of National Power

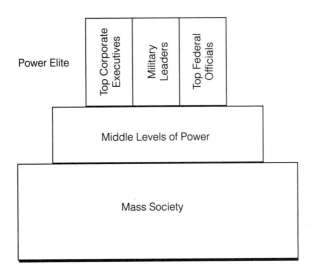

could be dominated by a handful of men if the major corporate, political, and military institutions they represented were in agreement. But even as Mills wrote, new issues were emerging, such as the civil rights question, which could not be easily contained at the elite level of power but would have to be fought out at the middle level, particularly in Congress, and sometimes in the streets.

MILLS, HIS CRITICS, AND THE PROBLEM OF ELITE COHESION

Much of the criticism directed at elite theory in the community-power literature is also relevant to Mills' work. For example, pluralists have argued that Mills' methodology, like Hunter's approach in Atlanta, is flawed by a failure to examine actual decisions as a test of the power of the elite. In this section, we will focus on the substantive issue that has provoked the most debate among Mills' readers: the problem of *elite cohesion—that is, the extent to which the members of a hypothesized elite hang together in pursuit of common objectives and in opposition to other groups.* (For other lines of criticism, see the community-power discussion above and Domhoff and Ballard 1968.)

The cohesion issue was effectively posed in a classic essay by Dahl, directed at both Hunter and Mills, which accused the elite theorists of "confusing a ruling elite with a group that has a high *potential for control*" (Dahl 1967: 28). To be politically effective, potential for control must be coupled with "potential for unity." The American military has the potential to impose a dictatorship on the nation, but that potential means nothing unless military leaders are in agreement on that objective. Mills defined an elite in terms of key positions in organizations that control important resources (potential for control). But he failed, according to Dahl, to demonstrate a political consensus among the members of his elite (potential for unity).

Mills' pluralist critics are clearly predisposed to the belief that unity among power contenders is difficult to achieve, especially in contemporary America. This inclination was strikingly illustrated in the best-known pluralist alternative, David Riesman's version of the American political system, which was developed in a more general book on American society and culture, *The Lonely Crowd* (1953). In the course of his treatment of national power, Riesman asked two questions: "Is there a ruling class left?" and "Who has the power?" His answers were, respectively, no and no one. As the first question implies, Riesman believed that the country had a ruling class in the past. Early in the history of the republic, the ruling class consisted of the landed gentry and mercantile interests that constituted the Federalist leadership and, later, of captains of industry. But by the 1950s, the ruling class had been supplanted by an amorphous constellation of "veto groups," organized representatives of specialized interests that included "business groups, large and small, the movie-censoring groups, the farm groups and the labor and professional groups, the major ethnic and major regional groups" (Riesman 1953: 246). The veto groups are distinguished from the powerful of previous eras by their inability to take positive initiatives to impose their own will. Feeling themselves powerless, chary of offending other groups, their function is largely defensive, "to neutralize those who might attack them" (Riesman 1953: 247).

How can any decision be taken in such a political context? Is anyone in charge? Riesman's reply was that leadership may be needed to initiate something new or halt something in progress, but little leadership is needed to maintain the status quo. To the extent that anyone exercises power, it is over very specific and narrow issues. Power that might be effective over a broad range of issues or in the face of big questions that affect the nation as a whole is smothered by the action of the veto groups.

Mills characterized Riesman's amorphous power structure as "a recognizable although a confused statement of the middle levels of power, especially as revealed in congressional districts and in Congress itself" (Mills 1956: 244). Power at that level is indeed a semi-organized stalemate. In Mills' view, Riesman was guilty of a mindless empiricism that equated all interest groups and all issues. Banks and organizations representing motorcycle riders are both concerned with national legislation. But to describe them as two veto groups misses the point. Some groups are more important than others because they have powerful resources at their command and because they deal with issues that are more vital to the nation. As we have seen, Mills was only interested in the big economic and foreign policy questions and regarded as trivial most of the other issues that consume the attention of the middle levels of power.

Cohesion presented a particular problem for Mills. He had to demonstrate both that the three distinct elites are internally cohesive and that they are drawn together into a single power elite. Mills did present a series of mechanisms through which elite unity might be achieved. They fall into two categories: social-psychological and structural. The social-psychological mechanisms include similarities in origins, education, career, and lifestyles, which produce "a similar social type" and contribute to ease in informal association (Mills 1956: 19). Mills presented evidence on these topics for each elite. He noted that elite men tend to be drawn from upper-class or upper-middle-class, urban, white, Protestant families and that they are likely to be educated in Ivy League schools. There is a significant overlap between the world of the power elite and upper-class "society," with its elaborate links among "proper" families, select prep schools, distinctive class values, and notions of style. (These commonalities, as Mills conceded, apply more to the civilian elites than to the military.)

In addition, members of the three elites have similar career experiences, even if they do not move through the same institutions. The experience of managing a large organization is shared by the corporate rich, the "political directorate," and the chiefs of the Pentagon. The character of modern bureaucratic life has tended to blur the distinction between leadership in a large corporation, a civilian department of government, and an army. Mills contended that these shared elements of background and careers and the considerable material rewards attached to elite position tend to make members of the power elite conscious of the differences between themselves and the great mass of the population and to draw them together; they develop a form of upper-class consciousness, which leads them to view the world from a similar perspective.

The structural mechanisms of cohesion examined by Mills concern the more-or-less-formal connections between institutions. One critical link is the interchange of personnel among the three institutions, especially the movement of representatives of the corporate world into and out of top political positions. Another tie between these two is the dependence of political candidates on financing from the corporate rich. The military is closely allied with the corporations, who are its suppliers, while the militaristic foreign policy pursued by the political directorate strengthens its ties to the generals and admirals. All three elites are, of course, compelled to consider each other by virtue of the inevitable interdependence of institutions operating on such a scale.

In sum, the pluralists are persuasive when they argue that Mills must prove that his power elite is cohesive. Otherwise, it is little more than a list of important people. In reply, Mills points to a series of factors that tend to unify the individual members of the power elite and draw the three major institutional sectors together: common social background, shared lifestyle and values, similar job experience, interchange of personnel, campaign financing, and institutional interdependence. In this chapter we will examine contemporary evidence on most of these topics.

PLURALISM, STRATEGIC ELITES, AND COHESION

American Elites, by Robert Lerner, Althea Nagai, and Stanley Rothman, is a recent pluralist contribution to the debate spawned forty years ago by *The Power Elite*. The book is based on a series of interview surveys of people in elite positions, conducted by Rothman and various associates in the 1980s. Note that the coauthors of *American Elites*, unlike Mills, refer to elites in the plural. Their work develops from a pluralist perspective that emphasizes a division of power in modern societies among various "strategic elites" with distinct functional responsibilities. The leaders of business, the media, and the military are examples. This brand of pluralism does not deny that power is concentrated in the hands of a few. Small groups of powerful men (and occasional women) may dominate the media, the military or the corporations, but none of these strategic elites carries much influence outside its own sector. There is, in this conception, no core elite like Mills' power elite (Lerner, Nagai, and Rothman, 1996: Ch. 1).

Although this functional differentiation of elites is, in some ways, desirable, it also carries risks for the country. The distinct elites will tend to develop separate viewpoints, even conflicting values, which can contribute to disunity in the larger society. Rothman and his coauthors share with many political conservatives the conviction that the influential elites associated with the cultural sectors in American society, such as television, journalism, and academia, tend to reject many traditional American values. The result has been a national struggle over values the authors characterize as a virtual "culture war" (Lerner, Nagai, and Rothman 1996: 140). Obviously, Rothman and his colleagues, like Mills, bring a political perspective to the study of elites.

Ironically, the authors show little interest in how strategic elites exercise power. They do not examine relationships among strategic elites, ask whether all are equally important, or follow the advice of other pluralists to study actual decisions as a test of elite power. They are, it turns out, more interested in the powerful than in power itself. *American Elites* focuses on the social backgrounds and viewpoints of members of the strategic elites. These topics are directly relevant to the question of elite cohesion raised in the preceding section. If the authors of *American Elites* can demonstrate that strategic elites differ in the backgrounds or opinions of their leaders, they will weaken the claims of Mills and other elite theorists.

Rothman and his coauthors define twelve distinct strategic elites, falling into three broad categories: economic, political, and cultural. They range from corporate executives, lawyers, and government officials, to union leaders, journalists, and film makers. Together they constitute a very

TABLE 8-1 Social Background of "American Elites"

	In Percent
Family Income Growing Up	
Above average	38
Average	33
Above average	29
Total	100
Father's Occupation	
Professional/Manager	54
Other White Collar	19
Blue Collar	27
Total	100
Completed College	91
Attended Elite College	31

SOURCE: Calculated from Tables 2.1 and 2.2 in Lerner, Nagai, and Rothman 1996.

broad conception of elite status, especially given the way the individual strategic elites were defined for the surveys. For example, the corporate leaders interviewed were drawn from both "upper and middle management." The government officials were "high-ranking bureaucrats," excluding officials appointed by the president—which is to say, excluding the top decision makers (Lerner, Nagai, and Rothman 1996: 141). Taken together the leaders studied by Rothman and his associates represent the top and middle levels of power in a broad array of key national institutions. Rothman's strategic leadership is not only more differentiated, but also larger and less selective than Mills' power elite.

Table 8-1 summarizes the survey findings on the social backgrounds of strategic leaders from a chapter the authors entitle "Room at the Top." In class terms, the surveys reveal a pattern of recruitment that favors the children of privilege, but not exclusively. Most respondents report that their families had average or above average incomes when they were children. More than half had fathers who were professionals or managers, well above the national proportion of men in these categories during the years the today's leaders were growing up. But many had fathers with less impressive jobs and below average incomes. Almost all the respondents completed college, but surprisingly few graduated from highly selective institutions. The authors also report that the majority of their strategic leaders are not Anglo-Saxon Protestants, though virtually all are white and male. More detailed tabulations in the book show that these generalizations hold up fairly well for the twelve elites taken individually. The biggest exceptions are labor leaders who are, as might be expected, largely from blue-collar backgrounds; and corporate lawyers and public interest group leaders, who come disproportionately from privileged backgrounds. (Public interest groups, often led by lawyers, include organizations that lobby on behalf of civil rights, consumer protection, the environment, and other civic causes.)

Rothman's findings tend to undercut Mills' portrait of an Anglo-Saxon elite drawn together by the exclusive social pedigree of its members. Mills' social conception of the elite—which was certainly more accurate in 1950s when he wrote the *Power Elite*—does not fit the social diversity of these strategic leaders in the 1980s. (It is, however, possible that the strategic elites would look more like the power elite if they were defined more exclusively.)

At the end of *American Elite*, the authors assert that they have demonstrated that there is a high level of ideological division among American strategic elites. But the data, at least as presented in the book, are not nearly as decisive as this suggests. Asked to classify themselves as conservative, moderate, or liberal, respondents differed in predictable ways: corporate executives and military officers generally took the conservative label; majorities of labor leaders, movie makers, journalists, bureaucrats, and religious leaders described themselves as liberals (Lerner, Nagai, and Rothman 1996: 50). But as the authors note, such labels mean different things to different people and can hide as much as they reveal. Answers to specific opinion questions reveal a more complicated picture. For example, responses to a series of questions about the desirability of activist government show the bureaucracy quite close to conservative thinking of business and the military (Lerner, Nagai, and Rothman 1996:52).[2] This is not what would be expected from the ideological labels, but close to what Mills' power elite thesis would predict. More liberal responses to the activist government questions came, as expected, from labor officials and public interest group leaders.

Some questions revealed an unexpected degree of inter-elite consensus. There was a consensus among strategic elites (with the notable exception of judges) that the legal system favors the wealthy. Virtually no one thought that corporations should be publicly owned. But strong majorities in the various elites (except labor and public interest) agreed that business is "fair to workers" (Lerner, Nagai, and Rothman 1996:106).

The biggest differences emerged from a series of life-style and values questions. Although the right to abortion was strongly supported by all but the religious elite, questions regarding gay rights revealed greater divergence in opinion. Seventy percent of military officers thought that gays should not be allowed to teach in public schools, a position rejected by virtually all members of the media, television, movie, and public interest elites. Business executives and bureaucrats were more equally divided on this question (Lerner, Nagai, and Rothman 1996:91).

---

[2] These items ask whether the government should reduce income differences, guarantee jobs, regulate business, and protect the environment.

In sum, *American Elites* argues for a pluralist conception of national power based on differentiated strategic elites. Rothman and his colleagues cannot demonstrate the sharp opinion differences among the strategic elites their theory implies. They do, however, document some social diversity in elite recruitment. The strategic leadership is drawn disproportionately, but by no means exclusively, from the privileged classes. This finding tends to weaken Mills' claim that shared social background contributes to elite cohesion.

POWER ELITE OR RULING CLASS?

If Rothman and other pluralists believe that Mills fails in his efforts to prove that his tripartite elite is cohesive, one Marxist critic contends that Mills is all too successful. In a perceptive essay on *The Power Elite*, Paul Sweezy contends that there is an unresolved tension in the book between two views of the elite. The first is based on social class: Mills provided evidence that "those who occupy the command posts do so as representatives or agents of a national ruling class which trains them, shapes their thought patterns, and selects them for their positions of high responsibility" (Sweezy 1968: 123). Much of the evidence Mills presented for elite unity seems to point in this direction, especially the material on social background and recruitment patterns. The second view emphasizes the bureaucratic elites at the top of three "major institutional orders"; here Mills treated the corporate, military, and political realms as distinctly separate domains with autonomous leadership, which comes together to form the power elite. Sweezy was highly skeptical of the second view, particularly given the evidence that Mills presented for the first. The American military, Sweezy contended, is firmly under civilian control, and the political elite is dependent on the class that rules the corporations; thus, the justification for thinking in terms of institutional elites collapses.

Sweezy's criticism suggests an alternative to both the Millsian and the pluralist views of national power: the identification of power with the class that controls income-producing wealth, which he calls the ruling class. This approach—the last of the three theoretical conceptions of power we mentioned at the beginning of the chapter—assumes that Mills' "corporate rich" have largely subordinated competing elites to their will. Mills never dealt at length with this line of criticism, although in a breezy reply to critics on the left, he commented, "They want to believe that the corporation and the state are identical... I don't believe it's quite that simple" (Mills 1968: 224).

It probably is not "quite that simple." But Mills, like pluralist Rothman, appears to underestimate the power of wealth because he wants to fit it into a broader conception of elite power. We will use much of the remainder of this chapter to explore the bases of capitalist-class power. (We prefer the term "capitalist class" to Sweezy's "ruling class" because it avoids the implicit assumption about political arrangements.)

THE NATIONAL CAPITALIST CLASS: ECONOMIC BASIS

We will define the capitalist class as consisting of people who receive more than half of their income from invested wealth. Although such people exist at virtually all income levels, only at very high levels—on the order of several hundred thousand dollars a year—does dependence on property income become the predominant pattern (see Table 4-3 on page 95). A basic division within this class is that between local and national capitalists. The national capitalists are those who own or manage major national corporations; Mills collectively dubbed them "the corporate rich." The locals are affluent, but community-oriented business people, such as local media owners, real estate investors, and large local retailers. Since we have touched on the local capitalist class in the section on community power and our interest here is national power, we will restrict most of the discussion that follows to the national sector of the capitalist class.

In Chapter 4 we found that households with annual incomes over $1 million receive most of their income from accumulated wealth, in such capitalist forms as interest, dividends, and rents (see Table 4-3). We discovered that wealth, especially corporate wealth, is highly concentrated in the United States. For example, the richest 10 percent of households own 84 percent of publicly traded stock most of it in the hands of the top 1 percent of households (Table 4-6 on page 104).

Here we take a closer look at the wealth of the very rich. The asset portfolios of the top 2.5 million wealth holders in the United States, about 1 percent of the population, are analyzed in Table 8-2. All have net worth in excess of $500,000. Twenty-five thousand were worth more than $10 million in 1986. The table reveals a significant difference between the merely comfortable and the truly wealthy. Those with the largest fortunes have the highest proportions of their total holdings in the form of corporate stock and other business assets. For people worth between $500,000 and $1 million, real estate (including the individual's own residence) is the largest asset category. For those worth more than $10 million, stock is the largest category, making up about half the value of total assets. The average wealth holder at this level owns over $11 million in corporate stock.

Even these data tell us relatively little about the largest American fortunes. For that purpose, we turn to the list compiled annually by *Forbes* magazine of the 400 wealthiest individuals in the United States. The net worth of the 400 individuals listed by *Forbes* in 1996 ranged from $415 million to $18.5 billion. As recently as 1991, the list ran from $275 million to $5.9 billion. Fortunes at the top have swelled in the 1990s. Table 8-3 contains a sampling of the *Forbes* 400, with information on the size and source of their fortunes. The list begins with Microsoft billionaire Bill Gates, according to *Forbes*, the richest man in the world. The second part of the table samples *Forbes'* intriguing (though less systematic) list of the 100 largest *family* fortunes. (Forbes *1991, 1996*)

TABLE 8-2 Assets of Top Wealth Holders by Net Worth, 1986

Size of Net Worth[a]	Number of Persons	Mean Value of Total Assets	Value as a Percentage of Total Assets								
			Corporate Stock	Real Estate	Cash	Noncorporate Business Assets[b]	Bonds	Mortgages and Notes	Life Insurance Equity	Other Assets[c]	Total
$500,000 to $1 million	1,548,300	$763,946	21.9	33.9	13.7	4.6	8.4	3.9	1.5	11.9	100
$1 to 2.5 million	710,000	$1,618,095	27.8	29.2	9.6	7.6	10.3	3.9	1.0	10.6	100
$2.5 to 5 million	150,300	$3,820,885	36.6	23.9	7.0	8.6	10.2	3.4	0.8	9.4	100
$5 to 10 million	55,500	$7,426,306	41.3	21.6	5.7	9.6	11.8	2.5	0.4	7.0	100
$10 million or more	25,000	$23,102,440	49.5	13.5	5.1	11.0	9.5	4.0	0.2	7.1	100
Total	2,489,100	$1,565,090	32.0	26.7	9.4	7.6	9.7	3.7	1.0	9.9	100

[a]Reflects debt not shown separately.

[b]Net value of sole proprietorships, farms, and share of partnerships.

[c]Includes intangible and depletable assets, annuities, pensions, and personal property.

SOURCE: Schwartz and Johnson 1990 (based on estate tax data).

TABLE 8-3 Large Fortunes

Personal Fortunes	Estimated Net Worth (Millions)	Primary Source of Wealth
William Gates III	18,500	Microsoft
John Werner Kluge	7,200	Metromedia Co.
Jim C. Walton	4,800	Inheritance (Wal-Mart Stores)
Donald E. Newhouse	4,500	Media
Barbara Cox Anthony	4,000	Inheritance
H. Ross Perot	3,300	Electronic data management
Jay A. Pritzker	3,000	Financier
Marvin Davis	2,000	Oil, real estate
Bennett Dorrance	1,800	Inheritance (Campbell Soup)
David Rockefeller, Sr.	1,400	Inheritance (oil)
Ralph Lauren	1,300	Fashion
Peter Haas	1,200	Inheritance (Levi Strauss)
Paul Mellon	1,200	Inheritance
Michael Bloomberg	1,000	Financial news service
John Sall	1,000	Software
Daniel M. Ziff	1,000	Ziff Brothers Investments
Whitney MacMillan	950	Inheritance (Cargill, Inc.)
Carl Icahn	900	Finance
Richard Mellon Scaife	870	Inheritance
Dwight D. Opperman	815	Publishing
Harold Honickman	800	Beverage distribution
George Kaiser	780	Oil and gas
John Orin Edson	700	Manufacturing
Edward Perry Bass	690	Oil Investments
Alan Gerry	665	Cable TV
Leon Hess	650	Amerada Hess
Betsey C.R. Whitney	625	Inheritance
Nelson Peltz	620	Leveraged buyouts
Michael Krasny	620	Computer Discount Warehouse
Harry Silverman	600	Investments
William Barron Hilton	600	Hotels
Michael Chowdry	600	Atlas Air
William Connell	560	Scrap metal
Elizabeth Turner Corn	560	Inheritance (Coca-Cola)
Zachary Fisher	500	Real estate
Phoebe Hearst Cooke	500	Inheritance
Thomas S. Monaghan	500	Pizza
Irenee du Pont & family	490	Inheritance (Du Pont Co.)
Viola Sommer	460	Inheritance (real estate)
Oprah Gail Winfrey	415	Television
Fortunes of Extended Families		
Upjohn	3,200	Inheritance (Upjohn Co.)
Bicardi	2,000	Liquor
Lilly	1,500	Pharmaceuticals
Clapp	1,400	Lumber
Jordon	950	Inheritance (media, retailing)
McClatchy	625	Newspapers
Miner	600	Oracle Corporation
Sulzberger	600	New York Times
Coors	600	Beer
Houghton	500	Corning Glass

SOURCE: Forbes 1996.

Forbes touts the 400 list as evidence of America's open class structure. The magazine claims that about half the people on the list are, like Gates, "self-made millionaires." About 30 percent of the people on the list are heirs to giant family fortunes, including those associated with Ford Motor Company, Levi Strauss, Hearst newspapers, Campbell Soup. Some 20 percent apparently benefited from more modest inheritances, which they built into 400-level fortunes. For example, the Newhouse brothers, owners of one of the country's major publishing empires, inherited the "largest privately held newspaper chain in the country," in addition to Condé Nast, publisher of *Vogue* and other magazines. (Father Samuel Newhouse bought Condé Nast for $5 million in 1959 as an anniversary present for his wife: "She asked for a fashion magazine," he later claimed, "and I went out and bought her *Vogue*") (*Forbes* 1991: 152).

By focusing on individual net worth, the *Forbes* 400 and similar listings underestimate the concentration of wealth at the top. Many, if not most, of the largest fortunes are held in common by members of extended families. These clans are based on descent from a founding ancestor, who typically accumulated the initial fortune during the late-nineteenth-century expansion of American capitalism. Some are represented among the 400. For example, the list includes multiple descendants of oil tycoon John D. Rockefeller, banker Thomas Mellon, auto manufacturer Henry Ford, and industrialist E.I. Dupont.

But in many cases, shares in these established family fortunes are sufficiently dispersed after a generation or more that no individual controls the minimum $415 million required for inclusion on the *Forbes* 400 list. In 1996, *Forbes* included the Sulzbergers, publishers of *The New York Times*, and the Coors brewery family on its list of wealthy families. Both had fortunes estimated at $600 million, but neither was represented on the 400 list. The Kennedys, whose collective fortune the magazine estimated at $350 million, in 1991, were not even on the family list five years later.

The concentration of personal wealth is paralleled by the concentration of wealth in the corporations themselves. The dominant position of the largest corporations seems to have been further consolidated since Mills wrote. In 1950, shortly before the publication of *The Power Elite*, the 100 largest U.S. industrial corporations (among nearly 200,000) already controlled approximately 40 percent of all industrial assets; by 1992 their share had grown to 75 percent. Finance capital was even more concentrated. The twenty-five largest banks controlled almost half of all bank assets (Dye 1995: 15, 19).

Mills observed that even the largest personal fortunes were small relative to the assets concentrated in large corporations. Economic power, he contended, belonged to those who controlled the corporations. But who controls the corporations? Mills's answer, as we have seen, was "the corporate rich": top corporate executives and extremely wealthy families like the Fords with substantial stakes in major corporations.

But as major corporations have grown larger, their stock has become more dispersed and their relationship to even the largest stockholders has grown more distant. The biggest corporations have hundreds of thousands of stockholders; individual holdings of even 5 percent (enough to give the owner significant influence over management) are relatively rare. At the same time, many wealthy families have sought to reduce the risks to their fortunes by spreading investments across different corporations and sectors of the economy. These developments appear to undermine Mills formulation of corporate power: the rich owners among his "corporate rich" are fading away, leaving hired corporate managers in control.

Small and medium-sized corporations are still likely to be controlled by owners—typically families or small groups of stockholders, including many of the Forbes 400 listees.[3] But by 1980, only 22 of the 100 largest industrial corporations were controlled by owners. Most were run by hired managers, remarkably free of external control (Herman 1981:61). Even before Mills wrote, observers of corporate life were speculating about a "managerial revolution"—a shift not simply in the staffing of the corporation, but also in its ultimate objectives (Berle and Means 1932; Burnham 1941; Galbraith 1967). These writers argued that a manager who is neither an owner nor subject to the dictates of owners is free to substitute personal goals for those of the stockholders. The traditional owner-entrepreneur, they believed, operated a firm with one consideration in mind: profit. The professional manager is not quite free to ignore profit but, with little direct stake in the corporation's earnings, might pursue a broader array of objectives, including corporate growth, technological progress, the welfare of employees and customers, and public esteem for self and company. One well-known writer was so moved by this image of corporate high-mindedness that he dedicated an essay to celebrating "The Soulful Corporation" (Kaysen 1957).

The trend toward professional corporate management has continued unabated. Even the Ford Motor Company is no longer managed by people named Ford. But the related broadening of corporate objectives that some anticipated has not materialized. Major corporations are, even more than they were twenty years ago, singularly focused on making money for their shareholders. Three factors appear responsible. First, the balance of power between managers and owners has tilted back toward the owners. The reason, according to sociologist Michael Useem (1996), is the triumph of "investor capitalism." Useem explains that stock ownership in large corporations remains highly dispersed, but individuals are increasingly likely to own their stock shares through institutions, such as mutual funds,

---

[3] *Forbes* also publishes an annual list of the 500 largest privately owned (not publically traded) companies. Most have few stockholders. Included are some large corporations such as Levi Strauss and UPS. In 1996 all had revenues in access of $400 million. Ninety-two were owned by individuals included in the Forbes list of the 400 richest Americans (Kichen, et al., 1996)

pension funds, and bank trust departments. Frequently, the largest blocks of stock in major corporations are concentrated in this form. The so-called "money managers" who invest for the owners of a mutual fund or pension are under constant pressure to produce ample returns for the shareholders and will, in turn, pressure corporate managers to maintain profit levels (reflected in healthy dividends and rising stock prices). When profits lag (and drag down stock prices and dividends), money mangers sometimes press corporations whose stock they control to replace their top executives, fire thousands of workers, or make sweeping changes in corporate strategy. Useem does not mean to suggest that corporate executives have become powerless but, rather, that they must share power with the money managers who represent millions of (profit hungry) investors.

A second factor reshaping corporate governance is the more competitive economic environment of recent years. Under these conditions, corporate managers fear that competing firms at home and abroad will gain on them if they do not remain strictly focused on maximizing profit. The third factor is the changing character of executive compensation. As we noted in Chapter 4, the earnings of top corporate executives have soared in recent years. In 1996, the average CEO of a major corporation earned $5.78 million—almost three times the average six years earlier (*Business Week* 1997, 1996). At the same time that executive compensation has reached unprecedented heights, it has become increasingly dependent on the financial performance of the corporation. Incentives, including yearly bonuses and long-term stock purchase plans (stock options), are added to base pay to reward executives who enrich their stockholders. In 1982, about 40 percent of CEO compensation came from such profit-oriented incentives; by 1995 the incentive proportion had jumped to 70 percent. Obviously, the fortunes of CEOs and stockholders are, more than ever, bound together. For senior corporate officers just below the rank of CEO the pattern is similar: high and rapidly climbing rewards, with total compensation increasingly dependent on return to investors (Useem 1996: 243–250).

The spectacular ascent of executive pay, coming in a period of generally stagnant wages and massive corporate layoffs, has inevitably attracted critical fire. A typical target was AT&T chairman Robert Allen, whose compensation jumped 143 percent to $15.9 million in the year AT&T laid off 40,000 people. But, as a prominent business school professor told *Business Week*, "Today, a CEO would be embarrassed to admit that he sacrificed profits to protect employees or a community" (*Business Week* 1996: 103). In the era of investor capitalism, the pressures on corporations and the incentives offered to senior executives, ensure that they will devote themselves singlemindedly to generating corporate profit.

The model of the American class structure that we described in Chapter 1 includes top corporate executives in the capitalist class. The preceding discussion of corporate ownership and governance explains why. Top executives are richly rewarded. Their compensation ties their interests directly to the stockholder's. Although the CEOs of the largest corporations might not hold large proportions of their company's stock, they are given the opportunity to buy shares advantageously. CEOs can easily accumulate substantial fortunes. Smaller corporations are generally owned and managed by families or small groups of wealthy investors—most of whom we would also include in the capitalist class.

THE NATIONAL CAPITALIST CLASS: SOCIAL BASIS

Parallel to the economic basis for a national capitalist class in the corporate economy, there is a social basis in an upper-class social world built on prestige and exclusive patterns of association. Among the institutions identified with this world are the select prep school, the *Social Register*, and the proper metropolitan men's club. These three have been widely used by researchers as formal indicators of membership of a socially defined upper-class (Domhoff 1970: 9–23).

We have already described the appearance of the *Social Register* in early industrial America, when the new rich were being socially merged with the established upper classes. Although occasionally capricious in its inclusions and exclusions, a century after its creation, the *Social Register* remains, as Mills' once described it "the only list of registered families...the nearest thing to an official status center that this country, with no aristocratic past, no court society, no truly capital city, possesses..." (1956: 57).

A small circle of prestigious prep schools, such as St. Paul's (New Hampshire), Hotchkiss (Connecticut), Foxcroft (Virginia), and Chapin (New York), draw their students from upper-class families (Domhoff 1970: 9–32). These day schools and boarding schools tend to be concentrated in the Northeast, but they draw many students from throughout the country. They are secular or nominally Episcopalian (the religious affiliation most common in the upper class) and traditionally single-sex institutions, although many have become coeducational in recent years. A few are older than the republic, but most were founded or experienced their major expansion around the time the *Social Register* appeared. They have, moreover, served a similar function: the integration of old prestige and new money (Baltzell 1958: 292–319).

Prep school graduates are informally referred to as "preppies," a term with mixed undertones of admiration and contempt, which is used more loosely to refer to various aspects of an upper-class lifestyle (Birnbach 1980). The extension of the term is not inappropriate. The style and values the prep schools inculcate in their students equip them for participation in an upper-class social community. The network of personal ties that develop among prep school students and their families will serve them well in their subsequent careers and social lives.

The prep schools contribute to a pattern of upper-class endogamy by bringing students and their siblings into contact with potential marriage partners, both directly and through debutante parties and other upper-class social functions to which prep school students are likely to be invited (Blumberg and Paul 1975).

Most major metropolitan areas have one or two men's clubs, such as New York's Knickerbocker, San Francisco's Pacific Union, or Philadelphia's Philadelphia Club, with a distinctly upper-class membership. The major clubs perform much the same function as the *Social Register* and the prep schools: They indicate who can be considered part of "proper society." They also provide an informal setting where upper-class associations can be developed and maintained and, on occasion, important business or political matters can be discussed free from outside scrutiny (Baltzell 1958: 336–354; Domhoff 1970, 1974). Thomas Dye (1995: 192) found that most corporate leaders belong to private clubs and about one third belong to one of few dozen exclusive metropolitan clubs.

The prep schools, top universities, and metropolitan men's clubs draw not only from their own regions but from a national upper-class population. For example, more than 30 percent of the members of San Francisco's Bohemian Club reside outside the San Francisco area (Domhoff 1974: 30). The national scope of the upper class is also evident in marriage patterns; the society page study referred to earlier found that 54 percent of the marriages reported were intercity (Blumberg and Paul 1975: 74). The *Social Register,* perhaps in recognition of the national character of the upper class, has recently abandoned its city-by-city editions and begun publishing a single national edition.

To what extent is the national social class we have been describing identical with the national wealth class examined in the previous section? We would be surprised if the two did not overlap substantially because our studies of wealth and prestige on a local level suggest an intimate connection. In Chapter 3, we noted that virtually all the founders of great fortunes in the late nineteenth or early twentieth centuries had traceable descendants listed in the *Social Register* by 1940. More recently, Blumberg and Paul (1975) tabulated occupational data on the fathers of brides and grooms in their society page study; nearly 60 percent of the men whose occupations were identified were corporate executives. Using *Social Register* listings, education at top prep schools, and membership in the exclusive metropolitan clubs as criteria, Domhoff (1967: 51) found that 53 percent of the directors of the top manufacturing and financial corporations could be classified as upper class. Using slightly broader criteria, Dye (1979: 170) concluded that only 30 percent of the presidents and directors of major corporations were of upper class *origins*. These (somewhat dated) studies suggest a significant overlap between the top wealth and prestige classes at the highest levels of American society.

For the student of national power arrangements, the significance of the link between upper-class society and corporate wealth is related to the problem of cohesion raised earlier. The achievement of consensus on specific policy issues and more generally the maintenance of class solidarity is made easier because those who own and control the major concentrations of national wealth encounter each other in a private sphere of informal relations. The schools and clubs are merely the outer manifestations of this realm whose deeper meaning resides in shared experience, intimacy, and the bonds of friendship and kinship that produce a consciousness of common identity and shared values. Domhoff points to group dynamics research in social psychology, which has established that physical proximity among the members of a group, frequent association, a group reputation of high prestige, and an informal atmosphere all contribute to group solidarity. These are characteristic features of the upper-class world we have been describing and prepare us for another basic conclusion of the same research: "Members of socially cohesive groups are more open to the opinions of other members and more likely to change their views to those of other members" (Domhoff 1974: 89–90, 96). Baltzell, in his earlier study of upper-class Philadelphia, made a related point, which he phrased as social control: An upper-class community inculcates and sustains "a mutually understood code of conduct" in its members. Upper-class people are especially subject to the "norms and sanctions of their peers. A man caught in an act of dishonesty or disloyalty fears, above all, the criticism of his class of lifelong friends" (Baltzell 1958: 61).

Unfortunately, most of the research in this area dates from the 1960s and 1970s. We cannot say how upper-class society was affected by the growing concentration of wealth and transformation of the corporate world in the 1980s and 1990s. Classical stratification theory would lead us to expect economic position and social prestige to align with one another in the long run.

MECHANISMS OF CAPITALIST-CLASS POWER: PARTICIPATION IN GOVERNMENT

In the next few pages, we will discuss the institutional mechanisms that extend the power of the capitalist class well beyond the economic sphere that is its immediate concern. Among the most potent of these mechanisms is direct participation in national government, especially at the top levels of the executive branch. Although few recent presidents have been drawn from the upper class, they have typically placed upper-class men in key cabinet positions and leaned on them for advice. Journalist David Halberstam captured the essence of this dependence in the opening pages of a perceptive book on the Kennedy-Johnson years. He records a meeting between John F. Kennedy, the month before he took office, and Robert Lovett, a Wall Street investment banker with impeccable social credentials and elaborate corporate connections—in Halberstam's words, "the very embodiment of the Establishment." Kennedy tried to persuade Lovett, a Republican who had voted for Kennedy's opponent, to accept a major cabinet post:

> Lovett declined regretfully…, explaining that he had been ill… Again Kennedy complained about his lack of knowledge of the right people, but Lovett told him not to worry, he and his friends would supply him with lists. Take Treasury, for instance—there Kennedy would want a man of national reputation, a skilled professional, well known and respected by the banking houses. There were Henry Alexander at Morgan, and Jack McCloy at Chase [Manhattan Bank], and Gene Black at the World Bank. Doug Dillon too. Lovett said he didn't know their politics. Well, he reconsidered, he knew McCloy was an independent Republican, and Dillon had served in a Republican administration, but, he added, he did not know the politics of Black and Alexander at all (their real politics of course being business). At State, Kennedy wanted someone who would reassure European governments: They discussed names, and Lovett pushed, as would Dean Acheson, the name of someone little known to the voters, a young fellow who had been a particular favorite of General Marshall's—Dean Rusk over at [the] Rockefeller [Foundation]. He handled himself very well, said Lovett. The atmosphere was not unlike a college faculty, but Rusk had stayed above it, handled the various cliques very well. A very sound man (Halberstam 1972: 16–17).

Ironically, the three top cabinet appointments made by Kennedy, a liberal Democrat, reflected the advice of Lovett and others like him. As secretary of the treasury, he chose C. Douglas Dillon, an investment banker connected with Dillon, Read, and Company, a major Wall Street firm started by Dillon's father; as secretary of defense, Robert McNamara, who had just been made president of Ford Motor Company; and as secretary of state, Dean Rusk, then president of the Rockefeller Foundation, a man with many admirers in corporate circles (Burch 1980: 175–177).

Cabinet recruitment studies show that Kennedy's appointments followed a pattern established by his predecessors and maintained by his successors. Although there is probably no one today who could speak for "the establishment" with the authority Robert Lovell carried in 1960, cabinet officers are still overwhelmingly drawn from the top of the class structure, most notably from the national capitalist class and the national prestige class we have been describing. Mintz (1975) examined the social and occupational backgrounds of people who served in the cabinet between 1897 and 1973. She found that nearly two-thirds of the fathers of cabinet officers had professional or managerial occupations. Seventy-eight percent of the cabinet officers themselves had been corporate officials or partners in corporate law firms; at least 60 percent could be classified as members of a national social upper class on the basis of *Social Register* listings, exclusive prep school attendance, or membership in any one of a long list of selective social clubs; and more than half of the cabinet members fit both the corporate and social upper-class categories.

Using slightly different criteria, Burch concluded that nearly two-thirds of cabinet officers and major diplomatic appointees from 1933 to 1980 were linked to the corporate world or sizable family fortunes. Prominently represented among these officials were corporate lawyers and investment bankers—two groups that appear to play a crucial mediating role between business and government, akin to the coordinating roles they play within the corporate world (Burch 1980: 378–384). Our own, more focused survey of the men who held the top cabinet posts (State, Defense, and Treasury) from Kennedy's inaugural cabinet in 1961 through Clinton's in 1993 reveals that 68 percent were drawn from major industrial corporations, financial institutions, or corporate law firms (*Who's Who in American Politics 1979; Who's Who in America 1980*, Dye 1995:81-85).

If business is well represented among top federal decision makers, there is no parallel representation of labor. Since 1913, only six men have served in the cabinet who were in any way connected with the labor movement. Most served for short terms and were Secretaries of Labor (Burch 1980: 377; *Statesman's Year-Book* 1979–1990).

What can be said of the class background of Congress? Approximately 75 percent of the senators and representatives who served in the 102nd Congress (1991–1992) came to elective office from positions in business and the professions, principally law (data from LTV 1990). Another 15 to 20 percent had spent their entire careers in politics. The representation of blue-collar workers or white-collar employees in the 102nd Congress was scanty indeed. Only two of the 535 members of the Congress had been union leaders—a background that is common for legislators in some Western democracies (data from LTV 1990). Evidence going back to 1906 shows that this pattern of recruitment to Congress has been consistent through the twentieth century (Nagle 1977).

According to the financial disclosure forms that members of Congress file annually, at least 28 percent of Senators and 14 percent of House members were worth $1 million or more in 1994. Unfortunately, the information required on the forms is sketchy and tends to understate the value of assets. For example, Representative Amo Houghton, member of a prominent industrial family, was worth $10 million according to his disclosure form, but $350 million according to *Forbes* (Lazerow 1995). Senator Daniel Patrick Moynihan was probably close to the truth when he observed, "At least half of the members of the Senate today are millionaires. That has changed the body. We've become a plutocracy . . . The Senate was meant to represent the states; instead it represents the interests of a class" (*New York Times*, Nov. 25, 1984).

Congress, then, is recruited from the upper levels of the class structure, although the typical senator or representative appears to be drawn from a stratum a notch below that of the typical cabinet officer. He or she is also more likely to be a career politician rather than a top executive or corporate lawyer on temporary assignment. The law and business backgrounds of many members of Congress suggest a smaller-scale, more localized version of the corporate world so richly represented in the cabinet. If the cabinet is recruited from the national capitalist class, the Congress draws on local upper-middle and capitalist classes.

What do cabinet and congressional recruitment patterns tell us about national politics? It certainly cannot be argued that class background allows us to predict the political behavior of individual decision makers. For example, Edward Kennedy, a liberal Democrat, and John Warner, a conservative Republican, who seldom vote together, are among the wealthiest members of the Senate. But we have already seen that the behavior and opinions of people in the aggregate are shaped in important ways by class position; in the next chapter, we will see that this generalization can be applied to politics. Moreover, we have noted how informal association shapes receptivity to opinions. The congressman whose personal associations are largely upper class or upper-middle class is likely to be more open to viewpoints common at the top of the class structure than to those prevalent toward the bottom.

The precise effect of these tendencies is not easy to gauge, but we can point to certain issues, such as health care and unemployment, that are of much greater concern to working-class people than to people of higher class rank. The failure of the United States to develop a national health care system or adopt policies designed to counter high unemployment rates may be related to the recruitment patterns of cabinets and congresses.

MONEY AND POLITICS

A month before the 1996 elections, a bill passed the U.S. Senate that was of particular interest to Frederick Smith, founder and chairman of Federal Express Corporation, a man with a fortune estimated at $415 million (*Forbes* 1996:354). Smith's fortune is likely to expand further as the result of a brief technical provision inserted in the bill specifically to benefit Federal Express. The controversial provision, for which Smith personally lobbied senators, will make it virtually impossible for the company's 100,000 workers to form a labor union. Federal Express, critics note, has long been generous to lawmakers and their campaign treasuries (Lewis 1996). A few weeks earlier President Clinton had signed a bill raising the minimum wage. The President had little to say about more than $20 billion in corporate tax breaks incongruously included to a law conceived to benefit the working poor. The business interests that benefited from the tax provisions made sizable campaign donations to members of Congress, especially those on the committees responsible for writing the legislation (Pianin 1997).

These two laws, passed during the 1996 campaign season, illustrate the power and limits of political money. Although labor unions and their friends in Congress lobbied hard to defeat the Federal Express provision, they were defenseless in the face of the company's well-financed efforts. On the other hand, business interests fought a long battle against the minimum wage bill. But they were unable to stop an extremely popular piece of legislation in the midst of an election campaign. The tax concessions added to the bill in its final stages were, at once, a reward for campaign contributions and a corporate consolation prize.

Money is not the only significant political resource. But it is the one political resource that the rich have in particular abundance. This fact has long excited popular suspicion and encouraged periodic attempts to regulate political money. In the wake of the Watergate scandals, which drove Richard Nixon from office in 1974, Congress passed reform legislation. The Nixon campaign had solicited enormous contributions from corporations, accepted illegal donations and, in at least one instance, changed federal policy in exchange for a contribution.

The Watergate-era reform laws limit individual campaign donations to $1,000 per candidate in each election and $25,000 annually to all federal candidates; restrict group donations, which must be made through registered Political Action Committees (PACs), to $5,000 per candidate per campaign; provide for federal financing of presidential campaigns on a matching basis; and require public disclosure of campaign donations and expenditures.

The major effect of the legislation was to constrain the influences of very rich individuals on candidates while increasing the political weight of certain organized groups, including corporations. In the 1972 campaign, some individuals gave hundreds of thousands of dollars to candidates for federal office. Direct contributions to candidates on that scale were no longer legal. As a result, the wealthy contributor became less important than the corporate PAC.

The weight of political action committees in federal campaigns had grown rapidly since the 1970s as has the proportion of PAC money from corporate sources. The law does not permit corporations or labor unions to contribute money from their treasuries to political candidates. But they can use their resources to run PACs that solicit donations from union members or corporate executives and stockholders. Unions pioneered this form of fundraising, but by 1988, corporate PACs were contributing nearly three times as much as labor PACs to federal candidates (Makinson 1990: 19).

The Watergate legislation did not sever the traditional connection between wealth and politics, but it altered the character of the relationship between them. The limit on individual campaign donations—still relatively high from the viewpoint of middle-income people—compelled political fund raisers to broaden their class focus. Candidates for federal office, who still depend on individual contributors for the greater part of their campaign financing, must spend long hours chatting with upper-middle-class and upper-class people at fund-raising cocktail parties.

Who makes campaign contributions? Studies going back to the 1920s indicate that contributors are, not surprisingly, better educated, higher in occupational status and richer than the average American (Brown, et al 1995: 7; Sorauf 1992: 3, 34). A survey of contributors to the 1988 primary campaigns of presidential candidates, indicates that most donors had incomes in excess of $100,000—a level superior to the incomes of 95 percent of the population (See Table 8-4).

TABLE 8-4 Family Incomes of Contributors to Presidential Primary Campaigns

	Percent
Under $50,000	18
$50,000 to $100,000	21
$100,000 to $250,000	30
Over $250,000	30
Total	100

NOTE: Refers to contributions of $201 to $1,000, the legal limit. Such contributions accounted for more than 70 percent of the aggregate.

SOURCE: Brown, et al. 1995: 41.

Although the Watergate reforms capped individual donations at $1,000 per candidate and total giving to federal candidates at $25,000, they did not succeed in wholly eliminating very large contributions. As interpreted by the courts, the legislation does not limit *independent* expenditures made by individuals or groups on behalf of a candidate, as long as they do not co-ordinate their spending with the candidate's campaign organization. Organizations as varied as the National Rifle Association and the AFL-CIO have run independent ad campaigns in federal elections. The law also does not restrict so-called "soft money" donations that escape federal regulation by being channeled through party organizations. Nor does it limit the amounts that candidates or their families can spend on their own campaigns, except in the case of presidential contenders who accept federal financing.

Self-financing is especially important for nonincumbent candidates. In 1988, for example, new members of the House of Representatives spent an average of $80,000 of their own money—more cash than the average citizen has on hand—in their successful campaigns (Makinson 1990: 3). In 1994, thirty-nine House candidates each put more than $200,000 of their own money into their campaigns. Some wealthy men appear to have bought their way into the U.S. Senate. Senators Jay Rockefeller, Herb Kohl, and Bill Frist all spent millions on their own initial campaigns. In 1996, Michael Huffington, a famously underqualified candidate, was narrowly defeated after spending $30 million of his own money in a California race against a well-financed incumbent (Alexander and Haggerty 1987; Eismeier and Pollock 1996; Makinson 1990).

Soft money is the biggest loophole in the Watergate legislation. Political parties can collect unlimited sums of soft money from individuals, corporations, and unions. Although these funds cannot be spent directly on the campaigns of federal candidates, they can be used for "party-building activities," such as get-out-the-vote drives, that benefit them. In the 1996 campaigns, soft money was spent on television advertising featuring candidates but not explicitly urging a vote for them. The lax regulation of soft-money permits the wealthy to escape the usual contribution limits. In 1988, some 249 individuals gave $100,000 or more to Republican soft-money accounts. A single individual gave $500,000. Two years earlier, Joan Kroc, heir to the McDonald's fast-food fortune, gave the Democrats a $1 million (Makinson 1990: 11). Although labor unions make soft money donations, their contributions are modest relative to the amounts flowing from executives and other business sources (*New York Times*, Oct. 18 and Dec. 25, 1996).

In the 1996 campaign, the regulatory regime created in the 1970s, was virtually swept away by a flood of soft money contributions. Democrats and Republicans collected three times as much soft money they had in 1992, accounting for 60 percent of total contributions to the national party committees. Pushing the legal envelope, the parties spent millions of dollars in soft money on television ads featuring their Presidential candidates, but stopping just short of asking people to vote for them. Throughout the campaign, both parties aggressively pursued large contributors. Donors of $250,000 or more became part of privileged circle with special access to Republican leaders. President Clinton invited major contributors to sleep at the White House. Expensive overnights in the Lincoln bedroom became an easy target for stand-up comedians.

Political money has long been a source of power for the wealthy and the corporations. By capping contributions, the Watergate reforms reduced the influence of individual donors—but probably increased the collective influence of the privileged. Instead of appealing to a few extremely wealthy individuals or major corporations, a candidate was forced to appeal to many donors willing to give $500 or $1000 and to PACs limited to $5,000. From the viewpoint of the capitalist class, the reorganization of political giving was a phenomenon parallel to the earlier reorganization of corporate ownership: Both contribute to upper-class cohesion by making class interests and power more collective and less individual. The expanding role of soft money is undermining this system, especially in Presidential races.

BUSINESS LOBBIES

We have examined the political influence of the capitalist class exercised through direct participation in government and through campaign finance. We now turn to a third channel of influence: persuading public officials to adopt (or abandon) particular policies. Especially since the emergence of the modern corporate economy, business representatives have actively lobbied the Congress and the federal departments and regulatory agencies that carry legislation into practice. At the turn of the century, corporate lobbies—backed by abundant flows of cash—moved Congress with dependable ease. For example, in 1913, a lobbyist for the National Association of Manufacturers (NAM) publicly acknowledged that he had bought legislative favors with bribes and influenced House leaders to appoint congressmen favorable to NAM to House committees and subcommittees

(*Congressional Quarterly* 1976: 654, 662). In recent decades, major business lobbies have generally employed more subtle methods, partly because competing interests are better organized and the possibilities of unfavorable publicity are greater.

Currently, the principal business lobby organizations in Washington are the U.S. Chamber of Commerce and the Business Roundtable. Also important are the National Association of Manufacturers, which speaks for smaller industrial corporations, and the increasingly influential National Federation of Independent Business, representing small business. The Chamber gains special strength from its ability to mobilize pressure on individual members of Congress through local affiliates. The local capitalist class, which dominates the affiliates, is likely to include important elements of the power structure in a legislator's home district, as well as people who belong to the same social networks as the legislator. When the Washington office wants to pressure senators or representatives on a vote, it can systematically mobilize letters, phone calls, and personal visits from a local business owner, a legislator's former law partner, or a fellow member of the local country club.

The Business Roundtable consists of the chief executive officers (CEOs) of approximately 200 of the largest corporations. Its power is based on the formidable resources controlled by these corporations and the prestige of those who lead them. The Roundtable typically operates more quietly than the Chamber. Its stock-in-trade is the personal visit from a CEO and the carefully crafted economic study or legal brief supporting its position. The leaders of the Roundtable have access to members of the House and Senate and even to the president—entree that no ordinary lobbyist could hope to duplicate. A congressional aide commented, "A visit from a CEO has an unbelievable impact, as perhaps it should. It shows a commitment" (Green 1979: 29).

All these business lobby organizations and many smaller business groups collaborated to defeat plans for a national health care insurance system introduced by the Clinton administration in 1993 (Johnson and Broder 1996: 194–224, 601–636). The administration hoped to counter the accelerating inflation in health care costs and extend coverage to the millions of Americans who had no health care insurance. Approximately 15 percent of the population—concentrated among the young and the working poor—had no health care protection. The figure had been rising in the 1980s partly because employers were eliminating benefits to reduce labor costs.

Bill Clinton's promise to establish universal coverage helped elect him in 1992. National polls consistently showed that approximately 70 percent of adults agreed with the proposition that health insurance should be guaranteed to all. Despite this remarkable consensus, the administration's plan never even came to vote in Congress and alternative plans to extend coverage were equally unsuccessful.

Organized business spent more than $100 million to fight the administration—enough to finance a national presidential campaign (Johnson and

Broder 1996: 212). In fact, the opponents hired political professional experience in electoral politics and employed the sophisticated tactics of a national election campaign. Their efforts included targeted campaign donations, a well-staffed lobbying effort on Capitol Hill, regular public opinion polling, a large-scale television advertising campaign, phone banks, and field operations in key states and Congressional districts. Supporters of national health care (including labor unions) simply did not have the resources to operate on the same scale as the opponents. Their efforts were weak and disorganized in comparison.

The administration's inept political strategy, divisions among Democrats in Congress, and controversial aspects of the Clinton plan itself undeniably contributed to the administration's defeat. But, given popular support for universal health care, a compromised plan might well have emerged from Congress without massive opposition from organized business.

There are, as pluralist writers would remind us, a multitude of lobby groups operating in Washington, many of them directly opposed to business lobbies on key issues. But as the defeat of national health care suggests, their mere existence is not confirmation of the pluralist image of the political order as an athletic league in which all major social interests are represented with more-or-less-equal strength. Some interests are not represented at all. Others cannot compete in the same league as the representatives of business. Business lobby efforts are certainly better financed and probably better organized than those of any other sector of the society.

The superior organization of business interests is in part a reflection of a more general phenomenon referred to earlier: People toward the top of the class structure have higher rates of social participation. Only a minority of blue-collar workers are members of labor unions, but virtually every business belongs to some representative organization.

The class element in U.S. politics becomes clearest when issues arise that pit a liberal lobby alliance often led by labor unions against a conservative lobby alliance led by business groups. Since the late-1970s, the business alliance has tended to win such confrontations.

POLICY-PLANNING GROUPS

A step removed from the conflictual world of political campaigns and legislative battles is a quieter and less visible realm of organizations dedicated to formulating and disseminating broad proposals for national policy. We have already noted the influence of one of them, the Urban Land Institute, which was set up to study urban renewal prospects. Like that institute, the most influential policy organizations—groups such as the Council on Foreign Relations, the Council for Economic Development, and the Business Council—have been created and financed by the corporate elite, which plays a prominent role in their activities. A social-network analysis of the boards of major corporations, memberships of exclusive men's clubs, and participants in key policy-planning organizations reveals an elaborate pattern of interconnections (Domhoff 1975).

Similar to the policy groups, both in their functions and their links to the national upper class, are the major charitable foundations and the policy research "think tanks." Foundations such as Rockefeller, Ford, Lilly, and Kellogg (all named for the wealthy families that endowed them) fund research and pilot projects to test policy ideas. Many of the best-known think tanks are clustered in Washington, where they can feed their research findings and policy recommendations to sympathetic politicians, lobbyists, and journalists. Among the best known think tanks are the Brookings Institution, which has particularly influenced Democratic policy makers, and the Heritage Foundation and the American Enterprise Institute (AEI), which are influential among Republicans.

One key function of the policy groups, foundations, and think tanks is to back the careers of appropriately conservative public policy intellectuals, many of whom are channeled into government positions. When Ronald Reagan became president, he filled thirty-four positions with people from AEI and thirty-six with people from Heritage. At the time, Walter Wriston, AEI board member and chairman of Citicorp bank, described AEI's influence with a line from a Barry Manilow song: "I write the songs the [whole] world sings" (Blumenthal 1986: 54).

Wriston himself had good reason to sing. The prominence of AEI reflected, as we will see later in this chapter, the success a long-term corporate strategy to enlarge business influence in national politics.

INDIRECT MECHANISMS OF CAPITALIST-CLASS INFLUENCE

Capitalist-class influence over government is not limited to the direct means we have been describing (recruitment to decision-making positions, campaign financing, lobbying, and domination of policy-planning institutions). The capitalist class can also affect government policy indirectly, through its control of the economy and the mass media. A defining characteristic of a capitalist society is the existence of a relatively small class that controls most productive wealth and therefore independently makes investment decisions that can decisively affect the welfare of other classes. Although governments in capitalist societies have limited control over what business leaders do, their political fortunes are closely linked to business decisions. The connection is often phrased as "business confidence." If business leaders lack confidence in a government or its policies, they are not likely to risk their capital in new investments. The resulting decline in aggregate level of investment will soon be reflected in a rising level of unemployment, which, in turn, will subject the government to pressure from an electorate dissatisfied with the state of the economy. If the government wants to rectify the situation without making basic changes in the capitalist economic order, it must find a way to regain the confidence of investors. What sorts of government policies are likely to alienate business confidence? Basically, any that threaten business profits, from "excessive" taxation of corporate income to the imposition of expensive regulations designed to reduce pollution or guarantee worker safety. The precise factors that lead to a loss of confidence are less important than the essential fact that this mechanism gives the capitalist class an indirect veto over government policy.

Two aspects of the business-confidence veto are particularly worth noting. One is that it can influence a government without actually curtailing investment. The mere risk of such action is enough to persuade decision makers to reconsider a proposed policy. The possible effect of government action on investor behavior is frequently raised as an issue in public policy debates. The other is that the veto mechanism does not require conscious, concerted action by members of the capitalist class to be effective. Isolated investment decisions made on the basis of an objective assessment of potential risk and profitability can collectively produce a downturn in business activity and subject a government to popular pressure (Bloch 1977).

Governments are also subject to the limits imposed by private control of the mass media, through which people receive information about public affairs. In the United States, virtually all significant media are owned by the local and national capitalist classes (though they might just as well be organized as cooperatives, like the respected French paper Le Monde, as semiautonomous public bodies, like the British Broadcasting Company, or as organs of political parties, like a number of European papers). Control of the media has become highly concentrated, and the principal media organizations are typically major corporations or owned by major corporations. The most notable exception to this rule is the Public Broadcasting System (PBS), which only has a small portion of the resources or audience of the commercial networks.

Four TV networks provide Americans with most of the news they hear: ABC, CBS, NBC, and CNN. ABC and CNN are owned by large media conglomerates. NBC and CBS are owned by General Electric and Westinghouse, respectively. The corporations that own the two most influential daily newspapers, the New York Times and the Washington Post, and the two major newsmagazines, Time and Newsweek are also among the country's largest corporations. Fifteen newspaper chains account for more than half of the country's daily newspaper circulation. Most American newspapers get their national and international news from a single source, Associated Press (AP), a cooperative owned by the media it serves (Dye 1995: 113).

Capitalist-class influence over the media is not limited to the power of ownership. Since the media are operated for private profit and most of their income comes from corporate advertising, media managers are sensitive to pressure from advertisers. The networks are also subject to the influence of the affiliated stations that broadcast their programs to local audiences. Early in the history of television broadcasting, Edward R. Murrow's brilliant and controversial current affairs program, "See It Now," carried by CBS, was forced off the air after its corporate sponsor withdrew and no regular replacement could be found. In the 1950s, programs that dealt with the issue of racial discrimination or employed black actors could not appear on network television because corporate sponsors refused to be associated with them, and some affiliates (particularly in the South) refused to carry them. Today, there are few confrontations between advertisers and networks. As an executive of a major advertising agency explained, direct interference by the advertiser is seldom necessary, "because the producers involved and the writers involved are normally pretty well aware of what might not be acceptable" (Barnouw 1978: 54; see also Tuchman 1974).

As the ad man's comment suggests, the main power that the capitalist class exercises over the media is the power to impose implicit limits on what is "acceptable." The media, in turn, operate on their audiences not by imposing specific ideas but by defining the subjects that are appropriate for consideration and delineating the range of reasonable opinion. In other words, they define the public agenda. (Their ability to do so is increased by the concentration of media control.) Thus, until the late 1950s, racial inequity was not a national issue, though it was most certainly a serious national problem. During the Cold War, the question that so troubled Mills—the possibility that our leaders were pursuing policies that could carry us toward nuclear annihilation—was never raised in a serious fashion by the major media.

THE CAPITALIST-CLASS RESURGENCE

Who has the power? This is the question that pluralists, elitists, and class theorists are trying to answer in the debate we examined at the beginning of this chapter. The question assumes that the distribution of power is stable—it does not vary over time. But this is probably not a safe assumption. It seems clear, for example, that the capitalist class and business interests in the United States have gained at the expense of other competitors for power during what we have called the Age of Growing Inequality (Blumenthal 1986; Blumenthal and Edsall 1988; Edsall 1984; Ginsberg and Shefter 1990).

In the early 1970s, there was a growing sense of vulnerability and declining power in capitalist circles. The economic system was changing in ways that seemed threatening and unpredictable. Wages had been rising; profits had been stagnating; productivity had been declining. The international economy, dominated by the United States since the end of World War II, was becoming much more competitive. The U.S. economy was twice shaken in the 1970s by abrupt leaps in the world price of oil. Increasing government regulation in areas from environmental practices to consumer protection and workplace safety seemed to be raising the cost of doing business. Many business leaders felt politically isolated.

John Harper, then chairman of Alcoa, recalled the period from the happier perspective of the 1980s:

> We [corporate leaders] were not effective. We were not involved. What we were doing wasn't working. All the polls showed business was in disfavor. We didn't think that people understood how the economic system works. We were getting short shrift from Congress. I thought we were powerless in spite of the stories of how we could manipulate everything (Blumenthal 1986: 77).

Since the early 1970s, business leaders have taken a more direct and aggressive role in national politics. Corporations, as we have seen, seized the opportunity presented by the 1974 PAC legislation. Business lobbyists perfected the upscale "grassroots" campaign—mobilizing local business leaders, stockholders, depositors, suppliers, or dealer networks to influence Congress. In 1972, Alcoa's Harper joined other top corporate leaders to found the Business Roundtable. About the same time, the Heritage Foundation was started, with the help of a $250,000 donation from Colorado brewer Joseph Coors, and the American Enterprise Institute began its transformation from an inconsequential research center into a key player in public policy. Both received financial backing from major corporations and foundations endowed by wealthy families such as the Mellons, the Pews, and the Olins (Blumenthal 1986; Edsall 1984: chapter 3).

These efforts to reassert the power of the privileged have paid off. In the late 1970s, business won a series of key legislative battles—for example, defeating both the consumer protection agency bill and labor reform legislation that would have made it easier for unions to organize workers. In earlier chapters, we saw that federal taxes on high incomes, inherited wealth, and corporate earnings have all been reduced since the early 1970s. The real incomes of workers have declined, and the distribution of income has become more concentrated. In the next chapter, we will see that the power of organized labor declined as the power of business grew. The power shift that began in the 1970s was one of the factors that contributed to three successive victories of Republican presidential candidates in the 1980s and the election of a Republican majority to both houses of Congress in 1994.

The last two decades have seen an unmistakable transformation of the structure of power. However, there is no reason to assume that the new power structure is unalterable. The changes that have taken place might themselves set off further change, in directions we cannot now imagine.

CONCLUSION

We began this chapter by reviewing studies of community power structures. Although we reached no definitive resolution of the debate between elite or class and pluralist views, we did learn that local decision makers are overwhelmingly drawn from the upper and upper-middle classes and that local affairs are strongly influenced by the decisions made by national elites, especially corporate elites.

The latter conclusion forced us to think about national power structures. We examined the work of C. Wright Mills, which alerted us to institutional developments that have tended to concentrate power in the leadership of huge organizations. Within Mills' power elite, the corporate sector appeared to be dominant over the political and military sectors. This led us to focus our attention on the national capitalist class, and we learned the following: A small class controls most corporate stock, and although major corporations are typically run by their top executives, there is no division of interest between these hired managers and stockholders. There is apparently—our information on this is somewhat dated—a significant overlap between the national capitalist class and the national upper class represented by such institutions as the *Social Register* and exclusive men's clubs and prep schools; the upper class provides some of the social glue that binds the members of the capitalist class together. Finally, the national capitalist class has powerful means to shape national politics; these include placement of its members in top decision-making positions, campaign financing, lobbying, creation of policy-planning organizations, exercise of the business-confidence veto, and control of the public agenda through the mass media. Looking back over the last two decades, we concluded that the power of the capitalist class has grown.

A pluralist would be quick to point out that neither the formidable array of political resources available to the capitalist class nor recent indications of growing capitalist-class power are definitive proof of domination by that class. Any such broad conclusion would have to be based on a more comprehensive analysis of the political system than is possible within the bounds of this book. We are, however, committed to examining how social classes participate in the political system and how they interact in the political arena. With those goals in mind, we will amplify the picture we have painted here in the next chapter, which deals with class consciousness and conflict between classes in electoral and industrial contexts.

SUGGESTED READINGS

Bottomore, Tom. 1966. *Elites in Modern Society.* New York: Pantheon.

> *Short, lucid survey of elite theory.*

Domhoff, G. William, and Hoyt B. Ballard, eds. 1968. *C. Wright Mills and the Power Elite.* Boston: Beacon.

> *Excellent set of critical essays on Mills' power elite thesis.*

Dye, Thomas R. 1995. *Who's Running America?: The Clinton Years.* 6th edition. Englewood Cliffs, N.J.: Prentice-Hall.

> *The national elite, sector by sector.*

Edsall, Thomas B. 1984. *The New Politics of Inequality.* New York: Norton.

> *The growing power of the affluent in American politics and the resulting shift in national policies.*

Ferguson, Thomas 1995. *Golden Rule: The Investment Theory of Party Competition and the Logic of Money Driven Political Systems.* Chicago: University of Chicago.

A theory of the role of money in American politics, from the nineteenth century to the 1990s.

Johnson, Haynes and David Broder 1996. *The System: The American Way of Politics at the Breaking Point.* Boston: Little Brown.

An engaging, thoughtful account of the Clinton administration's failed attempt to create a national system of universal health care.

Judis, John B. 1991. "Twilight of the Gods." *Wilson Quarterly* 5 (Autumn): 43–57.

Intriguing account of the rise and fall of the "American establishment" of bankers, corporate lawyers, and scholars who once made U.S. foreign policy. Helpful annotated bibliography.

Makinson, Larry and Joshua Goldstein 1996. *Open Secrets: the Encyclopedia of Congressional Money and Politics.* 4th edition. Washington, D.C.: Congressional Quarterly.

A treasure trove of data on campaign financing. Want to know where your member of Congress gets campaign money? Look here.

9

Class Consciousness and Class Conflict

I believe that leaders of the business community, with few exceptions, have chosen to wage a one-sided class war today in this country.

Douglas Fraser, president of the United Auto Workers (1978)

We owe the concept of class consciousness to Karl Marx. Its role within his theory was pivotal, joining individual experience to broad social structures and transforming alienated individual resentment of the capitalist present into decisive striving for the socialist future.

Class consciousness implies an *awareness* of membership in a group defined by a relationship to production, a sense that this shared identity creates *common interests* and a common fate, and, finally, a disposition to take *collective action* in pursuit of pursuit of class interests. At some points in his work, Marx implied that only a group whose members experience such a consciousness can be defined as a class. Elsewhere he carefully distinguished between a *class-in-itself* and a *class-for-itself*. The first is a class in a formal, definitional sense; its members share a social position (defined by the analyst) but are unaware of their common situation. The second is a class in an active, historical sense: Its members are aware of common interests, they engage in militant action focused on goals that they conceive as being in direct opposition to those of other classes. Thus, embodied in Marx's conception of class consciousness—especially in the notion of a class-for-itself—is the expectation of class conflict.

In its fullest sense, class consciousness is not just an aspect of public opinion ("What percentage of blue-collar workers supported Roosevelt?"), but an intense, collective involvement in the events of a critical historical juncture. It develops out of a long series of strikes against bosses who exploit workers and riots against authority that brutalizes the masses. It culminates in urban mobs roaming the streets and burning the buildings that symbolize upper-class domination and in peasants seizing the land they work, and it ends with a revolutionary seizure of power in the name of the oppressed: Paris in 1871, Mexico in 1910, Moscow in 1917, Peking in 1949, Havana in 1959.

Revolution is rare, but simmering class struggle is endemic. Slave revolts, violent strikes, local mobs on a rampage—these occur regularly in many societies. More institutionalized and controlled forms of class struggle, such as union organizing campaigns and political movements that seek legislative power to help the underprivileged, are considered a normal and healthy part of a democratic society. Historians study past revolutions; sociologists usually focus on the initial development of class consciousness, which could, under very specific historical circumstances, lead toward revolutionary consciousness but is more likely to result in peaceful change or historical stagnation. In this chapter, we will examine the extent to which people are aware of sharing a class identity and class interests, the social factors that advance or retard the development of this consciousness, and its relationship to political opinion and behavior. The final sections of the chapter will focus on class conflict as reflected in two arenas: electoral politics and labor relations.

MARX AND THE ORIGINS OF CLASS CONSCIOUSNESS

One of Marx's major objectives was to isolate the social forces that could produce the transformation of a class-in-itself into a class-for-itself; he hoped that by understanding the process, he could determine how to intervene and accelerate it. Specifically, he asked, What inherent tendencies of capitalist society are likely to produce a class-conscious proletariat? Here are the factors that he especially stressed:

1. *Concentration and communication.* The process of industrialization in capitalist society concentrates the proletariat in big cities, working-class neighborhoods, and large factories. This process promotes communication among workers, leading to a recognition of common problems and facilitating efforts at political organization.

2. *Deprivation.* Marx expected a progressive impoverishment of the proletariat, if not in an absolute sense, at least relative to the rising productive capacity of the industrial economy and the wealth of the bourgeoisie.

3. *Economic insecurity.* Marx was convinced that the proletariat's sense of deprivation would be exacerbated by the periodic experience of unemployment during the downturns in the capitalist economy, which, he observed, is quite subject to boom-and-bust cycles.

4. *Alienation at work.* Marx identified the mindless, repetitive, unsatisfying quality of factory-type labor with capitalism. (For a modern rendition of this argument, see Braverman 1974.) Such labor is fundamentally at variance with human nature as Marx understood it and is therefore a spur to the development of class consciousness.

5. *Polarization.* The swings of the capitalist economy drive smaller enterprises out of business; their owners are forced into the proletariat, and control of the economy becomes further concentrated at the top. The result is the steady depletion of the middle ranks and the corresponding development of a society polarized between a minuscule, affluent bourgeois minority and an impoverished proletarian majority.

6. *Homogenization.* Within the proletariat, Marx observed a lowering of skill levels and therefore an equalization of wage levels produced by adaptation to the simple requirements of machine tending in the modern factory. This tendency leads to a less stratified, more homogeneous proletariat, which, because of its shared condition, is more disposed to unified political action.

7. *Organization and struggle.* To defend itself, the proletariat is drawn increasingly into working-class parties and labor organizations. Marx believed that participation in such organizations and the experience of struggle against capitalist employers (backed by the bourgeois state, with its police and armies), would promote the development of a revolutionary class consciousness.

The revolutions that Marx's theory anticipated in the advanced industrial countries never came. However, class-based revolutions in industrializing agrarian states (Mexico, Russia, and China, for example) were a characteristic feature of twentieth-century history. In the industrial nations, working-class parties and labor movements reshaped political systems and economic life. In both cases, the seven factors listed have shaped events. Understood as variables operating under specific circumstances, they can even help us understand the failure of revolution in the advanced countries. For example, Marx expected parallel processes of class polarization and homogenization in capitalist society. Property relations did became polarized: productive property was increasingly concentrated in the hands of a small minority. But homogenization of the proletariat was undercut by occupational differentiation and the corresponding spread in incomes, even among manual workers (see Chapter 3). This process undercut the sense of common identity and the shared experience that are the bases of class consciousness. We might say that Marx was correct in identifying the key sociological processes, even if he was historically wrong in predicting their outcome. Much of this chapter will be devoted to applying his best insights to social and political circumstances he could not have imagined.

RICHARD CENTERS AND CLASS IDENTIFICATION

Contemporary interest in the concept of class consciousness is based on the idea that it provides a link between objective class position (measured by occupation, income, or wealth) and political behavior. That is, we (like Marx) assume that people who recognize and articulate their class position are more likely to promote their class interests. A systematic and sustained effort to investigate this linkage in American society grew out of the work of Richard Centers (1949), who focused on one aspect of class consciousness, *class identity,* the sense of belonging to a particular social class.

Centers began by noting that in previous public opinion surveys (such as the famous one conducted by Fortune magazine in 1940), about 80 percent of Americans called themselves middle class. Some popular writers seized these figures to proclaim that America was almost completely a middle-class country—that if Americans had any class consciousness at all, it simply meant that they mostly thought of themselves as belonging to the same big group. But Centers noticed that the figure quoted came from the following survey item, which offered only three alternatives:

> What social class do you consider that you belong to?
>   1. Upper class
>   2. Middle class
>   3. Lower class

He also noticed that when respondents were asked the question in open-ended form (without a specific list of answers from which to choose), many called themselves working class. Centers made a reputation for himself by adding *working class* to the reply alternatives offered by the *Fortune* survey. When he asked a nationally representative cross section of adult white men which of the four classes they belonged to, the responses were radically

TABLE 9-1 Class Identification

	In Percent	
	Centers (1945)	General Social Survey (1994)
Upper Class	3	3
Middle Class	43	46
Working Class	51	45
Lower Class	1	5
Don't know/Other	2	1
Total	100	100
N =	(1097)	(2992)

SOURCES: Centers 1949 and tabulation from GSS data.

different than those to the 3-choice question posed by *Fortune*. Now the majority of respondents chose the label working class (See Table 9-1).

Centers rightly concluded that Americans do not like the term *lower class* and that this attitude was the main conclusion to be drawn from the *Fortune* survey, not that most Americans thought of themselves as middle class. His confidence in the results of the surveys was strengthened because only a tiny minority of respondents refused to accept one of the labels suggested in the question. Even fewer were inclined to deny the existence of social classes. Centers was moved to say, "The authenticity of these class identifications seems unquestionable" (Centers 1949: 78).

Variants on Centers' class identification question have been used in numerous surveys over the years and gotten roughly similar results. Table 9-1 shows the responses to the Centers' item in the 1994 General Social Survey, a representative national survey of adults conducted annually. The most notable difference in 1994 was the higher, though still small, percentage of adults willing to accept the label "lower class"—reflecting, perhaps the growing number of Americans who feel themselves falling behind the mainstream.

Despite the remarkable durability of his class identification item, Centers' claim of "unquestionable authenticity" for these responses is not quite justified. Surveys that ask class identification without suggesting answer alternatives—so called "open-ended" questions—produce less orderly results. The responses are, predictably, much more varied, including some that deny the existence of class altogether. We cannot, then, use the answers to Centers' "forced choice" question as literal descriptions of the way people freely conceive of their own class positions. We can, however, assume that the forced-choice question is in some sense a measure of class consciousness because, as we will see, responses to it are related to both class position and political attitudes.

CORRELATES OF CLASS IDENTIFICATION

What types of persons chose the particular labels offered in class-identification surveys? Centers regarded occupation as the principal basis of class identification. When he sorted respondents by occupation, he found that 70 percent or more of professionals and businessmen considered themselves middle class, while more than 70 percent of manual workers considered themselves working class. A series of election-year surveys conducted from 1956 to 1968 by the University of Michigan's Survey Research Center (SRC) obtained similar results (Schreiber and Nygreen 1970). But as even these figures indicate, there was not complete consensus. Centers' data conform to a pattern that is by now familiar: The results were fairly clear-cut at the extremes of the class structure but somewhat ambiguous in the middle. In particular, more than a third of sales and office workers among Centers' respondents and about half of that group among the SRC respondents labeled themselves working class (Centers 1949: 86, Hamilton 1975).

Centers' view that occupation is the main determinant of identification is supported by an analysis of national survey data by Hodge and Treiman's (1968). They found that occupation (of the family's main earner) was a stronger predictor of class identification than either family income or respondent's education, but the three variables considered together left much of the variance in class identification unaccounted for. In other words, objective class position seemed to be a crucial but not decisive determinant of this aspect of class consciousness. Hodge and Treiman found one additional factor that was significantly and independently related to class identification: association—one of our basic class variables. The class positions of friends, neighbors, and kin were strong influences on the formation of class identification. People with largely high-status associations, whatever their own position, were more likely to identify as middle class. Likewise, those with predominantly low-status associations were more likely to think of themselves as working class. Put differently, your class identification depends partly on your objective class position and partly on whom you know.

MARRIED WOMEN AND CLASS IDENTIFICATION

Centers' original class-identification surveys referred only to men, but the SRC data cover both men and women. If we are interested in the relationship between objective class position and identification, the inclusion of women raises an intriguing problem: Does a working wife base her class identification on her husband's occupation, her own occupation, or some combination of the two? Hodge and Treiman implicitly allowed for the influence of the wife's job on the self-placement of both spouses when they measured earnings by *family* income. But by defining occupation for all respondents as "main earner's occupation"—the husband's in most households—they ducked these questions.

Recent research suggests that the class identification of working wives is influenced by both their own and their husband's occupations—though the influence of the husband's job is typically stronger (Sorensen 1994: 34-5). Beeghley and Cochran (1988) wondered why this might be true. Their analysis of data from several national surveys concludes that the key factor is the wife's attitude toward gender roles. The surveys asked respondents whether they supported the Equal Rights Amendment (ERA) and whether they believed that a woman should hold a job if her husband was able to support her. Using answers to these items, respondents' attitudes toward gender roles were classified as traditional or egalitarian. As expected, working wives with traditional gender views base their class identity on their husband's occupation. Those with egalitarian views considered their own and their husband's jobs, along with other class factors pertaining to both spouses.

However, the effect of wives' attitudes on the overall pattern of class identification for men and women is probably modest. The reason is that wives' and husbands' occupations are highly correlated. That is, men with better jobs tend to be married to women with better jobs. Few wives of white-collar men hold blue-collar jobs. Although the proportion of wives of working-class men holding white-collar positions is larger, it is likely that such women are concentrated in the lowest sorts of clerical or sales positions.

We can speculate that women and men answering the class identification question are not simply weighing two occupations but are thinking of a standard and style of living, a set of associations, and particular values and attitudes that are shared by the members of a household. Traditionally, all these things depended largely on the husband's job, and it was reasonable to assume that the class positions of most individuals were socially fixed by the characteristics of the (male) heads of their families. But rising female participation in the labor force, growing family dependence on dual incomes, and corresponding changes in gender norms undermine this conventional wisdom. Research on the class identification of working wives is consistent with the idea that family continues to be the relevant basis of class position, but the standing of families is no longer solely dependent on the activities of men.

CLASS IDENTIFICATION, POLITICAL OPINION, AND VOTING

Our interest in subjective class identification, like our concern with class consciousness generally, stems from the idea that consciousness is a critical link between objective class position and political attitudes or behavior. If there is anything to the notion of class identification, we would, for example, *expect manual workers who identify themselves as working class to take liberal positions on economic issues and manual workers who identify themselves as middle class to take more conservative position.* We would expect a similar pattern in candidate preferences in elections.

After Centers had satisfied himself that class identification was closely correlated with occupation, he explored its relationship to ideological differences. For this purpose, he asked respondents a series of questions designed to measure attitudes toward such matters as union activism and government social programs. On the basis of their answers, he placed respondents on ideological scale, ranging from "ultra-radical" to "ultra-conservative." Centers demonstrated that answers to the class identification question were independently predictive of position on the ideological scale. As he expected, the most conservative answers came from white-collar respondents who identified as middle class; the most radical, from blue-collar respondents who thought of themselves as working class. These results support his contention that class identification, which he used as an indicator of class consciousness, is a strong predictor of political orientation (Centers 1949).

Subjective class identification should also be predictive of voting behavior. Table 9-2 shows how class identification influenced choices in the 1992 presidential election at different income levels. Note the large gap, in the vote for Clinton, between low-income working-class identifiers and high-income middle-class identifiers. Clearly, income and consciousness of class position were both shaping voter preferences.

THE ORIGINS OF WORKING-CLASS CONSCIOUSNESS IN COMPARATIVE PERSPECTIVE

If class consciousness is a predictor of class-oriented voting behavior, perhaps we can turn the equation around and use voting behavior as an indirect way of getting at the origins of class consciousness. To do so, we need to answer this question: What are the social factors that predict class-oriented voting? Two sociologists, Lipset (1960) and Szymanski (1978), synthesized the relevant research on working-class voters in Western Democracies. Nearly all the studies they examined present data on working-class support for leftist or liberal parties—in the United States, the Democratic party, and in Europe, communist and socialist parties. We can summarize much of what Lipset and Szymanski found with the following list of social correlates of leftist or liberal voting among workers:

Lower skill level
Unemployment
Union membership
Larger city
Larger workplace
Economically advanced region
Minority ethnic or religious group
Specific occupations: miner, fisherman, sailor, longshoreman, forestry worker
Male
Working-class origin

TABLE 9-2 Class Identification and Presidential Preference, 1992

| | *Percent Voting for Clinton* | | |
| | Class Identification | | |
Income Level	*Working*	*Middle*	*Total*
Under $25,000	58	48	54
$25,000–$50,000	47	39	43
Over $50,000	47	36	38
Total	52	39	45
N =	(435)	(381)	

NOTE: Working includes lower; middle includes upper-class identification (see Table 9-1).
SOURCE: Tabulated from 1994 GSS survey.

These variables are, by and large, consistent with Marx's theorizing about class consciousness. For example, Marx's emphasis on "concentration and communication" anticipates the significance of larger workplaces, big cities, and occupations such as miner and longshoreman that tend to isolate workers in separate communities (Kerr and Siegal 1954: 190–191). Marx stressed the sense of economic insecurity produced by unemployment and the influence of organization, represented here by union membership. His emphasis on alienation at work is reflected in the significance of lower skill levels and larger workplaces. On the other hand, Marx did not attach great significance to gender (which has not, until very recently, played an important role in U.S. elections) or ethnicity.

BOTT: FRAMES OF REFERENCE

Class identification gives us an indirect way of measuring class consciousness and estimating its political importance. But class identification tells us little or nothing about how people actually turn their experience of the class structure into some notion of their own place within it. Elizabeth Bott has pointed out that "people do not experience their objective class position as a single clearly defined status." We might add that such clear definitions are the result of calculated decisions on the part of academic researchers; they create concepts such as "upper-middle class" or "bourgeoisie" and through hard thinking attach some specific empirical criteria for membership. (Granted, they may start with words in popular usage, but by the time they have finished their ratiocinations, the original words have taken on new meanings.) Naturally, they endow their concepts with connotations that derive from the researchers' own general philosophy; thus, Warner thought of prestige strata, and the Marxists thought of actual or latent conflict groups.

Then the researchers go into the field and try to discover the degree to which the populace thinks as the concepts suggest they should, and the investigators feel a growing sense of triumph the more closely they can fit the data to the concepts.

But, said Bott:

> When an individual talks about class, he is trying to say something, in a symbolic form, about his experiences of power and prestige in his actual membership groups both past and present. These membership groups—place of work, friends, neighbors, family, etc.—have little intrinsic connection with one another, especially in a large city, and each of the groups has its own pattern of organization. The psychological situation for the individual, therefore, is one of belonging to a number of segregated, unconnected groups, each with its own system of prestige and power. When he is comparing himself with other people or placing himself in the widest social context, he manufactures a notion of his general social position out of these segregated group memberships... The group memberships are not differentiated and related to one another; they are telescoped and condensed into one general notion (1954: 262).

The man or woman on the street is aided in conceptualizing by ideas and terms that have diffused into popular culture from intellectual debate. Thus, especially in Europe, there are many factory workers who have long been subjected to propaganda that stems from Marxist ideology. Naturally, they not only use class-conflict terminology; they perceive their own position and interpret their everyday experiences in terms of conflict. Similarly, the American middle classes have been bombarded with propaganda about our equality and absence of classes. Consequently, they tend to perceive as individual differences experiences that a European would see as common class experiences.

Therefore, any individual's self-perception in a stratification order is a combination of (1) actual experiences in a wide variety of contexts in many membership groups and (2) verbal theories about society, which are usually vague and somewhat contradictory common-sense notions that have filtered down from the theorizing of intellectuals and propagandists. Consequently, the social reality of identification that we are studying is complex rather than simple, and when we simplify it (as we must for certain purposes) into categories such as middle class or working class, we do violence to the original facts. The simpler and neater the scheme, the further it is from reality.

Let us complicate the picture even more. Bott went on to point out that

> The individual performs a telescoping procedure on other people as well as on himself. If they are people who have the same, or similar, group memberships as himself, he is likely to feel that they have the same general position and belong to his own class. If they are outsiders, his knowledge of them will be indirect and incomplete so that there is plenty of room for projection and distortion...

> The suggestion advanced here is that there are three steps in the creation of a class reference group: First, he internalizes the norms of his primary membership groups—place of work, colleagues, friends, neighborhood, family—together with some more hazy notions about the wider society; secondly, he performs an act of conceptualization in reducing these segregated norms to a common denominator; thirdly, he projects his conceptualization back on to society at large... The main point is that the individual himself is an active agent. He does not simply internalize the norms of class which have an independent external existence. He takes in the norms of certain actual groups, works them over, and constructs class reference groups out of them (1954: 263–265).

Finally, Bott reminded us that because the conceptualizations were both hazy and tied to a variety of actual experiences in the life history of an individual, they could shift in the course of an interview. Sometimes respondents would think of the people they knew as children in their hometown; sometimes of their workmates; sometimes of their dreams for their own children. These shifts, plus those of the forms of the questions being asked, would lead them to shift frames of reference, changing their conception of the class system or their own place within it.

ELECTIONS AND THE DEMOCRATIC CLASS STRUGGLE

Although no advanced industrial country has experienced the convulsive class revolution envisioned in the *Communist Manifesto,* most have passed through periods of bitter class confrontation and continue to experience less dramatic, institutionalized struggles over conflicting class interests. Class conflict is especially evident in two realms: electoral politics and labor relations. Most of the remainder of this chapter will be devoted to these areas.

Elections in modern democracies have been characterized as manifestations of "democratic class struggle" in recognition of the representative role of political parties (Anderson and Davidson 1943; Lipset 1960: chapter 7). Synthesizing the available evidence in 1960, Lipset wrote:

> Even though many parties renounce the principle of class conflict or loyalty, an analysis of their appeals and their support suggests that they do represent the interests of different classes. On a world scale, the principal generalization which can be made is that parties are primarily based on either lower classes or the middle and upper classes (p. 230).

In most parliamentary systems, parties can be arrayed on a spectrum from right to left, with the former upholding the interests of the privileged classes, and the latter attacking them on behalf of the less fortunate. In Great Britain, for example, the Labor Party has traditionally drawn working-class support, while the Conservatives have run strong among managers and professionals. In France, manual workers lean toward the

Socialists (and, in the recent past, the Communists); business owners, executives, and professionals and farmers tend toward the right-wing parties.

The relationship between classes and parties has traditionally been looser in the United States than in most Western democracies. Democratic political systems typically have at least one major party that identifies itself as socialist and presents itself as a partisan of the working class (for example, the Communist, Socialist, Social Democratic, and Labor parties of Western Europe). The Democratic Party in the United States has traditionally been regarded as the party of the "common man," but it has never called itself socialist, and it has become increasingly coy about appealing directly to working-class interests. Nevertheless, the Democrats have done better among working-class voters than among middle-class voters for as long as anyone has bothered to keep track (Lipset 1960: chapter 7; Abramson, Aldrich, and Rohde 1995: 153)

The percentage gap between professional-managerial and blue-collar voters in support for the Democrats has typically been modest (6 to 10 percent) and smaller than corresponding differences in European democracies.[1] The income data in Table 9-3 suggest greater polarization—though that impression is exaggerated by the difference between extreme categories (under $15,000 and over $100,000, each accounting for about 10 percent of voters).

Explanations for the "peculiarism" of the American party system often revolve around the role of ethnicity and religion in our politics. Blacks, Jews, and Catholics are more likely to vote Democratic, whatever their social class, than are white Protestants. Abramson and his associates (1995: 158-9) summarize evidence on Protestantism, Catholicism, and class as follows:

> [T]he effects of social class and religion are cumulative. In every [presidential] election from 1944 through 1992, working-class Catholics have been more likely to vote Democratic than any other class-religion combination. In all thirteen elections, white middle-class Protestants have been the least likely to vote Democratic. They are more consistent over time than any other group we have studied. An absolute majority voted Republican in every election from 1944 through 1988, and Bush won half of their total vote in the three-way contest of 1992.

The American party system is rooted in the relationships among class, ethnicity, and politics as they were worked out in the period 1928 to 1948. Samuel Lubell (1956) has interpreted this period in a classic study of electoral behavior, emphasizing the emergence of the urban ethnic working class. The generation preceding 1925 witnessed the greatest mass immigration that this country has ever experienced, and most of the newcomers

---

[1] *New York Times*, Nov. 6, 1986, and Nov. 8, 1990. For some comparisons see Frears 1988, Daalder and Koole 1988, Charlot 1985.

TABLE 9-3 Presidential Preference by Income

	1992				1996			
	Clinton	Bush	Perot	Total	Clinton	Dole	Perot	Total
Under $15,000	58	23	19	100	60	29	11	100
$15,000–$30,000	45	35	20	100	54	37	9	100
$30,000–$50,000	41	38	21	100	49	41	10	100
$50,000–$100,000	39	44	17	100	44	49	7	100
Over $100,000	36	48	16	100	39	55	6	100

House Vote 1994

	Democrat	Republican	Total
	62	38	100
	52	48	100
	49	51	100
	46	54	100
	45	55	100

NOTE: Family income in current dollars, not inflation adjusted.

SOURCE: *New York Times*, November 13, 1994, and November 10, 1996.

went to work in the factories of the big cities. Simultaneously, many farmers were streaming into the cities, where they could make more money in the new industries; during and after the First World War, this internal migration included hundreds of thousands of blacks. By the 1920s, these people and their children were becoming voters in great numbers. For the first time, the urban workers approached dominance in national politics—not until then did more people live in cities than on farms. And these urban workers were primarily Democrats; by 1928, the Democrats outvoted the Republicans in most of the big cities. Lubell's central thesis is that although economic is-sues have always been important in our politics, in earlier years they were more closely connected with regional interests and conflicts, but by 1928 the parties had taken on the color of class parties—with the Republicans repre-senting business (and successful farmers) and the Democrats representing workers (and less successful farmers). This one split tended to override re-gional differences and divisions based on noneconomic issues (though of course they were still significant):

> Never having known anything but city life, this new generation was bound to develop a different attitude toward the role of government from that of Americans born on farms or in small towns. To Herbert Hoover, "rugged individualism" evoked nostalgic memories of a rural self-suffi-ciency in which a thrifty, toiling farmer had to look to the marketplace for only the last fifth of his needs. The Iowa homestead on which Hoover grew up produced all of its own vegetables, its own soap, its own bread...

> In the city, though, the issue has always been man against man. What bowed the back of the factory worker prematurely were not hardships inflicted by Mother Nature but by human nature. He was completely dependent on a money wage . . . A philosophy that called for "leaving things alone" to work themselves out seemed either unreal or hypocritical in the cities, where nearly every condition of living groaned for reform . . . If only God could make a tree, only the government could make a park (Lubell 1956: 33–34).

Franklin Delano Roosevelt, who came to office in 1933 during the depths of the Depression, was the first president to take advantage of these developments. He built an electoral coalition of Catholic and Jewish immigrants and their sons and daughters, blacks (in the urban North where they had the vote), and working-class people generally. As we will see later in this chapter, the labor movement grew rapidly under Roosevelt, and unionized blue-collar workers were among his strongest supporters. Roosevelt also managed to hang on to white Southerners, who had traditionally supported the Democratic Party.

The Democratic Party that emerged from the New Deal period was a party with strong working-class support but was not a working-class party. Ethnicity blurred class lines. Most white Protestant workers supported the Democrats, but some of them (more than among Jewish or Catholic workers) favored the Republicans. The Democratic Party also had important upper-class backers, drawn from the "ethnic rich"—men such as John F. Kennedy's multimillionaire father Joseph, a Catholic, or Jewish financier Bernard Baruch. The ethnic rich bankrolled the party, and, although they were probably more ideologically flexible than their Protestant counterparts in the Republican party, they were nonetheless a conservatizing influence on their own party.

The New Deal coalition that Roosevelt built in the 1930s enabled the Democrats to dominate American politics for decades. But beginning in 1968, the Republicans aggressively challenged the political status quo. At the same time, the New Deal coalition was losing internal cohesion. The result, apparent by the 1980s, was not a decisive realignment in national politics such as occurred under Roosevelt but, rather, an inconclusive standoff. The Republicans won most presidential elections after 1968. Although the Democrats regained the White House under Bill Clinton in 1992 and retained it in 1996, they did so by blunting the party's appeals to working-class interests. In 1994, the Democrats lost control of both Houses of Congress for the first time in 40 years.

This remaking of national politics was connected with important changes in American society. In the prosperous years after World War II, many of the children and grandchildren of immigrants moved into the middle and upper-middle classes. Even if they retained their Democratic affiliation, these people often became more conservative and more open to the political message of the Republicans. They were increasingly likely, for example, to see themselves as beleaguered taxpayers rather than as beneficiaries of government programs. The class differentiation of the descendants of Catholic and Jewish immigrants convinced Democratic leaders to dilute their party's already weak working-class identity. As second- and third-generation Americans moved to the suburbs, the urban Democratic political machines they had supported went into decline, undercutting the party's ability to mobilize voters.

At the same time, Democratic positions on issues such as Vietnam, affirmative action, welfare, and abortion were straining the loyalty of working-class supporters. These issues, which were ably exploited by the Republicans, divided a large sector of blue-collar Democrats from the party's upper-middle-class activists. Affirmative action issues in the workplace split black and white working-class Democrats, especially under the unstable economic conditions that prevailed in the 1970s and 1980s. Of course, the Democrats lost much of their traditional white support in the South because of the party's identification with the cause of civil rights.

Finally, the relative weights of business and labor in national politics began to change in the 1970s. In response to an increasingly competitive economic environment, organized business interests began to lobby more aggressively, give more money to politicians, and confront the labor movement more directly. In a shifting economy, the proportion of unionized workers in the labor force was declining—a trend accelerated by the policies of Republican administrations. Organized labor had been one of the mainstays of the New Deal coalition. Now its ability to mobilize blue-collar voters for the Democrats, its capacity to finance political campaigns, and the strength of its voice in Washington were all waning. Under these circumstances, Democratic officeholders inevitably became less attentive to the needs of organized labor.

The class differentiation of the descendants of immigrants, the emergence of issues that divided traditional Democratic constituencies, the decline of labor, and the gains of organized business interests in national politics all contributed to the decline of the New Deal coalition and further diluted the role of class in American politics.

CLASS, ETHNICITY, AND POLITICS

Richard Hamilton's *Class and Politics in the United States* (1972) remains, many years after its publication, the most thorough analysis of class in American politics. Hamilton's work is especially valuable because it consistently takes both class and ethnicity into account. In this section, we will try to update Hamilton by comparing his analysis of the 1964 election with our own of the 1984 and by examining other, more recent, data on class and ethnicity. The 1964 and 1984 presidential elections make a neat comparison: In 1964, liberal Democrat Lyndon Johnson crushed conservative Republican

TABLE 9-4 Presidential Preference, by Class and Ethnicity:
Percent Voting for Democratic Presidential Candidate

	1964		1984	
Class	WASPS	Ethnics	WASPS	Ethnics
Upper middle	37	74	18	40
Lower middle	52	84	37	49
Upper working	68	76	27	43
Lower working	81	92	35	48

SOURCES: 1964 from Hamilton 1972: 437; 1984 computed for this table from Center from Political Studies 1984.

Barry Goldwater; in 1984, conservative Republican Ronald Reagan (the man who nominated Goldwater in 1964) subjected liberal Democrat Walter Mondale to an equally humiliating defeat.

Table 9-4 compares our 1984 analysis with Hamilton's results from 1964. Data in both cases were drawn from the National Election Survey series conducted by the University of Michigan's Survey Research Center. Following Hamilton, we divided respondents into four classes by distinguishing working from middle class (manual versus nonmanual occupations) and then splitting both into upper and lower groupings on the basis of income.[2] We also employed Hamilton's two ethnic categories, which we retitled for brevity's sake as WASPS and Ethnics. The first refers to white Protestants and the second to all others—notably, African-Americans, Catholics (including most Hispanics), and Jews. (The 1984 figures in the table may not be perfectly compatible with Hamilton's from 1964, but we can use the table to compare the general pattern of presidential preference in the two elections.)

The overall pattern in the table is easy to anticipate. In both contests, Ethnics and people toward the bottom of the class hierarchy were more likely to vote for a liberal Democrat. But note that preference is much more decisively structured by class among WASPS than among Ethnics. In fact, among Ethnic voters, class virtually washes out as an influence.

An apparent anomaly in the 1984 data is the relatively high Democratic vote among lower-middle-class WASPS. This grouping was more likely to favor Democrats than any another class among WASPS. This phenomenon may reflect the proliferation of routinized, low-wage office jobs, which

[2] Respondents constituting a national sample of adults were placed in occupational categories (derived from the 1980 census) based on the most recent job held by the head of the respondent's household. Managerial, professional, technical, sales, and clerical jobs were considered middle class. All others were considered working class, including farm occupations. The middle class was divided into two categories at $30,000 in 1983 income. The working class was divided at $20,000.

TABLE 9-5   Support for Increased Spending on Social Programs,
by Occupation (1984)

Occupation	Percent
Managerial	30
Professional	32
Sales	27
Technical	40
Clerical	46
Crafts	40
Operatives	52
Service	46
Laborers	48
Farm	28
All occupations	39

SOURCE: Center for Political Studies 1984 (computed for this table).

probably do not merit the label lower-middle class. (Our class model, introduced in Chapter 1, classifies such jobs as working class.)

The 1984 data reinforce one of Hamilton's basic points: that upper-middle-class WASPS are an extraordinarily conservative segment of the electorate. In 1964, they were the *only class* to give Goldwater majority support. By 1984, the entire electorate had moved to the right, but upper-middle-class WASPS still stood out in their support for a conservative Republican, separated from both upper-middle-class Ethnics and lower-middle-class WASPS by approximately 20 points. Hamilton demonstrated, using national survey data going back to the 1950s, that they were likely to take very conservative stands on national issues.

On the basis of the same data, Hamilton argued that the electorate was, on the whole, quite liberal on social and economic issues—certainly more liberal that anyone would judge from national policy. How did his claim hold up in the conservative 1980s? Table 9-5, based on the 1984 survey, shows considerable support for spending on domestic social programs. Of course, from 1964 to the mid-1970s, there was an enormous expansion of federal spending on such programs. But by 1984, the national political debate focused on which programs to cut and by how much. Ironically, there was more support for liberal programs than anyone could guess from politicians' speeches or media coverage.

In the 1984 survey, people were asked whether they favored cutting, maintaining, or increasing federal spending on social security, food stamps, public assistance (welfare), medical care, and job programs. The table indicates, by occupation, the percentage of respondents who favored expanding at least three of these five programs. About 40 percent of those interviewed ("all occupations") favored such increases, and fewer than

TABLE 9-6   Percent Voting by Family Income, 1994

	Percent
Under $5,000	19.9
$5,000–$10,000	23.3
$10,000–$15,000	32.7
$15,000–$25,000	39.9
$25,000–$35,000	44.4
$35,000–$50,000	49.8
Over $50,000	60.1
All Voters	44.6

SOURCE: Klinkner 1996: 75 (also http://www.census.gov/ftp/pub/population/socdemo/voting).

20 percent wanted reductions in three or more programs.[3] There is an obvious class difference in opinions on this issue, with a gap of approximately 20 percent separating upper-middle-class occupations and the lower-working-class categories.

CLASS AND POLITICAL PARTICIPATION

There is a paradox, of which Hamilton is well aware in his analysis of American political orientations. If, with the exception of an upper-middle-class minority, Americans tend to take liberal positions on bread-and-butter economic issues and identify themselves as Democrats, how is it that liberal Democrats are not consistently elected to office and liberal measures are not regularly enacted?

One reason is that the people toward the top of the class structure are better informed politically and more likely to participate in political activities than those toward the bottom. Surveys consistently show that voter participation is higher at higher income levels. There is also some evidence that this class gap in participation is slowly widening (Teixeira 1992: chapter. 3). In the November 1994 balloting, for example, people from families with incomes over $50,000 were about three times as likely to vote as those with incomes under $10,000 (see Table 9-6). Thus, middle-to upper-middle-income households are overrepresented. The result is a middle-class electorate.

Of course, astute politicians can read the numbers, and they tend to craft their message for the middle class. A logical alternative for the

---

[3] The second figure, from the same source, is not included in the table.

TABLE 9-7 Presidential Preference by Occupation

	Percent of Major Party Voters Voting Democratic			
	1956	*1968*	*1980*	*1992*
Professionals	26	40	43	61
Managers	41	38	29	49
Owners & Proprietors	44	49	32	36
Lower White Collar	39	50	42	63
Skilled Workers	47	56	61	60
Semi/Unskilled Workers	53	56	43	67
All Major Party Voters	43	50	42	58

NOTE: Refers to voters in experienced labor force.
SOURCE: Hout et al. 1995: 818.

Democrats, suggested by sometime presidential candidate Jesse Jackson among others, is to match a liberal platform appealing to working-class interests with a massive effort to increase the electoral participation of lower-income people, both black and white. But many Democratic leaders are reluctant to take this path, fearing that it will send more middle-class voters to the Republicans and exacerbate the party's problems with white voters generally (Ginsberg and Shefter 1990: 15).

TRENDS IN CLASS PARTISANSHIP

Is social class a declining influence in U.S. elections? Many students of American politics are convinced that it is. This conclusion is supported by comparisons between working-class (blue-collar) and middle-class (white-collar) voters. The gap between these two broad classes in the percentage voting for Democrats fluctuates from election to election. But over a period of several decades, it has tended to shrink, especially among white voters (Campbell, Gurin, and Miller 1954; Pomper 1989; Abramson 1995:152-154). By this measure, class voting is declining.

One problem with this analysis is the loose manner in which class is defined. It is probably a mistake to assume that the significance of a white or blue collar has been stable over a period of forty or fifty years. Using more detailed occupational categories, Hout and two associates find change, but not decline in class voting.

Change is most evident among the white collar categories (see Table 9-7). Traditionally Republican professionals and lower white-collar workers are becoming more Democratic, despite their white collars. This shift is consistent with what we learned in Chapter 3 about occupational change in late industrial and postindustrial societies. An increasing proportion of professionals are salaried employees in relatively modest positions. And lower white-collar workers have been moving closer to routine blue-collar workers in pay, prestige, and nature of their jobs. We should not be surprised that they are drawn to the Democrats. At the same time, normally Republican managers and owners are becoming even more Republican. The vote gap between these two business categories and the manual worker categories is actually growing, suggesting increased class polarization.

These trends and the steep income gradient in voter preferences that we noted earlier in the chapter prove that class remains a strong force in American electoral politics.

CLASS CONFLICT AND THE LABOR MOVEMENT

The preceding sections focused on electoral politics as an arena of "democratic class struggle." Here we shift our attention to industrial conflict, the confrontation between capitalists and laborers in the workplace. American labor history has been distinguished by an ironic combination of violent struggle and limited class consciousness. Violence has grown out of tenacious capitalist resistance—not so much to specific economic demands as to the very right of workers to organize labor unions. The class consciousness of American workers has been limited, in the sense that they and their leaders have typically sought circumscribed goals—basically union recognition, economic security, and decent working conditions. Labor unions, as we will see, have played a key role in supporting the passage of liberal welfare measures whose main benefits have flowed not simply to their own members but to working-class and lower-class people generally. But unions have rarely sought a fundamental reordering of economic and political relationships designed to benefit the working class—which is to say that they have not developed a socialist class consciousness. In both ways, the American experience differs from that of industrial democracies generally. Elsewhere, employer resistance has been less evident, and the labor movement has been more committed (in rhetoric at least) to a socialism. In regard to the first point, a leading historian observed on the eve of the New Deal:

> Employers in no other country, with the possible exception of those in the metal and machine trades of France, have so persistently, so vigorously, at such costs, and with such a conviction of serving a cause, opposed and fought trade unions as the American employing class. In no other Western country have employers been so much aided in their opposition to unions by the civil authorities, the armed forces of government, and their courts (Lorwin 1933: 355, cited in Greenstone 1977: 19).

Nearly a half century later, an American labor economist writing on industrial conflict in Europe noted, "The resistance of U.S. employers to unionizing efforts has no serious counterpart in Europe today" (Kassalow 1978: 97, cited in Brody 1980: 248).

In Chapter 3, we sketched the history of the labor movement to World War I. A brief review of the decades that followed will place contemporary developments in a meaningful framework. World War I was a period of rapid expansion for the unions, especially those belonging to the American Federation of Labor (AFL), which had formed a wartime political alliance with the Wilson administration. Free to organize workers without government or employer harassment, the AFL unions grew from 2 million to 4 million members in two years (Brooks 1971: 134).

Coasting on its wartime momentum, the AFL continued to grow and gained some important strike victories in the immediate postwar period. But a reaction soon set in, fed by a national "red scare." In 1919, 300,000 steelworkers nationwide went on strike, protesting twelve-hour workdays and bare-subsistence wages. The response of employers and the federal and local governments that supported them was decisive and often brutal (Brooks 1971: 139–144; Boyer and Morais 1975: 202–209).

The steel strike collapsed after three bitter months. Its fate was indicative of what was to come in the 1920s. In particular, it revealed many of the weaknesses of working-class consciousness and organizations, as well as the effectiveness of the techniques used by the capitalists to resist unionization. The strike showed a working class divided by differences in skill level, race, and ethnicity. Blacks were used as strikebreaking workers. Language diversity kept many immigrant workers from communicating with one another or with native-born members of their class. All these divisions were systematically exploited by employers.

Clearly, violence played a major role in the suppression of strikes. Anti-union violence was common because local and national authorities generally sided with the capitalists. In effect, civil liberties were routinely suspended in strike situations. Without the normal protection of the laws, workers could be physically intimidated (often by thugs hired for this purpose), union organizers harassed, and leaders jailed. Much of this activity was coordinated by special firms that sold "union-busting" services (Litwack 1962: 95–115).

Resistance to the labor movement was so effective in the 1920s that the proportion of the nonagricultural labor force organized in unions declined from approximately 20 to only 10 percent (Brooks 1971: 148). Most of the core that remained by 1930 was in the relatively conservative AFL craft unions, which showed little interest in organizing the masses of less-skilled mass-production workers employed in large-scale industry. At this late date, even the legal right of workers to bargain with their employers through labor unions remained in doubt. In short, there was little to indicate that the American labor movement was on the verge of an era of militant action, dynamic growth, and expanding national power.

The change, of course, came with the Great Depression and the concomitant shift in national politics, particularly during the years 1933 to 1937. A labor historian has described this period as "the highwater mark of class struggle in modern American history" (Davis 1980: 47). The transformation of labor relations that emerged from this period was the product of a clash between intense worker militancy and dogged capitalist resistance. When the United Textile Workers announced an industrywide strike in 1934, *Fibre and Fabric,* the New England trade journal, declared, "A few hundred funerals will have a quieting influence" (*Fortune* 1937: 122). Before this bitter, violent strike had completed its three-week run, thousands of National Guard troops had been mobilized in seven states, and twelve strikers and one deputy had been killed. The union lost.

But during these years, there were more labor victories than defeats. Some of the most significant were gained through a new tactic, the sitdown strike; workers forced concessions from employers by taking physical control of the workplace. First used in the rubber industry in 1936, the innovation, which appealed to the militant mood of workers at the time, spread rapidly. One labor official remembers 1937 as the year he received calls daily saying, "My name is Mary Jones; I'm a soda jerk at Liggett's; we've thrown the manager out, and we've got the keys. What do we do now?" (Brooks 1971: 180).

By 1938, the right to union representation had been written into law through the National Labor Relations Act, known as the Wagner Act, and unions had successfully established themselves in the mass-production industries, such as steel, automobiles, rubber, and electrical goods, which stood at the center of the American economy. A conjunction of social and political developments made these accomplishments possible. Of critical importance were the developments in overcoming the division within the working class and the labor movement that had plagued earlier unionization drives. The significance of ethnic differences declined as the sons and daughters of immigrants joined the labor force. No longer cut off from one another by language barriers, more confident of their place in American society than their parents had been, and more demanding of their rights, these second-generation Americans helped recast labor relations just as they contributed to the revamping of national electoral politics.

Differences between skilled and unskilled or semiskilled workers in manufacturing did not disappear, but a barrier to unionization was removed when the Committee for Industrial Organization (CIO) was formed within the AFL in 1935, with the explicit purpose of organizing workers on an industry-by-industry basis rather than on the craft basis that was typical of the AFL unions. The following year, the more aggressive CIO broke with the tradition-bound AFL, retitling itself the Congress of Industrial Organizations. The CIO strove to remove another source of weakness by organizing both black and white workers.

As CIO organizers set about their task, the American working class was in an extraordinarily militant mood. The story of the drugstore soda jerk may be apocryphal, but it suggests the atmosphere of the times. The rank and file frequently ran ahead of union organizers, who found themselves forced to restrain premature action they were not in a position to support.

Working-class solidarity grew to the extent that big strikes attracted workers from other industries and localities, who came to offer moral and even physical support (Greenstone 1977: 44).

A key to the worker militancy of the 1930s was the experience of the Depression. We have seen that economic insecurity feeds class consciousness. The insecurity that workers experienced during this period was connected to the breakdown of the entire economic system. The confidence that workers had in their employers was shattered by the recognition that even such powerful companies as U.S. Steel and General Motors were subject to the vagaries of the marketplace. When capitalists responded to the Depression by laying off workers, reducing benefits, and speeding up work, they lost the loyalty of many workers. An elderly Ford worker complained that he had devoted his working life to "helping to create a millionaire," and an unemployed machinist about to lose his home concluded, "The bankers and industrialists who have been running our country have proved their utter inability, or indifference, to put the country in a better condition" (Brody 1980: 77).

But neither the reduced factionalization nor the growing militancy of the working class could have accomplished the transformation of industrial relations of the 1930s without the changed political context represented by the New Deal. As the experience of the 1920s made clear, if government was hostile to labor, or at least so indifferent as to ignore patently illegal forms of employer resistance, unionization was impossible. The new state of affairs was clear from the passage of the Wagner Act in 1935. The act guaranteed the right of workers to form labor unions, prohibited employers from interfering with the exercise of that right, and set up a National Labor Relations Board (NLRB) thereafter, to ensure that these provisions were carried out in practice. Shortly thereafter, the LaFollette Civil Liberties Investigating Committee began hearings "which, through exposure, largely neutralized the repressive weapons heretofore used in the fight against trade unionism" (Brody 1980: 139).

Both the labor act and the LaFollette investigation would have been unthinkable under earlier administrations. Ironically, neither was encouraged by Roosevelt, whose own attitude toward labor was ambivalent (Brody 1980: 138–146). He did not initiate many of these reforms but was so dependent on working-class political support that he was forced to go along with them.

THE POSTWAR ARMISTICE

The organizational base that the labor movement had built in the 1930s was strengthened during World War II, through close cooperation between the government and the unions in support of the war effort. By 1945, approximately 36 percent of the nonagricultural labor force was organized (Dubofsky 1980: 9). A wave of strikes (which had been prohibited during the war) in 1945 and 1946 gained substantial wage increases for workers, in sharp contrast with the repression of unions in the wake of World War I.

Yet the labor movement was not, as some thought at the time, invulnerable. In 1947, capitalist-class interests gained passage of an important piece of antilabor legislation, the Taft-Hartley Act. In the next couple of years, as a new "red scare" developed, an upheaval within the labor movement resulted in the removal of Communists and other leftists from the positions of leadership they had held in some unions and the expulsion of several left-leaning unions from the CIO. The Taft-Hartley Act made it easier for employers to resist union organizing, and the loss of the left deprived the union movement of its most effective organizers, people who had played a major role in the union drives of the 1930s. Both contributed to the collapse of a major organizing effort in the South and severely weakened other efforts to extend union coverage.

Despite these defeats, the labor movement had for the first time established a secure place for itself. In the process, it had significantly reformulated the relationship between the capitalist and working classes in key areas of the economy. Unionization not only provided mechanisms to press for better wages and hours but also afforded some control over working conditions and protection from the traditional petty tyranny of the foreman. Though the majority of workers were never represented by unions, the unionization of some firms and industries posed an implicit threat that tended to constrain the behavior of capitalists in other sectors.

If the 1930s represented a period of explicit class conflict during which these changes were forced on the capitalist class, the decade after World War II was the time when the details of a class armistice were worked out. When leaders of the major labor organizations met in Washington with key business representatives in 1945, their purpose was to "lay the basis for peace with justice on the home front." The conference was convened by President Truman at the suggestion of conservative Senator Arthur Vandenberg, who told the president, "Responsible management knows that free collective bargaining is here to stay ... and that it must be wholeheartedly accepted" (Brody 1980: 175).

Although many business leaders were willing to accept the existence of unions, they were determined to preserve for the capitalist class what they termed the "right to manage." At stake was participation in decisions regarding such matters as investment (including plant openings and closings), product design and production methods, and the pricing of final products. Had labor gained a share in these decisions, as did some contemporary European unionists, the labor movement could have had meaningful influence on employment and other basic economic questions. Unions would thus have been able to represent, not just their own members but the working class more broadly. But management insisted on the notion of "property rights," refusing to even concede to unions the right to examine the books of the enterprises with which they bargained. The right to manage gradually ceased to be an issue (Brody 1980: 173–213).

By the late 1950s, the shape of the industrial peace was unmistakable. Unions were firmly established among blue-collar workers at the core of the economy in heavy industry. Here they could gain substantial benefits for their members as long as they did not interfere with management preroga-tives—benefits that would allow a large segment of the working class a life of relative affluence. "The labor movement," concluded auto union leader Walter Reuther, "is developing a whole new middle class" (Brody 1980: 192). Serious industrial conflict was banished to the periphery of the econo-my—to the smaller firms, weaker economic sectors, and backward regions (especially the South). If such conflict was relatively infrequent, that was mostly because the labor movement had grown satisfied and unaggressive. By the end of the decade, the giant AFL-CIO (the two had re-merged in 1955) was behaving, in the words of labor economist Richard Lester like a "sleepy monopoly" (Dubofsky 1980: 8).

Oddly enough, labor came to play a more dynamic role in national elec-toral and legislative politics than it did in the workplace. During the 1930s, the labor movement began to cast off its traditional "volunteerism," the deter-mination to steer clear of politics except when its own direct interests were in-volved. In the postwar period, labor emerged as a major supporter of the Democratic Party and a broad array of liberal social and economic programs. Writing in the late 1960s, David Greenstone (1977) concluded that the rela-tionship between labor and the Democratic Party, although unofficial, had come close to that between Social Democratic parties and labor unions in some Western European countries. The political goal of the activists within the labor movement was "to complete the transformation begun by the New Deal and make the Democrats the genuine party of the common man in America" (Brody 1980: 229). Having assumed this burden, labor came to rep-resent not just union members but the working and lower classes generally.

Labor was politically active in two broad arenas: electoral and legisla-tive. In the former, the labor movement promoted liberal candidates, both within and on behalf of the party, raised a substantial part of the party's campaign money, and fielded thousands of campaign workers. In areas such as Detroit, where unions were especially strong, the party and the union's political organization became virtually indistinguishable (Greenstone 1977: 119–140). In Washington, labor maintained a formidable lobbying apparatus. During the 1960s and early 1970s, the labor lobby played a major role in obtaining passage of liberal legislation in such areas as civil rights, health care, minimum-wage protection, public employment programs, nutrition programs for the poor, and occupational health and safety (Greenstone 1977: xvii–xxiii, 319–360). Clearly, the benefits of such legislation flowed toward the lower portions of the class structure. Ironically, union lobbyists could not match their successes in social legisla-tion with victories in labor legislation, such as the long-sought repeal of bothersome portions of the Taft-Hartley Act.

During these years, union lobbyists were frequently the leaders of broad liberal coalitions that confronted business lobbies and other conservative groups over critical pieces of legislation. In effect, a class cleavage ran through the center of national legislative politics. Although that cleavage disappeared from view over many issues (such as the Vietnam war), it was still apparent during the 1990s in the struggles over budget and tax policies and issues such as the minimum wage, parental leave, and health care.

LABOR IN DECLINE

The 1980s were years of devastating decline for the American labor movement. The clearest measure of labor's fate was the drop in the proportion of workers who belonged to unions. Membership rates, which had been slowly eroding since the mid-1950s, plunged in the 1980s. By 1994, less than 11 percent of private sector workers were union members, the lowest since the early 1930s (U.S. Census 1995a: 443; U.S. Labor 1980: 412).

Labor's problems stemmed to some extent from basic shifts in the economy and occupational structure, which we discussed in Chapter 3. Manufacturing, the bastion of union power since the 1930s, was itself in decline. Increasing foreign competition and distorted monetary exchange rates bolstered the sales of imports in the U.S. market and priced American goods out of markets abroad. American manufacturers responded by closing older U.S. plants and moving operations to low-wage havens abroad or to the anti-union states of the Sunbelt. In the course of the decade, one of every three jobs in heavy industry disappeared (Berman 1991: 7).

Employment growth in the 1970s and 1980s was strongest among those categories of workers who are the most difficult to organize: white-collar employees, service workers, employees of small establishments, and female workers. (Most of these groups are among those we have found to be the least class conscious.) The only area in which the unions have made significant gains in recent years is in the public sector, whose future employment growth is uncertain at best. In short, labor's natural economic base has been crumbling.

Labor's organizational decline was soon reflected in political weakness. The 1978 battle over the Labor Reform Bill in Congress gave an early warning of what was to come. Both labor and business regarded the measure, which sought to restore the effectiveness of the 1935 Wagner Act in protecting workers' rights, as a critical test of strength. The legislation was all-important to union hopes of regaining lost ground. Over the years, employers had discovered that they could stave off union organizing efforts by legal maneuvering and other tactics designed to discourage union activists and postpone representation elections. The longer the delays stretched out, the greater the probability that a union drive would collapse. The basic provisions of the bill were designed to guarantee speedy elections to determine whether workers wanted union representation, to ensure prompt decisions from the NLRB in unfair-labor-practice cases, and to stiffen penalties for violation of existing labor laws, such as the firing of union sympathizers.

The bill evoked one of the most extensive and expensive lobby campaigns of the era. The AFL-CIO spent $3 million on its campaign in support of the bill. A coalition of business groups, including the Business Roundtable, National Association of Manufacturers, U.S. Chamber of Commerce, and National Federation of Independent Businessmen, spent approximately $5 million to secure its defeat. Some 20 million pieces of mail were generated by the two campaigns (Cameron 1978: 80). Both made extensive use of state and local affiliates to pressure members of Congress, who found themselves deluged by communications from businessmen and labor leaders in their states or districts (Cameron 1978; Green 1979: 51–53).

The reform bill glided through the House and found strong majority support in the Senate. However, senators sympathetic to business mounted a determined filibuster to keep the matter from coming to vote (an ironic parallel with the problems the bill was intended to remedy), and business lobbies applied strong pressures to sustain the filibuster. Labor supporters failed by two votes to gain the 60 percent of the Senate needed to close off debate and force a vote on the bill.

Unionists were stunned by both the undemocratic character of their defeat and by the composition of the coalition that had confronted them. They were used to the idea that the lesser capitalists represented by the Independent Businessmen, the Chamber of Commerce, and the NAM harbored strong anti-union sentiments. But the fact that the Business Roundtable—representative of the largest American corporations, the firms with which the unions had made their peace in the 1950s—would support initiatives designed to thwart union organizing came as a terrible revelation. It opened old wounds. In the midst of the conflict over the bill, Lane Kirkland, a ranking AFL-CIO official who later became president of the federation, had charged the corporations with embarking on "a campaign to kill the hopes of the most oppressed and deserving workers in this country. It is class warfare..." (Zeitlin 1980: 32). Shortly after the Senate defeat, Douglas Fraser (1978), president of the United Auto Workers, resigned from the semiofficial Labor–Management Group in Washington with a direct attack on his Business Roundtable colleagues in the group: "I believe leaders of the business community, with few exceptions, have chosen to wage a one-sided class war today in this country... The leaders of industry, commerce, and finance in the United States have broken and discarded the fragile, unwritten compact previously existing during a past period of growth and progress."

Ronald Reagan, elected in 1980, proved to be the most anti-union president in decades. Reagan signaled his intentions early in his presidency by firing thousands of striking air traffic controllers and putting their union out of business. Under his administration, the protections of labor law were further diluted by pro-management appointments to the NLRB and the Labor Department. One experienced union lawyer commented that the labor board had come to operate "like a bloodless bureaucratic death squad" (Geoghegan 1991, quoted in Berman 1991: 7).

In the 1978 elections, for the first time, business PACs outspent labor PACs (*New York Times*, Aug. 4, 1981). In Congress, the ability of labor lobbyists to move (or block) legislation was waning.[4] The AFL-CIO's attempt to recoup political losses by throwing the confederation behind a pro-union presidential candidate in 1984 ended in embarrassing failure.

Business, encouraged by the new atmosphere in Washington and often threatened by competition, assumed a more aggressive stance in labor relations. Many firms forced unions to yield "givebacks" of favorable wage rates and working conditions won in earlier negotiations. Others sought to destroy their unions by forcing decertification elections or provoking a strike and hiring nonunion replacement workers. In the Sunbelt, corporations found a legal and political atmosphere in which union organizing was extremely difficult. Even large, established corporations were taking advantage of the services of a militant new breed of labor consulting firms.

The new consultants pose an intriguing contrast with the union-busting specialists of the 1920s and 1930s. The firms that once supplied tear gas, small weapons, and strikebreakers have been supplanted by masters of manipulation, sophisticated in their application of social scientific knowledge. For example, the consultants show management how to convey to employees an artificial sense of participation in company decisions and how to use systems of subtle rewards and punishments to influence employee attitudes or create anxieties that undercut pro-union sentiment. Employers are advised to avoid certain categories of workers, because they are more likely to be receptive to union appeals. Consultants also instruct corporations in the advantages they can glean from calculated violations of poorly enforced labor laws (Langerfeld 1981).

Organized labor contributed to its own problems. After the purges of leftist activists in the late 1940s, the well-paid officers of many unions had grown complacent, in a few cases corrupt, and distant from the problems of ordinary workers. Such leaders might have been qualified to conduct the daily business of established unions in quiet times, but they were less effective against the rising tide of political and economic change.

Labor's defeats in recent years have real implications for the dynamics of the American class system. The unions always spoke for a much wider constituency than their own members. In the workplace and in national politics, however indirectly and imperfectly, organized labor has represented the interests of working-class Americans against those of the capitalist class. At work, the unions set a standard that even the employers of nonunion workers had to acknowledge. In politics, union voters, activists,

---

[4] For example, during the inflation-ravaged decade of the 1980s, labor was unable to obtain an increase in the minimum wage, whose purchasing power steadily declined. Also, in 1991, the unions lost a major legislative battle with business over the U.S.–Mexico free-trade agreement. The union regarded the agreement as a serious threat to wage rates and union organization in the United States.

and campaign contributions backed liberal candidates at all levels. In Washington, labor lobbyists helped usher liberal social and economic measures through Congress. The labor movement remains active on all these fronts, but with much diminished effectiveness.

In the 1990s, there were some signs of a revival of union political influence, but few observers expected the unions to regain anything like their former strength. Labor's long decline altered the balance of power among social classes in ways that will affect this country for years to come.

CONCLUSION

Our concerns in this chapter have been class consciousness and class conflict. Marx first raised the issue of class consciousness because he believed that subjective awareness of one's objective position in the property and occupational system would lead to a sense of belonging with others of similar position and promote conflict with those above or below. Those experiences in turn would generate political beliefs or ideologies that interpreted the world and suggested appropriate forms of organized action to advance class interests.

Contemporary sociologists have used Marx's views to create specific hypotheses that could be tested in empirical research. They wanted to explain the origins of class consciousness as indicated in various forms of verbal response, including use of class terms of self-identification, and they wanted to pinpoint the consequences of identification for political belief and action. They have consistently found, in the United States and other countries, that objective position in the occupational system is an important influence on identity, with manual workers usually considering themselves to be working class and nonmanual workers calling themselves middle class. These terms carry connotations much broader than just a classification of jobs: They imply a way of life, a set of friends, and a system of values. And they cover all members of a family, not just the principal breadwinner, whose job is usually used to define class membership. In the United States, very small percentages of people consider themselves either upper class or lower class (or poor). In response to forced-choice surveys, the vast majority of Americans consider themselves working class or middle class in roughly equal proportions.

But identification flows from other causes besides occupation, so it reflects more than one social fact of life. Particularly important is racial or ethnic membership. When people belong to a minority group suffering from discrimination that leads to economic insecurity and social stigma, they respond by identifying with other underdogs in the system. And underdogs in general develop ideologies of liberal or leftist tone that demand changes in society to promote more equality and justice. These ideologies are even more pronounced among individuals and groups who have had these additional experiences: economic insecurity; a high rate of interaction within their own group and isolation from other groups; two or three generations of similar social status; and membership in organizations, especially unions, that promote class conflict. All these specific indicators, except perhaps for aspects of ethnicity, can be deduced from Marx.

In the Western democratic countries, class consciousness has not expressed itself in the sort of revolutionary upheaval that Marx anticipated. Instead, class conflict has been channeled into electoral competition and labor politics. The democracies typically have a right–left spectrum of parties, with the left parties tending to draw their members from the bottom of the class structure and the right parties from among the privileged.

The United States fits this general pattern, but the class identities of our major parties have always been somewhat blurred by ethnicity and regional traditions, and the class gap in party support has never been as great as the corresponding difference in European systems. From the 1930s until the 1970s, the Democratic Party dominated national politics. The party was built on a working-class base, reinforced by traditional ethnic and regional loyalties. In recent years, the strength of this so-called New Deal (or Democratic) coalition has eroded, changing the shape of national politics.

In labor politics, the American labor movement consolidated its strength at the same time that the New Deal coalition emerged under Roosevelt, in the early 1930s. It became one of the pillars of the Democratic Party. Through the 1930s, American labor history was marked by violence and management resistance to the right of workers to unionize. The post–World War II period produced an "armistice" in labor relations, as the major corporations accepted unions as a fact of life. But that has come undone in recent years. Business, often supported by conservative Republican administrations in Washington, has become much more aggressive in its confrontations with labor. Union strength has declined sharply, contributing in turn to the decline of the Democratic Party. These developments have shifted the class balance in American politics in favor of the capitalist class.

SUGGESTED READINGS

Brody, David. 1980. *Workers in Industrial America: Essays on the 20th Century Struggle.* New York: Oxford University Press.

   A lively introduction to the history and historiographic literature of the labor movement in the twentieth century.

Croteau, David 1995. *Politics and the Class Divide: Working People and the Middle-Class Left.* Philadelphia, Temple University Press.

   Class and political participation: upper-middle class liberal activism vs. working-class fatalism. Based on extensive interviews and participant observation.

Edsall, Thomas, and Mary Edsall. 1991. *Chain Reaction: The Impact of Race, Rights and Taxes in American Politics.* New York: Norton.

   The "wedge" issues that splintered the Democratic class-ethnic coalition.

Ehrenreich, Barbara. 1989. *Fear of Falling: The Inner Life of the Middle Class.* New York: Pantheon Books.

The evolving consciousness of the upper-middle class and its consequences for American politics and culture. Sources range from Hollywood to Dr. Spock.

Geoghegan, Thomas. 1991. *Which Side Are You On: Trying to Be for Labor When It's Flat on Its Back.* New York: Farrar, Strauss & Giroux.

A labor lawyer's personal account of the 1970s and 1980s.

Rieder, Jonathan. 1985. *Canarsie: The Jews and Italians of Brooklyn against Liberalism.* Cambridge, Mass.: Harvard University Press.

Political ethnography. The disintegration of the Democratic coalition from the bottom up.

Thompson, E.P. 1963. *The Making of the English Working Class.* New York: Vintage Press.

Classic historical portrayal of the development of working-class consciousness.

10

The Poor, the Underclass, and Public Policy

We stand at the edge of the greatest era in the life of any nation... Even the greatest of all past civilizations existed on the exploitation of the misery of the many. This nation, this people, this generation, has man's first chance to create a Great Society; a society of success without squalor, beauty without barrenness, works of genius without the wretchedness of poverty.

Lyndon B. Johnson, 1964

We have decided to abjure a glitzy, splashy, high-profile announcement of new [poverty] programs and a grand new strategy. We concluded that there were no obvious things we should be doing that we weren't doing that would work. Keep playing with the same toys. But let's paint them a little shinier.

Bush administration officials, 1991

[T]his legislation will end welfare as we know it.

Bill Clinton, 1996

During the era we have titled the Age of Growing Inequality, Americans added a new word to their political lexicon: *homelessness*. The term was not a recent addition to the dictionary, but it was being used in a new way— to name a social problem. And although homelessness had once referred to people without fixed residence, who drifted from place to place, it was now being applied to people who literally had no access to conventional housing, people who slept in doorways, packing crates, bus stations, and shelters. No one knew exactly how many there were. National estimates suggested that there might be 500,000 homeless on an average night and that some 1.5 million people would find themselves homeless sometime in the course of a year (Rossi 1989: 20–70; Wright 1989: 19–32).

Although 500,000 people could populate a city the size of New Orleans, they constitute only a fraction of a percent of the nation's population. Yet the homeless attracted intense national attention. Homelessness had been little known since the Great Depression of the 1930s. In the 1980s, it suddenly became visible on city streets across the country. There was general recognition that the homeless population was growing and a broader sense that the homeless were simply the ominous tip of the iceberg of poverty.

In fact, poverty rates were rising in the 1980s and early 1990s to levels not seen since the 1960s. In 1995, 36 million Americans were considered poor by the government's official standard, approximately 14 percent of the population (U.S. Census 1996d). And the poor were becoming poorer. In a book on the homeless, sociologist Peter Rossi calculated that there were as many as 7 million unmarried, working-age adults who could be described as "extremely poor"—people with incomes so low that they were one step from homelessness. This was, Rossi found, more than twice as many as there had been in 1970 (Rossi 1989: 72–81). Against a background of rising childhood poverty rates, the nonpartisan National Commission on Children (1991: 124) offered the following observation on hunger in its 1991 report to the president and the Congress:

> While there is debate over the prevalence of childhood hunger in America, there is no doubt that the problem has increased over the past decade. Recent estimates of the number of children who experience hunger range from 2 million to 5.5 million. The increase is closely related to the high rates of childhood poverty and may become even more severe . . . if poverty among families with children is not reduced.

In the 1990s, most Americans and their leaders in Washington were aware of poverty but little inclined to do anything about it. They were more concerned about the erratic performance of the national economy and their own (in many cases) shrinking or stagnant incomes. Americans were not as confident as they had once been in the capacity of government to tackle big social problems. Some were convinced that the poor were responsible for their own situation. As we will see in this chapter, the national will to fight poverty has waxed and waned in recent American history.

In this chapter, we will deal with several key questions. How is poverty defined and measured? How many Americans are poor? Who are the poor? What are the long-term trends in poverty? What are the causes of poverty? How has government policy responded to poverty? We will begin by going back to the 1930s, when poverty was first recognized as a problem requiring federal action, and the modern system of social programs was born under Franklin Roosevelt's New Deal.

THE BEGINNINGS OF WELFARE: ROOSEVELT

The Great Depression hit the nation with devastating effect. The unemployment rate was above 20 percent from 1932 through 1935 and did not go below 15 percent until the eve of war in 1940. Those who had jobs saw their wages fall. Frightened, angry citizens joined widespread and often violent protests. Private charities were overwhelmed, and the attempts of local governments to provide assistance were totally inadequate. With the election of Franklin D. Roosevelt in 1932, the federal government moved quickly and devised entirely new approaches to the problems of unemployment and poverty. The first step was direct "relief" in the form of cash payments to any family in desperate need, financed by "emergency" grants from the federal to the state governments, which made up their own rules of distribution within broad federal guidelines. In 1934, one sixth of the population was receiving assistance. In addition, the Civilian Conservation Corps provided some jobs for unemployed young men in reforestation and similar projects (Piven and Cloward 1971: chapters 2 and 3).

By 1935, it was decided to expand the job programs and phase out direct relief. The Works Progress Administration (WPA) was set up to finance all sorts of construction projects and other activities, mainly organized by local governments. From 2 to 3 million people (mostly men) were employed in the later years of the decade by the WPA and another million people in other projects (U.S. Census 1975: vol I, 339). The main responsibility for direct relief was turned back to the states and localities, but with the federal government sharing the costs of supporting certain categories of people considered unemployable: the blind, the physically or mentally disabled, the elderly, and mothers who had small children but no husbands to provide for them. (The small program for mothers eventually grew into AFDC, Aid to Families with Dependent Children.) It was assumed that these programs would gradually shrink as the new programs of the federal social security system took hold and as the country worked its way out of the Depression.

The core of that new system of social security was the 1935 Social Security Act, which established a national social insurance system. Enrolled individuals (especially in the earlier years, not everyone was included) would receive full coverage after ten years of contributions made by employees and employers to the trust funds. Retired persons would get permanent pensions, as would those who were disabled to the point of not being able to work. Widows and orphans (survivors) of insured workers would get benefits. A system of unemployment insurance was also established, that would give temporary payments to insured workers during periods of layoff, usually up to twenty-six weeks.

These programs, and others that would be added later, came to be called "entitlements," because all who meet specified prerequisites are entitled to receive them and the government is committed to providing the necessary funding. Some entitlements, such as AFDC, are "means-tested": they are only available to people with incomes below a certain threshold. Others, such as old age benefits under Social Security, are available to people all income levels.

These features would, in the long run, have significant political consequences. Entitlement spending, because of its open-ended character, proved difficult to contain. When budget deficits became a national concern in the 1990s, means-tested entitlements were the natural target because of their narrower and relatively powerless constituency, the poor.

REDISCOVERY OF POVERTY: KENNEDY AND JOHNSON

During the Age of Shared Prosperity after World War II, the scene had changed. The Social Security system was beginning to pay out large sums to the elderly, and because it was viewed by the public as an insurance program that returned to them in the form of pensions money that they had earlier contributed in the form of taxes on wages, it was not stigmatized as "relief" and was popular among all segments of society. (Actually, the program taxes current workers to pay retirees, who generally receive much more than they pay in, but most people do not think of it that way.) The "make-work," or special public jobs, had disappeared. The unemployment compensation system was working smoothly and was taken for granted. Although these programs were created by New Deal Democrats over Republican opposition, a national, bipartisan consensus developed around them. When the Republicans regained the White House in the 1950s, they did not attempt to undo what the New Deal had done. A benign mood had settled on the generally prosperous country, and neither party had much taste for innovations in social policy.

Only a few critics seemed worried about the state of domestic economic affairs. One was John Kenneth Galbraith, an iconoclast among the economists, who wrote a book ironically titled *The Affluent Society* (1958). Galbraith did not want to celebrate prosperity but to demonstrate that the new affluence was creating new problems that were not being noticed by public policy: a built-in, tendency toward inflation; old cities decaying without sufficient investment for proper renewal of public services; and a new type of poverty that was not caused by economic depression and persisted despite economic prosperity. The new poverty, according to Galbraith, affected two types of people: individuals who suffered from one or another personal handicap (bad health, inadequate education, low intelligence) and who should be helped by social security or social welfare but were not being adequately supported; and individuals living in islands of poverty in areas of the country untouched by the new prosperity—for example, the Appalachians, peopled with subsistence farmers and underemployed coal miners, or the inner-city slums to which the rural poor were moving in large numbers. These people and their children needed help from deliberate government action; economic growth by itself was leaving them behind.

Then came a stronger statement: Michael Harrington's *The Other America: Poverty in the United States* (1962). Harrington estimated that more than 40 million Americans were poor and that the new prosperity was not helping them but, in many instances, actually making their situation worse.

> One might summarize the newness of contemporary poverty by saying: These are the people who are immune to progress. But then the facts are even more cruel. The other Americans are the victims of the very inventions and machines that have provided a higher living standard for the rest of the society. They are upside-down in the economy, and for them greater productivity often means worse jobs; agricultural advance becomes hunger (1962: 12).

These two books reminded Americans that a large sector of the population remained poor in the midst of general prosperity. Harrington's book was widely read by members of the new Kennedy administration, including the president and his influential brother, Robert, who had been shocked by the misery he had seen in West Virginia during the electoral campaign. Galbraith became an advisor to the president. The new generation of Democrats who came to Washington with Kennedy in 1961 were determined to take on the problems of poverty and the social pathologies associated with it. They saw themselves as heirs to the New Deal, and, like Roosevelt in the 1930s, found popular backing for their anti-poverty initiatives.

The Kennedy and Johnson (1961–1968) administrations made important improvements in both the social security and welfare systems. They added three new programs: food stamps (poor families could receive script that was redeemable at the grocery store for food); Medicare (insurance to cover hospital and doctor bills for the elderly); and Medicaid (a medical insurance program for low-income people, that came to serve both children on public assistance and seniors receiving long-term care in nursing homes). Through court challenges, often brought by government legal-aid lawyers, access to welfare benefits became more dependable, open to all who met clearly codified criteria. Civil rights legislation passed in this era opened new opportunities to the minority poor.

Lyndon Johnson hopefully called his program for the poor the "War on Poverty." "We stand at the edge of the greatest era in the life of any nation,"

Johnson told his countrymen in 1965. "For the first time in human history, we have the abundance and the ability to free every man from hopeless want, and to free every person to find fulfillment in the works of his mind or the labor of his hands." (The rest of the passage is reprinted at the beginning of this chapter.)

From the cold perspective of the 1990s, Johnson's grandiloquent rhetoric sounds quaint, and his optimism appears misplaced. The United States made steady progress against poverty in the 1960s and early 1970s, but lost ground after 1975. More serious, Americans have lost the hope and commitment that Johnson and many of his contemporaries displayed.

THE OFFICIAL DEFINITION OF POVERTY

The Kennedy administration came to office in 1961 with the problem of persistent poverty on its policy agenda. Government studies, supported by academic research, gave substance to Harrington's position that there were still millions of Americans living in misery—indeed, some in hunger. General living standards were rising steadily, so it seemed to many that the nation could finally afford to eliminate poverty completely.

To accurately assess the problem and develop remedial measures, an official definition of poverty was needed—something the government had never before enunciated. Using the definition, the government could measure the number and characteristics of the poor and analyze the causes of their poverty, and later gauge the effectiveness of anti-poverty measures.

The poverty standard finally adopted by the Kennedy administration was designed by Mollie Orshansky (1974), an economist at the Social Security Administration. Orshansky put two pieces of information together from government surveys: the cost of a minimum nutritious diet for a typical family of four and the proportion of income (approximately one-third) that the average family then spent on food. Multiplying the price of the food budget by three to allow for nonfood costs, Orshansky calculated an income "poverty line" of approximately $3,000 for a family of four. If you were a member of a four-person family with income below $3,000, you were poor by this standard. On the assumption that family needs vary with size, somewhat higher and lower poverty lines were computed, on a similar basis, for households that were larger or smaller than the typical four.

Orshansky's poverty line became the official federal standard. Each year, the Census Bureau uses it in conjunction with the annual survey of household income to estimate the number of poor people in the country. (The survey, by the way, measures pretax income, including income from such sources as Social Security and welfare payments.) By the new official standard, 22 percent of Americans were poor in 1960 (U.S. Census 1991b).

Because prices change, the poverty line must be regularly adjusted for inflation; each year, it is increased in proportion to the annual increase in the Consumer Price Index, the government's general measure of inflation.

By 1995, the poverty line for a family of four was about $15,600. Note that the poverty line is adjusted for changes in prices ($15,600 bought in 1995 what $3,000 bought in 1960) but not for changes in the general standard of living. It does not take into account the fact that, on average, Americans lived at a higher material standard in 1995 than they had in 1960. By using this measure, the government is saying something about the meaning of poverty. It is telling us that what makes some people poor is their material deprivation in an absolute sense rather than the relative gap between their standard of living and the standard typical of people in the same society. We will come back to this question shortly.

The Orshansky standard was as reasonable a poverty measure as anyone could come up with at the time. But it was problematic from the beginning and remains a target of political controversy to this day. The trouble started when the proposed measure was being considered by Kennedy's Council of Economic Advisors. Government nutritionists had come up with not one, but two food plans. (Both were based on the dubious assumption that the family cook was a sophisticated dietitian who never wasted a penny.) The first food plan was an emergency diet, suitable to maintain a family for a short period. The second plan, which cost 25 percent more, was designed to provide the nutrition necessary for long-term family health. It was left to the Council of Economic Advisors to decide which food plan to use as the basis for measuring poverty. Adopting a standard based on the second plan would result in a higher poverty line and, as it turned out, a much higher estimate of the number of impoverished Americans. That was apparently unacceptable to the Kennedy administration. The Council of Economic Advisors chose to base the official poverty line on the emergency food plan, thus opting for a more restrictive definition of poverty. The decision had little to do with nutrition and nothing to do with economics; it was essentially political.

The higher standard, now referred to as "125 percent of the poverty line," is used in some government statistical reports as a broader measure of poverty. The population that falls between the official poverty line and 125 percent of the poverty line includes many of those who belong to the class we call the working poor.

In principle, the poverty line was based on an objective, scientific standard: minimum nutritional need, plus a proportional allowance for rent and other essentials. But, in retrospect, it appears that there is no wholly objective way to specify minimum requirements. Even when political considerations don't get in the way, contemporary cultural values always intercede and help define what is appropriate. When the general standard of living goes up, so do ideas about the minimum needed to maintain subsistence. If there was any doubt about this, it was removed by the research of Oscar Ornati and his colleagues (1966), done about the same time the government was defining the poverty standard.

Ornati collected the subsistence budgets that had been used by local social service agencies, going back to 1905. He found that these budgets, which were designed to provide temporary relief for family survival, were typically based on the same notion of a "basic-diet-plus" that was adopted for the federal poverty standard. Ornati discovered that the conception of a subsistence budget tended to expand as national living standards improved. For instance, a 1908 budget for a family of four included 22 pounds of high-protein foods. A 1960 budget assumed that a family of this size needed 55 pounds. The 1908 budget makers supposed that a family needed four rooms but did not require a bath; by 1960, the standard was five rooms and a full bath. And we now assume that electricity, a refrigerator, and a high school education for children are included in a minimum standard of living, but they were not in 1908.

Sociologist Lee Rainwater (1974) further analyzed Ornati's data. He found that, year after year, the price of what the agencies considered subsistence budgets stayed close to 40 percent of the current average disposable income for a four-person family. By no small coincidence, the same was true of the federal standard when it was adopted in the early 1960s. Rainwater went a step further and looked at public opinion data from Gallup polls since 1946. In that year, Gallup started asking people this question: "What is the smallest amount of money a family of four needs to get along in this community?" The average response, from 1946 to 1969, was usually in the range of 50 to 55 percent of the average family disposable income for the year the question was asked.

The conclusion that can be drawn from the research of Ornati and Rainwater is that conceptions of a minimum living standard are inevitably *relative*. They depend on income levels and lifestyles in the societies where they are made. This generalization apparently applies with as much force to the experts as it does to popular opinion. Thus, what we now consider a poverty-line existence would have qualified as middle-class comfort at the beginning of the century and might still be regarded as such in some small town in the Peruvian Andes.[1]

This brings us back to the federal standard. If we could somehow summon the Kennedy administration experts, erase their collective memory of the poverty line they created in the 1960s, and make them do it again in the 1990s, they would almost certainly come up with something closer to our current living standards. But for thirty years, the government has treated the poverty line as an *absolute* standard, one that needs to be adjusted for increasing prices, but not for changing lifestyles.

Most sociologists believe that the government should use a relative standard to measure poverty. They reason that what makes people think of themselves as poor and causes others to regard them as poor is the comparison between their lives and the mainstream lifestyle in their community.

---

[1] Presumably, there is some nutritional level at which people starve or at least become weakened and highly susceptible to disease. Life at that bare level represents an absolute standard of poverty, without regard to community standards. But we don't know exactly how to define that standard; in any event, it is not very relevant to a society like our own.

The relative standard most frequently proposed is half the median family income. In 1960, the median family income was around $6,000, which meant that the poverty line of $3,000 for a four-person family was, in fact, close to half the median income when it was first minted. By 1995, the basic poverty line had fallen to 38 percent of the median family income—more precisely, the median had increased, leaving the poverty line behind (U.S. Census 1996d, 1996e).

The federal poverty standard has been subject to a different line of criticism—from conservatives. They argue that the official standard, because of a series of technical deficiencies, overstates the amount of poverty (Anderson 1978; Stockman 1984). For example, they claim that people tend to underestimate their income when they respond to government surveys. And they recall that the annual income survey, from which poverty is estimated, records only *money* income. This was not a problem when the poverty standard was established, but since then there has been a vast increase in the amount of *in-kind* aid given to the poor in such forms as food stamps, subsidized housing, and health care. These critics probably have a reasonable argument with regard to food stamps, which function as cash and might well be considered part of income. Unfortunately, they do not mention another change since the 1960s: Rising taxes, especially payroll and sales taxes, have cut into the real buying power of poverty incomes, but taxes are not reflected in the income survey data or the official poverty statistics.

HOW MANY POOR?

According to the various measures of poverty we have been discussing, there are somewhere between 29 and 59 million poor people in the United States (see Table 10-1). By the official standard, there are 36.4 million poor. This figure is based, as we have seen, on Orshansky's minimum subsistence family budget. The other statistics in Table 10-1 are refinements of the official measure, using the same data from the Census Bureau's income survey but reaching rather different conclusions. All are based on Orshansky's poverty-line concept and begin with a basic 4-person poverty threshold which is adjusted for family size.

The low-end figure of 29 million takes the conservatives' criticisms of the official standard to heart. The Census Bureau produced this estimate by adding to money incomes the estimated dollar value of noncash benefits such as food stamps and Medicare.[2] But the bureau also considered the income lost to all local and federal taxes.

---

[2] Assessing the value of Medicare and Medicaid coverage is especially problematic for estimates of this sort. The Census Bureau took into account only what it considered the fungible proportion of Medicare and Medicaid. If a family's income did not exceed the amount needed for food and housing, it was assumed that health coverage did not free up money for other purposes and should not be considered part of their income.

TABLE 10-1. Four Measures of Poverty in the United States, 1995

Poverty Measure	Poverty Line (family of 4)	Percent Poor	Number Poor (millions)
Official	$15,600	13.8	36.4
Adjusted official	$15,600	10.3	29.0
125% poverty line	$19,500	18.5	48.8
½ median income	$20,600	22.2	58.6

NOTE: "Adjusted official" reflects effects of in-kind benefits and taxes (Census definition #14). Poverty statistics based on 1/2 median income are estimates.
SOURCES: U.S. Census 1996d, 1996e; Michel et al. 1997: 303.

By the standard of 125 percent of the official poverty line, there are nearly 49 million poor. This number would, of course, have been the official statistic if the Kennedy administration had not chosen to base the poverty line on an *emergency* food budget, but rather on a diet that provides good health over a longer period of time.

A relative standard defines poverty in relation to mainstream standards of living. By the most frequently cited relative standard there are about 59 million poor people in the United States — a statistic equivalent to the combined population of all the states from the Rocky Mountains to the Pacific Ocean. As we noted earlier, when the official poverty line was first established, it was close to one half the median income. By this measure, the United States has made no progress against poverty: One-fifth of the population was poor in 1960, and one-fifth remains poor today.

WHO ARE THE POOR?

The debate about how to measure poverty is endless, and it is not difficult to imagine why. Any serious discussion of poverty in an affluent society, which also regards itself as democratic, inevitably stirs political emotions. Just below the surface, the technical dispute about measurement is a debate about the fairness of our political and economic institutions.

To examine poverty, we will have to settle on some measure. For most purposes in the remainder of this chapter, we will employ the official standard. If we keep its specific defects in mind, the federal standard is not an unreasonable way to measure poverty. It happens also to be the basis of most of the detailed poverty statistics published by government.

Figure 10-1 uses official data to answer a key question: Who are the poor? It starts with the total number of poor people, as measured in government surveys (about 36.4 million people) and breaks this figure down in various ways. Here are some of the things we learn:

FIGURE 10-1   Distribution of Poverty, 1995. Persons in poverty (in millions)

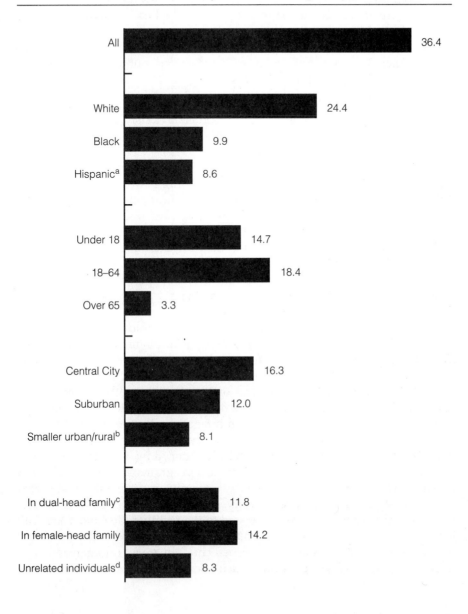

^a Hispanic can be any race.

^b Outside metropolitan areas. Generally less than 100,000 population.

^c Includes small number of members of male-headed families.

^d Living alone or with nonrelatives.

SOURCE: U.S. Census 1996d.

- There are more than twice as many poor whites as poor blacks.
- A high proportion of the poor are children. Few are elderly.
- There are almost as many poor in families headed by married couples as in families headed by single women.
- Only a minority of the poor live in the central cities of metropolitan areas. The majority are spread out among suburbs, small towns (under 100,000), and rural areas.

There is enough information here to contradict some popular stereotypes. In particular, the typical poor person is obviously *not* a member of a black, female-headed family, living in a big city. As it turns out, only 1 in 10 of the poor fit that description (U.S. Census 1991b).

Figure 10-2 answers a different kind of question: If you belong to a certain social group, what are your chances of being poor? This life chance is called the *risk* of poverty (or poverty rate). It is simply the percentage of the people in a group that fall below the poverty line. Just a quick glance at the graph reveals that there are very large differences in the risk of poverty. Blacks, although they are a minority of the poor, are three times more likely than whites to be poor. Children are at a much greater risk of poverty than adults. Approximately two out of five black and Hispanic children are poor. And whatever their ethnicity, female-headed families have poverty rates far above those of families headed by couples.

As we noted earlier, the measure of poverty employed here counts all money income—including, of course, the transfer payments people receive from the government for welfare, social security, veterans benefits, etc. These payments make a difference. Without them, the poverty rate in 1995 would have been 22 percent instead of 14 percent. But the effect of cash transfers is very uneven, as Table 10-2 demonstrates. Note that seniors and female-headed families with children have virtually the same high, pre-transfer poverty rate. But a dramatic gap opens between them after transfers. The poverty rate of children is hardly touched by transfers. Whatever advantages poor children receive from cash programs, they are clearly not intended to lift families out of poverty. The only government transfer program that significantly reduces poverty is social security for the elderly. The success of that program is not unrelated to the fact that the elderly are well organized in political pressure groups and vote in substantial numbers, while the adult relatives of impoverished children are an unorganized population with low rates of political participation.

TRENDS IN POVERTY

When Lyndon Johnson became president and promised to build a society free from poverty, ignorance, and exploitation, the official poverty rate was already falling. It would continue to fall during the presidency of Johnson's successor, Richard Nixon. But the poverty rate more or less flattened out in

FIGURE 10-2   Risk of Poverty for Selected Groups, 1995. Percent in Poverty

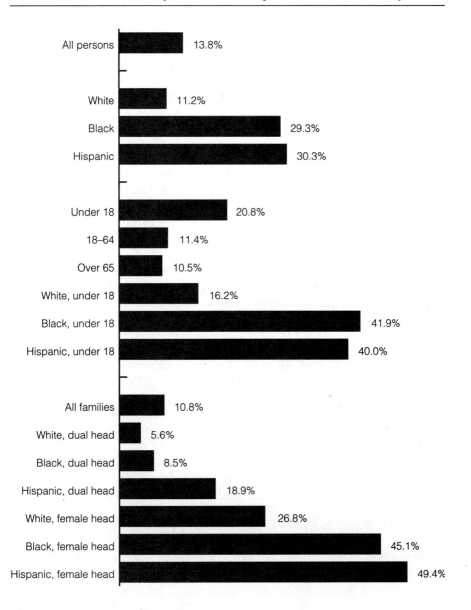

All persons — 13.8%
White — 11.2%
Black — 29.3%
Hispanic — 30.3%
Under 18 — 20.8%
18–64 — 11.4%
Over 65 — 10.5%
White, under 18 — 16.2%
Black, under 18 — 41.9%
Hispanic, under 18 — 40.0%
All families — 10.8%
White, dual head — 5.6%
Black, dual head — 8.5%
Hispanic, dual head — 18.9%
White, female head — 26.8%
Black, female head — 45.1%
Hispanic, female head — 49.4%

SOURCE: U.S. Census 1996d.

TABLE 10-2   Poverty Rate Before and After Cash Transfers, 1995

	Before Cash Transfers	*After Cash Transfers*
All persons	21.9	13.8
Under 18	24.2	20.8
Over 65	49.9	10.5
Female head w/children	51.7	44.8

SOURCE: U.S. Census 1996d.

the mid-1970s and actually climbed in 1980s and early 1990s. Figure 10-3, based on the official poverty standard, traces these shifts. Although the rate has recovered somewhat from the peak it reached during the recession of the early 1980s, the 1995 rate was well above the low point reached in 1973.

Comparisons with the years before 1970 are problematic, for reasons suggested in our discussion of poverty measures. As we noted earlier, poverty in 1995 was at the 1960 level by a relative measure. On the other hand, if we had retrospective statistics accounting for noncash transfers and taxes, they might show greater historical progress. But we can be confident of the general picture: *declining poverty in the 1960s, stagnation in the 1970s, and increased poverty in the 1980s and early 1990s.*

Over the years, as the poverty rate has changed, the composition of the poor population has shifted. Since 1960, the poor population has become somewhat more black, much more Hispanic, and more concentrated in big cities (Jencks and Peterson 1991: 7; U.S. Census 1995d). But the most dramatic changes have come in the age distribution and family structure of the poor. Between 1960 and 1995, the proportion of the poor living in female-headed households doubled (U.S. Census 1995d). Figure 10-4 illustrates the striking reversal that has taken place in the poverty rates of children and seniors, to the advantage of the latter. This shift was, as we will see, the result of federal policies that favored the elderly over families with children.

THE UNDERCLASS AND PERSISTENT POVERTY

Knowing how many people are poor from year to year does not tell us how persistent poverty is on an individual level. Many people fall below the poverty line during a given year as a result of temporary circumstances: loss of a job, physical injury, or divorce. When they get back on their feet, they are no longer poor. Others remain poor for many years. They might be disabled, living on meager retirement incomes, or working at a low-wage job while supporting a large family.[3] Some people are just incapable of

---

[3] Of course, under the federal standard, large families are more likely to be classified as poor.

FIGURE 10-3   Poverty Rates, 1960–1995

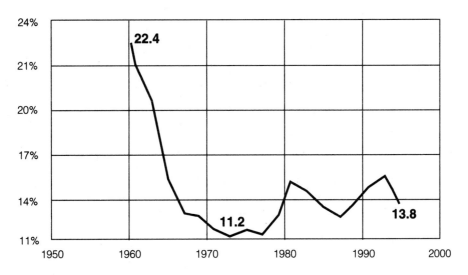

SOURCE: U.S. Census 1996d.

holding a job for long, and some have developed a long-term dependence on government benefits that allow them to survive but never approach the mainstream.

In recent years, interest in the persistence of poverty has turned into concern that the United States might be developing a permanent under-class—a growing class of people who are impoverished and mired in habits and circumstances that prevent them from ever joining the mainstream. This preoccupation has been fed by the increase in poverty in recent years and by the growing concentration of the poor in large cities, where they are most visible to the national media. We will return to the question of the un-derclass after we examine the few existing studies of turnover in the poor population.

Accurately measuring the *persistence* of individual poverty is difficult. It requires a "panel study"—finding the same people to ask the same ques-tions, year after year. The Census Bureau has a continuing survey that tracks people for two years. Data from the 1980s show that approximately 25 percent of the people who are poor in a given year are not poor the fol-lowing year. At this rate, there could be a complete turnover in the poverty population in four years. But since the bureau only follows people for two years, we have no way of knowing how many people fell back into poverty shortly after climbing out (U.S. Census 1989a, 1990e, 1991d).

The University of Michigan's Panel Study of Income Dynamics (PSID) has been tracking a large national sample of families since 1969. The PSID

FIGURE 10-4 Poverty Rates for Children and Seniors

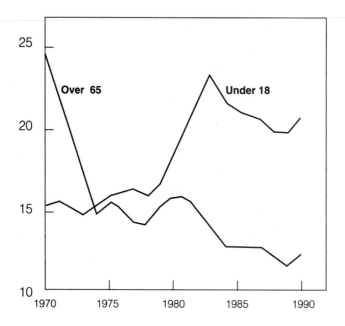

SOURCE: U.S. Census 1996d.

data indicate that brief spells of poverty are common: About a quarter of all families will be poor at least once in a ten-year period, but only a fraction of these families will remain poor for an extended period (Corcoran 1995: 241; Duncan, et al, 1984: 41-2).

One PSID study looked at rates of long-term poverty among the families of women aged 25–44 with children—both female-headed and married-couple families. A woman's family was considered long-term poor if its average income over the six years was below the poverty line. By this standard, only five percent of such families in the early 1980s were long-term poor (Duncan and Rodgers 1989, as reported in Jencks 1991: 35).

Studies of the welfare caseload produce a similar impression of continual turnover. AFDC, the main public assistance program, largely benefits impoverished single mothers and their children. Mary Jo Bane and David Ellwood (1994) studied the AFDC population using two decades of PSID data. Their two major conclusions seem contradictory: (1) Most of the women who start AFDC will receive only transitory aid, typically three years or less in total, but (2) most of the women receiving AFDC at any point in time are long-term recipients, who will remain on the rolls for ten years or more.

Bane and Ellwood explain this paradox by comparing the welfare system to a hospital. If we watched the admitting desk for a while, we would generally see people entering the hospital with acute but transitory conditions; the vast majority will be gone in a matter of days. But if we visited the wards, we would be surprised to find that the vast majority of beds are occupied by people with chronic conditions, hospitalized for extended periods. Although only a small fraction of the people admitted, chronic patients account for most of the bed-days and absorb most of the hospital's budget. Likewise, *a minority of the people who ever receive AFDC account for most of the caseload on a given day and consume most of its resources.* Bane and Ellwood estimate that women who will receive 10 years or more of AFDC account for 22 percent of those who enter the system, but 57 percent of those in the system at a given moment (1994:39)[4]

Who are the long-term recipients? Aside from the children, according to Bane and Ellwood, they are typically women who combine some of the following characteristics: entered the system as teenagers, high school dropouts, never-married mothers, women with no recent work experience. Many minority women share at least some of these characteristics, and they are overrepresented among long-term recipients.

All these studies demonstrate that there is rapid turnover in the poor population. *Only a minority of poor families, probably around 25 percent, are locked into a pattern of long-term poverty.* An even smaller minority of poor families are long-term dependents on welfare.

Can we equate this 25 percent with "the underclass"? That depends on how the term is understood. In the class model we presented at the beginning of this book, *underclass* was used in a broad economic sense to refer to the poor who are loosely connected to or wholly disconnected from the labor market—encompassing most of the people below the federal poverty line. But in the debate over the underclass, the term is applied to a smaller subset of the poor who are bound to their impoverishment by personal characteristics or structural circumstance. (See Auletta 1982; Jencks and Peterson 1991; Murray 1984; Wilson 1987.)

The underclass label often implies flawed character. The "conventional portrait" of the underclass, according to a *Washington Post* writer, links "extreme poverty, chronic joblessness, welfare dependency, out of wedlock births, female-headed households, [and] high dropout rates ..." (April 17, 1991). Many would have added crime to this list. Sociologist William J. Wilson (1987, 1991), one of the most influential writers on the underclass, more or less accepts this grim portrait, but defines the underclass as having marginal economic position coupled with geographic isolation. Wilson has in

---

[4] These estimates, based on PSID data, refer to the total time a person can expect to receive AFDC over a hypothetical twenty-five-year period. Some recipients will surpass the ten-year mark by repeatedly cycling on and off AFDC.

TABLE 10-3.   Intergenerational Transmission of Poverty

	Poverty Rate as Adults (in percent)	Family Income as Adults (in 1990 dollars)
Whites		
Non-poor as children	1.2	$53,400
Poor as children	9.3	$35,100
Blacks		
Non-poor as children	9.6	$36,100
Poor as children	24.9	$26,900

NOTE: Refers to adults age 27 to 35 in 1988. Based on stringent poverty measure that counts food stamps as income and considers average family income level over several years.

SOURCE: Corcoran 1995: 247.

mind the inner-city poor, living in areas with extremely high concentrations of poverty and caught in a postindustrial economic trap of shrinking job opportunities.

How big is the underclass? Wilson, in his contribution to a collection of papers on the topic, appears to accept an estimate of 2.4 million people in 1980, of whom 65 percent were black and 22 percent Hispanic (Wilson in Jencks and Peterson 1991: 464). This would mean that the underclass was roughly 1 percent of the national population, 8 percent of the poor, 20 percent of the black poor, and 15 percent of the Hispanic poor. Isabel Sawhill (1989) and her colleagues at the Urban Institute estimate the underclass is even smaller, about 1.1 million. They produced this figure by adding all the poor who live in "underclass areas"—defined as neighborhoods (or census tracts) with very high rates of school dropout, single motherhood, welfare dependency, and joblessness. The underclass, thus defined, accounted for just 4 percent of the poor in 1980. Sawhill reports that 59 percent of the residents of underclass areas were black, and 12 percent were Hispanic. She notes that less than 20 percent of the black poor and fewer than 6 percent of all African-Americans live in such areas. If there is an underclass, it does not, according to these estimates, exceed 1 percent of the population and only accounts for a small minority of the poor.

Implicit in most discussions of the underclass is the notion of a "cycle of poverty"—of social pathologies so severe that poverty is inevitably passed from one generation to the next. Given what we know about social mobility, we would be surprised if there were not at least some truth to this (seldom examined) assumption. But how much? According to PSID data for the years 1968 to 1988, *the vast majority of children raised in poverty are not poor as young adults.* As Table 10-3 reveals, this generalization holds for blacks as well as whites. Only one in four (24.9 percent) poor black children grows up to be a poor adult; for whites, the figure is less than one in ten. Childhood poverty does, of course, raise the probability of adult poverty and depress adult earnings, as the table indicates. But it is not the grim reaper of life chances that our public debates often suggest.

THE MYSTERY OF RISING POVERTY RATES

Here is the mystery facing students of poverty: Why did the 1960s progress in reducing poverty falter in the 1970s and 1980s? No one has the definitive answer to this question. But at least three factors are involved: (1) economic change, (2) a societywide revolution in family patterns, and (3) shifts in federal policy.

Economic Trends. In early September 1991, the national news media reported a fire in a North Carolina chicken-processing plant that killed twenty-five workers and injured more than fifty others. According to the stories, exits that had been locked to keep employees from stealing chicken had contributed to the death toll. In eleven years of operation, the plant had never been inspected by state or federal safety officials. The workers in the plant were typically single women with children, working eight hours a day for $5 an hour or less—a wage insufficient to raise a family of three above the poverty line (*New York Times* and *Washington Post*, September 4–11, 1991).

The late 1970s and 1980s saw a proliferation of such low-wage jobs and the loss of many of the better-paying blue-collar jobs. High unemployment rates and low skills compel many of the poor and near-poor to accept any available job. Relevant trends in employment and earnings are outlined in Figure 10-5. Unemployment rates (decade averages are used, to smooth out yearly gyrations) have increased steadily since the 1950s, but especially since 1970. Unemployment was 60 percent higher in the 1980s than it was in the 1950s.

Average weekly earnings, adjusted for inflation, rose in the 1960s, but by the late 1980s, earnings had slid back to the 1960 level. Of course, the weekly average includes high-paid executives and the workers at chicken-processing plants. The minimum hourly wage, a figure set by federal legislation, gives an indication of what was happening at the bottom. Wages at the lower end of the labor market, even for people who earn a little more than the federal minimum, tend to move up or down with the minimum. In the 1960s, the real value of the minimum wage rose faster than average earnings, but by 1990, the minimum wage had sunk to 86 percent of its 1960 value.

In short, the labor market turned sour. Unemployment rose. Wages went down. These developments affected the poor to the extent that they worked or were looking for work.

Just how much do the poor work? According to government data, about half of poor adults 25 to 64 worked in 1995. For diverse reasons, many of them worked part time or only part of the year, but 15 percent of poor adults worked full-time all year without rising above the poverty line. Among poor heads of households, the proportion is even higher: 25 percent worked full-time year round (U.S. Census 1996d).

Poor families headed by couples have particularly high rates of labor force participation. About 45 percent of non-elderly husbands in such families

FIGURE 10-5 Trends in Unemployment and Real Earnings, 1960–1990

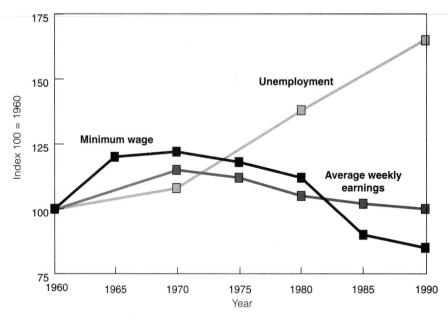

NOTE: All statistics are indexed to a common scale, with 1960 values equal to 100. Unemployment figures represented are averages for the preceding decade. An index of 150 means an increase of 50 percent over the base period; 80 percent indicates a 20 percent decrease. Weekly earings are for all nonagriculture wage and salary workers. Earnings and minimum wage are adjusted for inflation. Minimum wage for 1990 reflects a raise to $3.80 on April 1.
SOURCE: U.S. Census 1990c: 409, 412; Council of Economic Advisors 1990: Table C-32.

worked full time year round in 1995 (U.S. Census 1996d). Ellwood examined the work experience of parents in dual-headed families with pre-transfer incomes below the poverty line. He found that 44 percent of such families had the equivalent of at least one full-time, year-round worker in 1984. "The poverty of two-parent families," concluded Ellwood, "is the poverty of the working poor." (Ellwood 1988: 81, 89).

It is a sad commentary on our economy that many poor people work all year at full-time jobs and do not make enough to lift their families out of poverty. In the 1960s, the number of heads of poor households in this situation was falling fairly rapidly. More recently the number has been rising (Levitan 1990: 20).

About half of poor adults do not work because they are retired, ill or disabled, caring for children, or attending school. In other words, they are not, for the moment, potential workers. The other half are unemployed, underemployed, or working at jobs that pay substandard wages. These people and their dependents might escape poverty under different economic conditions.

In the 1980s, the capacity of a rising economy to lift people out of poverty was tested in the city of Boston. By 1988, the so-called Massachusetts Miracle had reduced the statewide unemployment rate to 3.3 percent, well below the national average (U.S. Census 1991c: 399). In such a tight labor market, jobs are easier to find, and both wages and working conditions improve as employers compete for workers. What effect did the economic boom have on poverty in Boston? In 1980, the city's poverty rate was higher than the average for U.S. central cities. In the decade that followed, poverty rates climbed in the average American city but fell in Boston. Among Boston families headed by nonstudents under age 60, the rate dropped from 18 percent to 11.5 percent. Among black families in this group, the rate plunged from 29 percent to 13 percent (Osterman 1991: 126). In Boston, it was proved that even tough big-city poverty responds to a healthy labor market.

Using national data for the years 1960–1984, Ellwood (1988: 97, 150) showed that it was possible to predict the annual poverty rate of families with children with remarkable accuracy using only two indicators: the median earnings of full-time, year-round workers and the unemployment rate. Ellwood's analysis again demonstrates how dependent the poor and the near-poor are on labor market conditions.

In the mid-1990s, unemployment rates came down. According to economic theory, wages should have risen in response. Remarkably, this did not happen, and many Americans continued to toil for a poverty wage.

Changing Family Patterns. The pressure of economic stagnation on poverty rates has been reinforced by sweeping changes in family patterns. Figure 10-6 summarizes the key trends. Relative to the recent past, Americans are now less likely to marry and more likely to divorce. Children are five times more likely to be born to unwed parents and almost three times more likely to live in female-headed households than they were in 1960. These trends are pervasive in American society. They affect the poor and the nonpoor, blacks and whites, teenagers, their parents, and grandparents. Together, they amount to a revolution in American family life.

The key change in family patterns that sums up the others and has contributed to the rise in the poor population is the increase in the number of female-headed families. Since 1960, the proportion of the poor population living in female-headed families has jumped from roughly 20 to 40 percent (U.S. Census 1996d). This development is transforming the lives of children. It is probable, according to one expert, that the majority of the children born today will spend part of their childhood in a female-headed family—and that most of the children who do will experience a period of poverty (Ellwood 1988: 45–47, 67).

Why the rapid increase in families headed by women? In part because in the course of the last generation, Americans have become much more tolerant of both unwed motherhood and divorce. Currently, about one-third of births in the United States are to unmarried women (U.S. Census 1996f).

FIGURE 10-6   Changing Patterns of Family Life

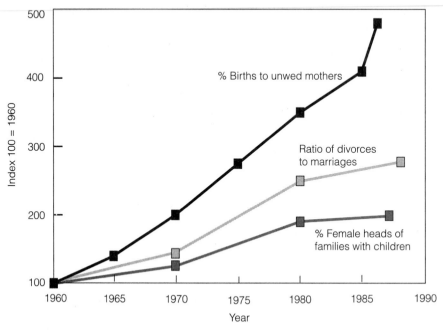

NOTE: Regarding index, see the notes for Figure 10-5.
SOURCE: U.S. Census 1986a, 1990c.

The popular conception of these mothers as irresponsible teenagers is dated and misleading. Girls who become mothers still face tough, poverty-ridden futures; they and their children will be overrepresented among the long-term welfare-dependent families. But teen birth rates have declined, probably as the result of greater use of contraceptives and access to abortion. There has been some increase in the number of *single* teen mothers because these girls are less likely to marry when they become pregnant. But the birth rate of single women in their twenties is higher than that for teens and is growing faster. Women in their twenties account for the majority of unwed births (U.S. Census 1996f; Ellwood 1988: 71–72; Jencks 1991: 83–93).

One reason for the shift in behavior and attitudes toward divorce, marriage, and single motherhood is that the choices facing women have changed. Job opportunities for women have expanded, and women are now expected to be capable of supporting themselves. At the same time, the falling wages and erratic unemployment rates of men without special skills or advanced education make them less dependable meal tickets. Sociologists William Wilson and Kathryn Nickerman (1986) present evidence to suggest that these trends have powerfully affected African Americans, among whom the majority of births are now to unwed women and nearly half of families are headed by women (U.S. Census 1996f). They find that the marriageable pool of employed young black males has contracted sharply, especially since the economic slump of the 1970s. Finally, public assistance became a more dependable source of family support beginning in the 1960s. Although the value of welfare benefits declined after 1970, access to public assistance was assured for families that qualified.

If the array of choices facing young women has expanded, the prospects for single mothers remain unenviable. The majority of absent fathers do not pay child support, and those who do contribute pay very little.[5] Employed women, of course, make considerably less on average than men. The conflicts between the nurturer and provider roles, which all working mothers face, are even tougher for employed single mothers, who bear the burdens of childrearing alone. Nonetheless, single mothers are more likely to work full-time than married mothers (Ellwood 1988: 145). Public assistance does not generally provide enough to place a family above the poverty line, and rule changes imposed under the Reagan administration made it impossible for women on AFDC to increase their families' incomes by working—unless they work "off the books" and hide their wages from welfare authorities. The most common way for a single mother to escape AFDC is through marriage; few manage to do so through their own earnings (Ellwood 1988: 148; Ellwood and Bane 1994: 57).

Given the difficulties that female heads of households face, it is scarcely surprising that their families are several times more likely to be poor than dual-headed families (see Figure 10-2) or that the growing number of female heads is associated with increasing poverty rates.

Government Policy. How much can we blame changes in government policy for the increase in poverty since the mid-1970s? The highly publicized cuts in social programs made under President Reagan in the 1980s certainly made life harder for the poor—especially for the working poor. A study by the Urban Institute estimated that the program cuts and worsening economic conditions were about equally responsible for the jump in poverty rates in the early 1980s (Palmer and Sawhill 1984: 14). But poverty rates began to rise before Reagan was elected. What we have already learned is that long-term changes in our economy and patterns of family life have been pushing poverty rates upward. The government has some power over the economy; it has virtually none over family life.

Liberals typically have great faith in the capacity of the government to change things. Ironically, in recent years, conservative writers have advanced the strongest claims for the power of federal poverty policy. For example, Charles Murray's influential book *Losing Ground* (1984) contended that rising welfare benefits encouraged the formation of dependent female-headed

---

[5] Child support is discussed more fully in Chapter 4.

households and were thus responsible for the increase in poverty rates. If Murray were correct, the proportion of female-headed families with children and the poverty rate should both have tumbled in the period from 1972 to 1984, when the value of welfare benefits declined 25 percent, back to 1960 levels (Ellwood 1988:58–59). Exactly the opposite happened. The growth in female-headed families accelerated, and the poverty rate climbed steeply. Public assistance is probably the least potent of the forces responsible for the change in family patterns that is affecting the entire society—not just the minority of the poor who receive welfare benefits. Subsequent research has not supported Murray's extravagant claims. (See Garfinkel and McLanahan 1986: 55–63; Ellwood and Bane 1985.)

In recent years, the federal policies that have most influenced poverty rates are not explicitly poverty policies. For example, the dramatic reduction in poverty among the elderly, visible in Figure 10-4 on page 266, is the direct result of real increases in social security benefits (Levitan 1990: 45). The subsequent leveling off of benefits contributed to the reversal of the long-term downward trend in the general poverty rate.

The failure of the government to uphold the value of the minimum wage and maintain the protections of the unemployment insurance system pushed some working poor families below the poverty line, though it is difficult to determine how many. In 1970, the minimum wage was high enough that a steady full-time worker could keep a family of three out of poverty; by 1990, the same worker could just manage to maintain a one-person household above the poverty level. Unemployment insurance can help keep low-income families from falling below the poverty line. But the proportion of the labor force covered, the value of benefits, and the duration of benefits have all fallen significantly since 1975 (Levitan 1990: 62–65; *Washington Post*, August 29, 1991).

National macroeconomic policies also influence poverty rates. In recent years national policy has focused on reducing inflation rather than maintaining employment and on increasing profits rather than upholding wages. These goals were reflected in tax and spending policies, as well as monetary policy. International trade policy opened the United States to increasing foreign competition, reducing both employment and wages in the affected industries, especially for workers at the lower end of the labor market. It is difficult to disentangle the results of these policies from the effects of the broad changes in the U.S. economy we discussed in Chapter 3. This much is certain: In the 1980s and early 1990s, federal policy did contribute to rising unemployment and declining wages; we have already seen how closely poverty rates are tied to these conditions in the labor market.

Toward the end of 1996, major legislation was passed restructuring national poverty programs. The full effects of these changes, which we discuss below, will not be felt for several years, but they are sure to push poverty rates further upward.

WELFARE REFORM: FROM NIXON TO BUSH

As we have seen, the expansion of social programs, made a big dent in the poverty rate before the early 1970s. But as these programs developed, so did opposition to them, particularly the means-tested ones that were defined by the public as "welfare." Opponents said that welfare was "growing out of control" and "full of fraud and abuse."

Aid to Families with Dependent Children became a particular target. AFDC grew very rapidly in the decade after 1965. Legal challenges opened the program to people who had been denied benefits through unfair, often discriminatory administrative procedures. But expansion was also a response to mounting social protest—especially black protest. Policy makers used AFDC to buy social peace. As protest waned, so did support for AFDC and related programs (Piven and Cloward 1971).

Presidents from Nixon to Clinton promised to "reform" the welfare system. What did they achieve? Nixon introduced an ambitious program to transform AFDC into a universal income support system with built-in work incentives (a "negative income tax" program). Carter backed a similar proposal, tied to federal program to create new jobs. Both died in Congress. Reagan changed the rules to constrain growth in welfare spending but left the basic system intact.[6] The Bush administration conducted a high profile review of poverty programs, but decided that nothing could be done. Clinton brought several of the country's leading welfare experts into his government but the bill they cobbled together in 1994, after a difficult struggle within the administration, got a hostile reception in a Congress controlled by the president's own party.

Officials involved in the Bush administration review had concluded "that there were no obvious things that we should be doing that we weren't doing that would work. . . . We have decided to abjure glitzy, spashy, high profile announcement of new programs . . . Keep playing with the same toys. But let's paint them shinier" (*New York Times,* July 6, 1991). These cynical sounding remarks about reform reflect hard political realities.

Americans and their presidents would like a welfare reform that would (1) provide an adequate standard of living for families, and especially for children, in need, (2) contain strong work incentives,[7] (3) encourage family stability and responsible parenthood, and (4) reduce federal spending. Politicians have encouraged voters to think that they can have all four at

---

[6] Reagan era welfare legislation tightened requirements, trimmed benefits, and effectively eliminated work incentives from AFDC. The Family Support Act (1988), a long-term plan passed in the last year of Reagan's presidency with bipartisan support encouraged states to place increasing emphasis on work in their welfare programs. But because neither the federal government nor the states provided adequate funding for work programs under the act, it had very limited impact.

[7] Work can be incentivized by providing a low support level for those who don't work and by limiting the reduction of support as job income expands—in other words, by allowing recipients of public assistance to keep most of what they earn.

once. Unfortunately these goals are often in conflict. For example, government support of minimum living standards for needy families undercuts work incentives. This support/incentive conflict can be attenuated by spending more money—for example, on graduated income subsidies for low-wage workers and child care for working parents. But, of course, increased spending compromises the fourth goal. Such contradictions defeated welfare reform efforts from Nixon's in 1969 to Clinton's in 1994 (Anderson 1978).

Nonetheless in August 1996, Congress and the president agreed to sweeping legislation restructuring the country's poverty programs. Three key factors made the new law possible and shaped its content. First, the national debt had doubled in real terms under President Reagan, imposing a continuing strain on federal finances (U.S. Census 1995a).[8] Reagan's fiscal legacy increased pressure to subordinate the social goals of reform (#1–3) to the overriding goal of reducing federal spending (#4). Second, candidate Bill Clinton had courted votes in 1992 with repeated promises to "end welfare as we know it." Exactly what this meant was unclear, but it was certainly a pledge to action. Third, a Republican majority took control of both houses of Congress in November 1994, for the first time in four decades.

The Republican Congress produced a poverty bill far more conservative than the administration's failed proposal two years earlier. But Bill Clinton was trapped between his own rhetoric regarding welfare reform and impending national elections in November 1996. Against the advice of his administration's welfare experts, who anticipated increased poverty under the legislation, Clinton signed the grandly titled Personal Responsibility and Work Opportunity Act (DeParle 1996; Edelman 1997).

RESTRUCTURING POVERTY PROGRAMS: THE 1996 LAW

The 1996 law fundamentally altered the social safety net that protected the poor by scaling back three of the major means-tested programs: AFDC, Supplemental Security Income (SSI), and food stamps.[9] (Policy-makers carefully steered clear of Medicaid, the largest and fastest growing means-tested program, whose beneficiaries include many seniors in nursing homes). The three affected programs had long operated as entitlement programs, providing guaranteed assistance to anyone who qualified (under clearly stated rules), backed by open-ended funding. This meant that

---

[8] Cavalier in his management of federal finances, Reagan simultaneously cut taxes and raised military spending, driving budget deficits to unprecedented levels. Interest on the accumulating national debt began to consume a big piece of the national budget. The rising cost of medical programs and the anticipated retirement of babyboomer workers added to the perceived need to control federal spending.

[9] This section draws on Burtless et al, 1997; CQ 1996; DeParle, 1996; and Edelman 1997.

spending automatically swelled with increasing need in economic bad times and shrank during boom periods.

AFDC, which provided cash-assistance to poor families with children, was a joint federal-state program. The states set their own benefit levels, which varied widely, but states had an incentive to be generous because the federal funding a state received for AFDC depended on its own spending level. SSI was a cash assistance program for low-income elderly, blind, or disabled persons, serving both adults and children. Food stamps provided the broadest safety net protection. The program assisted, in varying amounts, almost anyone whose income did not assure adequate nutrition, including indigent adults, the working poor, the blind and disabled, low-income seniors, and needy families with children.

Under the 1996 law, AFDC lost its entitlement status and became Temporary Assistance for Needy Families (TANF). *For the first time since the passage of the Social Security Act in 1935, there is no guarantee of income assistance for impoverished children and their families.* States will receive federal funding in lump sums known as block grants, whose value is projected to decline over time. The states are given wide latitude in designing new programs with block grant funds, subject to two key limitations: (1) Families may not receive more than five years of assistance, whether consecutive or nonconsecutive (*"Temporary" is clearly the operative term in the program's new title*) and (2) most adults benefiting from TANF will be required to begin work of some sort within two years of receiving assistance.

SSI and food stamps remain entitlements, available to all who qualify. But the new law reduces spending on both programs. The act stiffens SSI eligibility requirements for disabled children and makes most legal immigrants ineligible for SSI assistance, however great their needs. The new disability standard will halt benefits for some 300,000 children, nearly a quarter of those currently assisted. Legal immigrants, including people currently living and working in the country, will be refused benefits under most means-tested programs. Within a few years food stamp spending will be reduced by about 25 percent by excluding immigrants, limiting unemployed adults without children to three months of assistance in three years, and reducing general benefit levels. Even families with children will see their food stamp allotment shrink.

The Personal Responsibility and Work Opportunity Act offers more responsibility than opportunity. The act reduces or ends benefits for many, but does not provide transitional training, educational, or job opportunities. It will compel many single mothers with young children to take minimum wage jobs, but does not contain additional funding for child care. Forced to chose among the competing reform goals listed earlier, federal policymakers opted to save money, approximately $54 billion by 2002, a large bite out of the budget deficit.

A study conducted by the Urban Institute estimates that the 1996 law will push an additional 2.6 million people into poverty. More than 10 million families, most of them families with children, will lose income under its

provisions—on average, $1,300 per family.[10] These are conservative esti-
mates. The study, originally funded by the Clinton administration, is based
on two impossibly optimistic assumptions: Most long-term welfare recipients will find jobs, and the states will maintain current levels of funding for
poverty programs.

No one can say precisely what the long-term effects of this law will be.
Much will depend on how states respond to the considerable flexibility they
have gained. The act allows states to cut their current funding of poverty
programs. The states no longer have the incentive, created by the federal
matching provisions of AFDC, to maintain benefit levels. In fact, the states
have no obligation to offer any cash assistance. States may find creative
ways to help poor people become self-sufficient, but how many will be willing to increase spending for this purpose as federal funding declines?

Economic conditions will also play an important role. What will happen
during recessionary periods? Unlike AFDC, TANF is not an entitlement and
will not automatically expand in a downturn. At the time the act was passed,
the economy was growing and unemployment was falling, but wages for
low-skilled workers remained at the anemic levels. Can we expect those who
lose welfare benefits to find jobs that will enable them to support their families? The addition of millions of former welfare recipients to the labor force—
as the new law demands—may push unemployment up and wages down at
the bottom of the labor market. At the same time, the reduction in food
stamp benefits will make life tougher for the working poor.

The time limits and eligibility changes in the new law mean that many
families and individuals will see their benefits terminated—whether or not
they are capable of supporting themselves. We know from studies we reviewed earlier in this chapter that most people who receive benefits require
only transitional help to get them past a period of crisis, but a majority of
those receiving benefits at any one moment have received assistance for extended periods. People in the second group will have the hardest adjustment to the new regime. They typically have limited education, few skills,
and little work experience. Many have disabling physical or mental conditions or are caring for disabled family members. Under the pressure of
need, many former welfare recipients will—perhaps to their own surprise—
learn to provide for themselves. Some will continue to receive help in states
that care to provide it.[11] But, as a result of the 1996 law, we are likely to see
more poverty and crueler poverty, reflected in increased homelessness and

---

[10] These estimates are based on a poverty measure similar to the "adjusted official" standard in Table 10-1, which counts food stamps as income and adjusts for certain taxes.

[11] The act permits states to exempt 20 percent of TANF cases from the five-year, lifetime
limit, but it provides no additional funding to provide for these cases. Moreover, the evidence
does not suggest that anything like 80 percent of current recipients can become self-supporting
on a continuing basis. (States also have the option of adopting a time limit). The new food
stamp program has no similar time-limit exception to the cut off provisions for able-bodied
adults without dependents.

hunger, and growing numbers of children forced into the foster care system because their families cannot provide for them.

CONCLUSION

This chapter on poverty has, perhaps inevitably, revolved around a series of contentious issues. The first of these concerned proper the definition of poverty. Some writers favor an absolute definition: Poverty means not having enough food, proper housing, and so forth. Critics see absolute definitions as subjective in practice and largely irrelevant to an affluent society like the United States. They prefer a relative definition: Poverty means a living standard far below the mainstream material standard of the society. The continuing debate over the technical details of definition reflects differences over deeper political questions. As a practical matter, the choice of definition shapes statistical conclusions about the size of the poverty problem and the direction of trends.

The official definition adopted by the federal government in the 1960s and widely used today is an absolute definition. By this standard, some 36 million Americans were poor in 1995, about 14 percent of the population. (Figure 10-2 on page 263 tells who they are). The same data show that the poverty rate declined significantly during the Age of Shared Prosperity and has climbed in the Age of Growing Inequality.

A second issue concerned the idea of a poverty "underclass." We adopted the term as a neutral class label in the model introduced in Chapter 1. But some authors use underclass to refer to a new "dangerous class" supposedly emerging at the bottom of American society. From this perspective, poverty is increasingly a problem of misbehavior: out-of-wedlock births, welfare dependency, joblessness, violent crime—all concentrated in high poverty, often minority neighborhoods. Although the behaviors referred to are certainly part of the larger poverty problem, the neighborhoods where they are concentrated do not account for a very large proportion of the poor, even the minority poor, according to careful studies. Further, the notion of an inevitable cycle of poverty implicit in discussions of the underclass is not supported by the available data. People who grow up poor are not, by and large, fated to remain poor. In fact, there is a large turnover in the poverty population from year to year. On the other hand, a large minority of the poor do seem to be locked into long-term poverty and continuing dependency on government aid.

The question of the underclass is related to a third issue: the mystery of rising poverty rates. Why is the poverty rate climbing after falling for so many years? We did not find a simple answer. Economic trends have certainly played a critical role. For reasons we explored in detail in Chapter 3, the labor market has turned sour for those at the bottom. Even when unemployment rates come down, low-end wages do not rise. A surprising number of poor Americans are full-time workers. Changing family patterns have also influenced poverty rates. Increasing divorce and out-of-wedlock births, along with the growing economic independence of women, have brought a sharp increase in the number of female-headed families.

As we observed in Chapter 4, most female-headed families are at the lower end of the income distribution, and many are poor. The truncated educations of many young mothers, the typically lower wages paid to female workers, the widespread failure of absent fathers to pay child support, and the competing demands on single mothers all contribute to this result. But the growth in low-income female-headed households is, at the same time, a reflection of the changing economic conditions. Many women remain single who might not have done so in the past because the men in their lives are unemployed or working at low-wage, unstable jobs; they do not seem likely to become dependable family breadwinners. Government policy has also contributed to rising poverty rates—though probably less so than economic and social factors. Poverty programs have been cut back. The purchasing power of the minimum wage has been allowed to decline. More generally, the broad economic policies of the federal government do not reflect a priority concern with reducing unemployment and raising wages.

The chapter repeatedly returns to the issues surrounding federal poverty programs. The federal government first took responsibility for the poor during the 1930s under the pressures of the sudden mass poverty brought on by the depression. A series of protective programs from Social Security to AFDC were created under Roosevelt's New Deal. In the 1960s and early 1970s, an era of prosperity and protest, poverty returned to the national agenda. The existing programs were expanded and new programs were created—some of them were broad insurance programs like Medicare and others were means-tested programs like food stamps. Rising Social Security benefits cut traditionally high poverty rates among the elderly to below average levels. The food stamp program reduced hunger and malnutrition.

But by the 1970s, support for the continued expansion of spending on social programs was waning. Means-tested programs became particular targets of criticism. Politicians found that they could draw votes with promises to fight the abuses of "welfare queens" or "end welfare as we know it." In the 1980s, social spending was trimmed under Ronald Reagan, a conservative Republican. But repeated attempts to "reform" welfare failed, until 1996, when a Democrat, Bill Clinton, reached agreement with a Republican Congress to radically restructure three major means-test programs. Under the law they enacted, poor families with children have no guarantee of cash support, the food stamp program will be reduced, and some children and most immigrants will lose SSI benefits. The legislation is based on the premise that the poor can and should provide for themselves, but it makes no effort to help them do so. The consequences of this law remain uncertain and will not be felt for a number of years, until its full provisions come into effect. Much will depend on the separate responses of the states, which have gained increased flexibility and must assume broad new responsibilities. But 1996 law is quite likely to produce increased poverty, hunger, and homelessness.

SUGGESTED READINGS

Bane, Mary Jo and Ellwood 1994. *Welfare Realities: From Rhetoric to Reform* Cambridge: Harvard University; David T. Ellwood, 1988. *Poor Support: Poverty in the American Family.* New York: Basic Books.

> *Carefully reasoned, well-researched books on family poverty and public policy. After writing them, the authors took positions in the Clinton administration to work on welfare reform. Both were unhappy with the 1996 reform.*

Burtless, Gary, et al. 1997. "The Future of the Social Safety Net." In Robert Reishauer, ed. *Setting National Priorities: Budget Choices for the Next Century.* Washington , D.C.: Brookings Institution.

> *Best account of the provisions of the 1996 welfare reform. Also summary of social safety net programs.*

Jencks, Christopher, and Paul Peterson. 1991. *The Urban Underclass.* Washington, D.C.: Brookings Institution.

> *Excellent collection of social science research on poverty generally and the underclass specifically, inspired by (but often critical of) Wilson (see below).*

Levitan, Sar A. 1990. *Programs in Aid of the Poor.* 6th ed. Baltimore: Johns Hopkins University Press.

> *Brief, authoritative description of poverty programs. On this topic, also see U.S. House 1992 (and subsequent editions), the so-called Green Book.*

Piven, Frances Fox and Richard A. Cloward. 1971. *Regulating the Poor: The Functions of Public Welfare.* New York: Pantheon.

> *Stimulating analysis of the development of welfare programs in the context of the social and political forces that spawned them.*

Snow, David and Leon Anderson 1993. *Down on their Luck: Homeless Street People* Berkeley: University of California Press.

> *Ethnography of the daily lives of homeless people in Austin, Texas.*

U.S. Bureau of the Census. *Poverty in the United States.* Series P-60.

> *Basic source of poverty statistics, published annually.*

Wilson, William Julius. 1987. *The Truly Disadvantaged: The Inner City, the Underclass, and Public Policy.* Chicago: University of Chicago Press; 1996. *When Work Disappears: The World of the New Urban Poor.* New York: Knopf.

> *Two influential books on the origins and nature of ghetto poverty.*

11

The American Class Structure and Growing Inequality

The 1980s were the triumph of upper America—an ostentatious celebration of wealth, the political ascendancy of the richest third of the population and the glorification of capitalism, free markets and finance. But while money, greed and luxury had become the stuff of popular culture, hardly anyone asked why such great wealth had concentrated at the top...

Kevin Phillips (1990)

In this final chapter, we will synthesize what we have learned about the American class system and how it is changing. We will review the evidence we have found of increasing inequality and reconsider the possible reasons for this trend. Readers may notice a shift of tone. In the preceding chapters our overriding objective was to present the existing data and research as precisely and faithfully as possible. Here we will be less constrained. We will generalize broadly, emphasizing our own interpretations. We will, by and large, dispense with citations, numbers, and tables—to the relief, no doubt, of many readers.

HOW MANY CLASSES ARE THERE?

Those with good memories will recall our answer to this question from the first chapter: six, but it all depends on your viewpoint. There is no irrefutable answer. As we noted in Chapter 1, defining classes and specifying the dividing lines between them is as much art as science. The "teardrop" class model we developed (see Figure 1-1, p. 18) reflects what we have learned researching this book, but it inevitably imposes a simplified pattern on a complicated, even chaotic reality. Let's take a closer look at the model and how it was derived.

Drawing on Marx and Weber, we began with the assumption that the class structure develops out of the economic system. Our classes are based on economic distinctions, rather than the prestige distinctions that Warner and his successors employed to develop their class models. But because prestige is generally rooted in economic differences, our map of the class system is broadly similar to Warner's prestige model for Yankee City or Coleman and Rainwater's model for Boston and Kansas City. There are two major differences: we do not employ their new money/old money distinction at the top nor do we treat blue collar/white collar as the essential class distinction in the middle (Table 2-2, p. 38 compares the three models).

As we learned in Chapter 4, three basic sources of income are available to households in this country: capitalist property, job earnings, and government transfers. The first source allows us to distinguish a top class that is largely dependent on income from property. The last source points to a class that has virtually no property and limited labor force participation, but depends on government transfers. Those who fall between these class extremes rely on their jobs—ranked by the skill or education required, the judgment and authority exercised at work, and the level of earnings.

Taken together these factors suggest a structure of six classes:

1. A *Capitalist Class*, subdivided into nationals and locals, whose income is derived largely from return on assets.
2. An *Upper-Middle Class* of college-trained professionals and managers (a few of whom ascend to such heights of bureaucratic dominance or accumulated wealth that they become part of the capitalist class).

3. A *Middle Class* whose members have significant skills and perform varied tasks at work, under loose supervision. They earn enough to afford a comfortable, mainstream lifestyle. Most wear white collars, but some wear blue.

4. A *Working Class* of people who are less skilled than members of the middle class and work at highly routinized, closely supervised manual and clerical jobs. Their work provides them with a relatively stable income sufficient to maintain a living standard just below the mainstream.

5. A *Working-Poor Class*, consisting of people employed in low-skill jobs, often at marginal firms. The members of this class are typically laborers, service workers, or low-paid operators. Their incomes leave them well below mainstream living standards. Moreover, they cannot depend on steady employment.

6. An *Underclass*, whose members have limited participation in the labor force and do not have wealth to fall back on. Many depend on government transfers.

The cutting points suggested by this schema are not equally significant or salient. The capitalist class is clearly separated from other classes by a crucial distinction: wealth as a primary source of income. The upper-middle class is set off from the classes below it by valuable credentials and the rewards that flow to them. The underclass is isolated from the classes above it by its separation from the world of work. But the other class boundaries are not so neatly defined.

The hazy distinction between the middle and working classes, the two great classes at the center of the class structure, has long perplexed students of stratification. We have ignored the traditional blue-collar/white-collar distinction and emphasized differences of education, skill, and autonomy connected with particular occupations. Thus, the electrician is middle class and the clerical worker doing routinized office tasks is working class.

In summary, we are suggesting a model of the class structure based primarily on a series of qualitative economic distinctions. From top to bottom, they are ownership of income-producing assets; possession of sophisticated educational credentials; a combination of self-direction and freedom from routinization at work; entrapment in the marginal sector of the labor market, and limited labor force participation.

Our schema is summarized Table 11-1, which provides fuller detail than the teardrop graphic in Chapter 1. As the table indicates, we think of the six classes as divided into three broader categories: the *privileged* classes (capitalist and upper middle), the *majority* classes (middle and working), and the *lower* classes (working-poor and poor). The percentages in the table are rough estimates of the proportion of households in each class, based on the available occupation, income, and wealth data. (Our rationale for conceiving of classes as groupings of households was presented in Chapter 1.) The classes are defined by the sources of income and occupations listed in

TABLE 11-1   Model of the American Class Structure: Classes by Typical Situations

Class, Percent of Households	Source of Income, Occupation of Main Earner	Typical Education	Typical Household Income, 1995
Privileged Classes			
Capitalist 1%	Investors, heirs, executives	Selective college or university	$1.5 million
Upper-Middle 14%	Upper managers and professionals, medium-sized business owners	College, often post-graduate study	$80,000
Majority Classes			
Middle 30%	Lower managers, semi-professional, nonretail sales workers	At least high school, often some college	$45,000
Working 30%	Operatives, low-paid craftsmen, clerical workers, retail sales workers	High school	$30,000
Lower Classes			
Working Poor 13%	Most service workers, laborers, low-paid operatives, and clerical workers	Some high school	$20,000
Underclass 12%	Unemployed or part-time workers, many dependent on public assistance and other government transfers	Some high school	$10,000

the third column. Occupation here refers to the work of the highest earning member of the household. The last two columns list education levels and incomes typical of each class. Of course, both will vary considerably in practice. Household income, as we learned in Chapter 4, is very dependent on the number of family members in the labor force. Educational levels have risen in successive generations, so that a younger worker will tend to have more years of school than an older worker in the same class.

Let's take a closer look at the six classes.

THE CAPITALIST CLASS

The members of the tiny capitalist class at the top of the hierarchy have an influence on economy and society far beyond their numbers. They make investment decisions that open or close employment opportunities for millions of others. They contribute money to political parties, and they often own media enterprises that allow them influence over the thinking of other classes.

The capitalist class strives to perpetuate itself: Assets, lifestyles, values, and social networks (which are themselves a vital source of influence) are all passed from one generation to the next. The maintenance of family fortunes and cohesion becomes more difficult as kin multiply and holdings are divided. Families attempt to instill a sense of lineage in the young. Extended families are drawn together by regular contact and by mutual dependence on their shared estate. Members of this class are active supporters of preparatory schools and colleges for their children and for carefully selected newcomers whom they hope to see socialized into their world view.

The richest, most powerful members of the capitalist class operate on the national and international scene. They control major corporations and are the main source of the huge "soft money" donations in national politics. They fund foundations and public policy "think tanks." These national capitalists have less prominent counterparts in communities across the country—the people who own the local car dealerships, real estate empires, media outlets, banks, and department stores. They fund community nonprofit institutions and influence local politics. Their collective influence over national politics is considerable because they are likely to have easy access to their own Members of Congress (whose campaigns they help finance) and belong to local chapters of politically potent organizations like the Chamber of Commerce and the National Federation of Independent Business. At both the national and local levels, the political power of the capitalist class has grown since the late 1970s.

The capitalist class is defined by dependence on income-producing wealth. We include owners of substantial enterprises, investors with diversified wealth, heirs to family fortunes, and top executives of major corporations. Top executives are included because their multimillion dollar compensation typically includes a stake in the company they manage and permits the accumulation of personal fortunes.

The organization of wealth in this country has been changing for decades. Family-controlled enterprises still account for a large share of capitalist class wealth and income. Among the largest corporations, however, family control has, by and large, given way to a system of control by professional executives—increasingly subject to the influence of the money fund managers who invest in their stock. As this happens, members of the capitalist class are diversifying their holdings. At the national level, families are less likely to be identified with a single enterprise. Locally, families are selling the enterprises with which they established their fortunes to national corporations. The younger generation inherits diversified stock and bond portfolios. Local wealth becomes national wealth.

These changes create the basis for a more cohesive capitalist class, whose members are free from parochial identification with a particular firm, economic sector, or locality. In politics, this tendency has been reinforced since the 1970s by the business-PACs, business lobby groups, and policy research organizations that defend the interests of the capitalist class as a whole rather than those of individual capitalists.

THE UPPER-MIDDLE CLASS

Since the early decades of the twentieth century, the upper-middle class has grown in numbers and importance and its composition has changed. Increasingly, salaried managers and professionals have replaced individual business owners and independent professionals. The key to the success of the upper-middle class is the growing importance of educational certification in a society dominated by complex technology and large-scale organization. Weber, who died in 1920, saw this coming. He observed the spread of the bureaucratic form of organization, with its characteristic preference for formal credentials, in business, government, and elsewhere. Weber compared the "preferential social opportunities" available to the university-educated in modern societies to the privileges of the well born in aristocratic societies (Weber 1946: 301, 241).

The upper-middle class exercises enormous and growing influence in American society. Because the upper-middle class embodies American achievement ideas and also possesses vast purchasing power, its lifestyles and opinions are becoming increasingly normative for the whole society. The members of this class vote, volunteer, make campaign donations, and run for office. Their high level of political activity amplifies the force of their political views, which have become increasingly conservative.

We have described the upper-middle class as one of the two "privileged classes." It is, in fact, a fairly porous class, open to people of modest origins who manage to earn the right credentials. But at the same time, it is increasingly separated from the rest of society. The income gap between the upper-middle class and the non-privileged classes has widened. Upper-middle class families are more likely to live in class-segregated neighborhoods and send their kids to class-segregated schools. We are convinced that the gap between this class and the rest of the population has replaced the traditional blue-collar/white-collar division as the most important cleavage in the American class structure.

THE MIDDLE CLASS

The stratification hierarchy, as we have repeatedly observed, is clearest (and incidentally, mobility the weakest) at the extremes. Toward the center, distinctions become blurred, people move more often during their lives from one level to another, and status becomes ambiguous. This is particularly true at the point where the middle class and the working class intersect—or better, overlap—so the reader should not expect precision of classification.

The changing character of work has largely eliminated the traditional differences between blue-collar and white-collar employment. The declining income differential between them, the increasing routinization of clerical tasks, and the corresponding drop in the prestige value of a white collar per se have all helped close the gap between shop and office.

Viewed in terms of major occupational groupings, the problem of distinguishing the two majority classes centers on the sales, clerical, and craft categories. We distinguish jobs that require little preparation, are highly routinized, closely supervised, and, typically, not well paid from jobs that demand significant skill or knowledge, are fairly varied, autonomous, and better rewarded. On this basis, semiprofessionals (teachers or social workers, for example), low-level managers, craft workers, and foremen are all middle class. Operatives, belong in the working class, along with routine clerical workers, whose jobs often have an assembly line character and are not well rewarded.

Government statistics divide sales occupations into retail and nonretail jobs, which we categorize as working and middle class respectively. The nonretail group includes insurance salesmen, real estate agents, and manufacturers' representatives. A few highly paid sales workers—stock brokers, in particular—should be considered upper-middle class. On the other hand, we place the lowest paid sales, clerical, operative positions in the working-poor class.

By most definitions, the middle class is large and diverse. The first characteristic makes it a natural target of political campaigns; it is difficult, in most districts, to win elections without middle class support. But the diversity of the middle class makes for ambiguous, if not conflicting, political interests. Although its numbers are larger, the middle class probably has less political influence than the upper-middle class, which is much more active politically and has a clearer sense of political direction.

"The disappearing middle class" has become a recurrent theme for political observers, journalists, and pop sociologists. In fact, the middle class is probably growing and, given the occupational trends associated with postindustrial society, is likely to grow in the future. Of course, the middle class can be made to disappear and reappear by manipulating the way it is defined. From our perspective, it is not the middle class but rather the middle-income group that is shrinking because of declining earnings of many working-class positions and the growth of family incomes toward the upper end of the distribution.

THE WORKING CLASS

The traditional core of the working class is easy to identify: semiskilled machine operatives, in factories and elsewhere, whose proportion of the labor force has been shrinking since the 1950s. Workers in routine white collar jobs are also part of this class and their share has been growing. Working class households commonly combine white and blue collar workers.

The loss of many goods-producing jobs has powerfully affected the working class and the country as a whole. New high school graduates can no longer be assured of finding jobs that will enable them to support a family. The contraction of employment in goods producing industries has helped reduce union membership to a small fraction of the labor force. This trend has, in turn, reduced the political influence of the working class and reinforced the conservative trend in national politics.

Members of the working class have low rates of political participation. With the decline of the urban political machines and labor unions that once mobilized them, they are less likely to vote than they were as recently as the 1960s.

THE WORKING POOR

The working poor class includes most service workers[1] and the lowest paid operatives and sales and clerical workers. Aside from low pay, their jobs generally have other disadvantages, such as unpleasant or dangerous work, few benefits, and uncertain employment. The distinctions between this class and the classes below and above it are both problematic. Many jobs near the boundary between the working class and the working poor could be placed on either side. But the gap between average jobs in the two classes is quite large. Lifestyles also differ. The working poor tend to have less stable work histories—sometimes for reasons beyond their control—and more personal and family problems. The lower boundary of this class, defined by commitment to employment, is blurred by the tendency of some individuals to move back and forth across it—a pattern which might be described as oscillating mobility.

Many of the working poor are young workers who, with further training and experience, will be able to move up in the class hierarchy. But income studies that follow families for extended periods reveal that a growing proportion of low income families are experiencing declining fortunes. This tendency will be exacerbated by the provisions of the 1996 act restructuring federal poverty programs—in particular, the food stamp provisions, which will cut into the disposable income of many working poor families.

Like the underclass beneath them, the working poor are generally alienated from political life and have minimal influence on the political process.

THE UNDERCLASS

Low-income households whose members have little participation in the labor force form the underclass. They may work erratically or at part-time jobs, but their lack of skills, incomplete education, spotty employment records, and, in many cases, disabilities make it difficult for them to find regular, full-time positions. Many are single mothers, receiving limited help

---

[1] Among the exceptions are police and fire fighters, classified as service workers by the Census Bureau, but as middle class by our criteria.

from absent fathers, balancing nurturer and provider roles, and facing a job market that offers little to women with limited education. A significant proportion of the underclass depends on government transfer programs, including public assistance, Social Security, Supplemental Security Income (SSI), and veteran's benefits.

The 1996 poverty legislation will have a decisive impact on the underclass. The most affected will be families with children, trapped in long-term poverty, with small prospects for self-support. The new time limits on public assistance will leave many such families in desperate straits. Least affected will be the larger number of families whose impoverishment is the result of some transitory situation.

Although the underclass as we have defined it is not identical with the federal poverty standard, the rising official poverty rates of recent years are certainly indicative of growth in the size of the underclass. Economic change and continuing high rates of unwed parenthood will continue to feed this growth.

GROWING INEQUALITY

A key theme in this book, announced in the subtitle, has been the expansion of class disparities in recent decades. The U-shaped curves we saw in Chapter 1 trace a momentous shift, somewhere in the early 1970s, from increasing equality to rising inequality. On this basis we distinguished two periods: the Age of Shared Prosperity (1946–1975) and the Age of Growing Inequality (since 1975). These dates are rough approximations because the precise timing depends on the indicator chosen, but the general pattern is unmistakable in the data on wealth, income, earnings, poverty, and other measures we have examined (See Figures 1-2 on pages 21–21 for a graphic overview).

In the remaining pages of the book, we will summarize the trends we found and return to the question we have asked repeatedly: Why is this happening?

Occupational Structure. Around 1970, the United States completed the transition from an industrial to a postindustrial society—that is, from a society in which most workers were employed in predominantly goods-producing sectors of the economy to one in which most were employed in service-producing sectors (see Table 3-5 p67). We found that the postindustrial economy tends toward occupational polarization. It requires engineers, accountants, and physicians, but it also needs janitors, cashiers, and hospital orderlies. The demand for operatives, miners, and other direct goods producers shrinks in the postindustrial economy, thereby eliminating some of the better paying positions available to workers without sophisticated educational credentials.

Earnings. Job earnings provide one of the best measures of the trends we have been discussing. During the Age of Shared Prosperity rapid economic growth produced a steady rise in the average wage. But since the early 1970s, wages have more or less stagnated and the distribution of wages has become increasingly unequal. The change has been most dramatic at the top and bottom: In the 1980s and early 1990s, the real compensation of CEOs almost quadrupled, while the proportion of men receiving poverty earnings for full-time work doubled.

The trend toward earnings inequality is remarkably pervasive. Inequality has increased between the college and high school educated, between the skilled and the unskilled, between younger workers and older workers, but also *among* the college educated, among the high school educated, among doctors, among carpenters, among people employed in manufacturing, among those employed in the service sectors, and so on. Even among corporate executives inequality has increased, as CEO earnings surged ahead of the rewards available to vice presidents and middle managers. There appears to be only one notable exception to this pattern: the earnings gap between men and women has narrowed.

Wealth. Inequality of wealth, measured by the concentration of net worth in the top 1 percent of the population, declined during the 1960s, only to rise sharply after the mid 1970s. In a sense, wealth represents a deeper inequality than earnings or income: it can be passed from generation to generation, it can provide such life fundamentals as a college education or home ownership, and, in capitalist form, it can generate new income. The increased concentration of wealth suggests mounting inequalities in the future.

Poverty. Measured by the federal poverty standard, the poverty rate plunged in the 1960s, stabilized in the 1970s and rose again in the 1980s and early 1990s. Over these decades the make up of the poverty population has also changed. The poor are less likely to be older than 65 and more likely to be younger than 18 than they were in 1960. They are also much more likely to live in female-headed families.

Income. Income provides the broadest and most continuous measure we have of economic inequality. Family income trends unambiguously reveal the differences between the Age of Shared Prosperity and the Age of Growing Inequality. During the first period, incomes at all levels rose swiftly and the distribution gradually moved toward greater equality. During the second period, incomes declined slightly at the bottom, stagnated in the middle, and soared at the top. Both the poverty rate and the percentage of families with incomes over $75,000—opposed measures of economic welfare rose during the Age of Growing Inequality. Change was most dramatic at the very top: The incomes of the upper 1 percent of households—which had stagnated in the 1970s—almost doubled in the 1980s. These same households were blessed with a significant reduction in federal taxes.

Social Life. Wealthy Americans have always been able to insulate themselves from the grimy realities of life at lower levels of the class structure. Has the increased economic inequality of recent decades meant increased social inequality? One important piece of evidence suggests that it has. The decennial census shows rising residential segregation by income level from 1970 to 1990. In particular, high income families are increasingly likely to live in separate neighborhoods. (In contrast, earlier studies indicate that residential segregation by class was decreasing in the 1960s). This residential separation means more schools, malls, and soccer leagues that are class-segregated, and fewer settings like the fictional neighborhood bar in "Cheers," where people of different classes encounter each other as equals.

Social Mobility. Social mobility slowed in the Age of Growing Inequality. There was *less upward and more downward mobility* in the 1980s than in the two preceding decades. The decline was most evident among younger workers. Nonetheless, even among younger workers, upward mobility was still more common than downward mobility. And even at the extremes of the class structure, position is not fixed at birth. For example, most young adults who were poor at birth are no longer poor (though their incomes typically remain below average). And a large percentage of the superrich Americans on the *Forbes* 400 list were not wealthy as children (though their families may have been well off).

In short, in the Age of Growing Inequality, it's harder to get ahead, but upward mobility is still pretty common. Does slowing mobility produce increasing inequality? Not necessarily. For example, if everyone moved down a notch (or up a notch), all would remain in the same relative position—inequality would not be effected. Recent studies do not indicate that a changing pattern of intergenerational mobility is contributing to increased inequality.

Political Power. In the Age of Growing Inequality, power has shifted away from the working class and working poor and toward the privileged classes—in particular, the capitalist class. Of course, power cannot be measured like wealth or income, and classes are not conscious political actors—they are abstractions. Yet we conclude that the class balance of power has shifted because the relative influence of institutions representing the interests of different social classes has changed and because national decision making has become more favorable to the privileged.

In the early 1970s, national business leaders initiated a concerted drive to expand their influence in national affairs. With money from corporations and wealthy individuals they organized to protect business interests, promote conservative ideas and back conservative candidates. Frequently, these candidates represented upper-middle class constituencies, thus forging an implicit political link between the two privileged classes. During the same period, labor unions were in a steep decline, losing members, money, and political power. The unions had long spoken for a wider working-class constituency than their own members. The business political initiative and the decline of organized labor were among the factors that contributed to the election of Ronald Reagan in 1980 and the election of a conservative Republican Congress in 1994. Political developments from reductions in the tax rates on very high incomes to the defeat of efforts to create a national health care system reflect the shifting class balance of power.

WHY?

How, then, can we account for the Age of Growing Inequality? No one knows exactly, but here's a short, provisional answer: There were some big changes in the economy. The effects of economic change were amplified by the decisions of corporations, families, and government.

The onset of the Age of Growing Inequality roughly coincided with the transition from an industrial to a postindustrial society. As we have seen, wage disparities of all sorts have widened. One reason is that disparities are greater in the growing service sectors of the economy, such as restaurants, health care, and law, than in the shrinking goods-producing sectors, such as manufacturing, mining, and transportation. The new economy (in both good-producing and service-producing sectors) makes winners out of workers with advanced education and skills, and losers out of those who lack such training. In part, this is because of the effects of technological change and globalization. Technology increases the demand for engineers, scientists, technicians, and those who manage technology, while reducing demand for craftsmen, operatives, and laborers. The ease with which capital and goods now move around the globe favors the investor, the accountant, the aeronautical engineer, and the systems analyst, but undercuts the factory worker. Shoes, textiles, clothing, and consumer electronics products can be made by low-wage labor in Mexico or China.

A variety of institutional mechanisms that once constrained wage differentials have weakened since the early 1970s. For example, fewer workers are protected by labor unions and collective bargaining agreements. The value of the minimum wage has eroded. Fewer positions are filled through promotion from each firm's "internal labor market." Federal deregulation seems to have resulted in increased pay differentials in industries such as airlines and trucking.

Facing more competitive markets at home and abroad, corporations look for ways to cut labor costs. In the Age of Growing Inequality, corporations "downsize," "outsource," and move production abroad. They eliminate benefits, freeze wages, and create new classes of workers, including part-timers, temps, second-tier new hires, and leased workers. They develop union-avoidance strategies. Although managers and professionals have not escaped the layoffs and wage reductions, these "meaner, leaner" labor practices have had their biggest impact at lower occupational levels. Corporations still offer generous competitive rewards to those whose talents are important to their success, from software designers to store managers—and, of course, CEOs.

Wage earners are family members and the effects of economic change are refracted through the prism of changing family life. In the new era, Americans are less likely to marry, more likely to divorce, more likely to have children out of wedlock, and, as a result, much more likely to live in female-headed families. Because absent fathers typically pay little child support and women generally earn less than men, female-headed families are concentrated on the low end of the income distribution. Among lower-class men and women, the failure to marry may itself reflect the reduced capacity of young men with limited education to provide for families as they might have in the past.

In the Age of Growing Inequality, more women are employed, they work longer hours, and earn higher wages. Women's rising earnings enabled families to overcome the effects of erosion in men's earnings. But this trend is also contributing to disparities in family incomes. Families depending on one worker—male or female—typically fall well behind two-paycheck families. High-earning women tend to be married to high-earning men and to experience faster growth in earnings. The net result of women's increased employment is greater inequality among families. (However, as we have noted, among married-couple families considered separately, wives' earnings have probably reduced income inequality.)

Relative to the giant consequences of change in the economy and family life, the effects of government policy have probably been modest, but not insignificant. We have already referred to the failure of the federal government to maintain the value of the minimum wage and to the effects of deregulation. The decline of the labor movement is, in part, the result of weakened federal protection of union rights and of anti-union laws in states that provide a refuge for firms searching for cheap labor. Recent legislation has weakened the social safety net protecting the poor. Presidents of both parties have promoted the elimination of barriers to international trade, a policy which has probably been injurious to the lowest skilled workers, whatever its advantages to the economy as a whole.

All these policies allow greater freedom to market forces, which tend to produce unequal outcomes: rising rewards for some who are talented, well financed, or just lucky, and, stagnant or declining rewards for many others.

In broad terms, public policy in the Age of Growing Inequality has become more responsive to the privileged classes and less sympathetic to other classes. Tax policy, for example, has zigged and zagged over the last two decades, but the most significant net result has been sharply reduced federal taxation of wealthy households. Macroeconomic policies reflect the same tendency. Federal policy makers have been more concerned with reducing inflation than maintaining employment or wage levels. During the early 1990s, as unemployment declined, senior federal officials strained to reassure worried investors that wages were not rising as a consequence. No one bothered to reassure low-income workers about depressed wages. Government might have cushioned the effects of economic change for those most adversely affected. It might have created new opportunities for people who lacked the education or skills to compete successfully in the new economy. But it has, by and large, failed to do so.

STRAWS IN THE WIND

As this edition of the *American Class Structure* was written, new federal statistics on poverty, wages, income, and employment were being released. The new numbers, referring to the mid-1990s, were encouraging. Poverty was down, and real median income was up. The unemployment rate had dropped to levels seldom seen since the 1960s, but wages hardly budged. What did all this mean? Clearly more Americans, especially lower-income Americans, had jobs and their earnings were pushing family incomes up—even though the average wage had remained stagnant. The new statistics also showed that the country had climbed out of the economic recession of the early 1990s.

Only time will tell whether the new trends are just a squiggle on the economic charts or something grander: the end of the Age of Growing Inequality. The poverty rate is still well above what it was in the early 1970s. The real median wage for men is about what it was twenty-five years ago—a clear indication that the bottom 50 percent has fallen far behind the rapidly advancing elite at the high end of the labor market. Only the remarkably low unemployment rates of the mid-1990s mark a break with recent history and a cause for optimism.

The fundamental changes in the economy and family life that produced an Age of Growing Inequality are still in place. And public policy retains its cold indifference to the widening disparities in American society.

Bibliography

Abramson, Paul R., John H. Aldrich, and David W. Rohde. 1995. *Change and Continuity in the 1992 Elections.* Revised ed. Washington, D.C.: Congressional Quarterly Press.

Aiken, Michael, and Paul E. Mott, eds. 1970. *The Structure of Community Power.* New York: Random House.

Aldrich, Nelson W., Jr. 1988. *Old Money: The Mythology of America's Upper Class.* New York: Vintage.

Alexander, Herbert E., and Brian A. Haggerty. 1987. *Financing the 1984 Elections.* Lexington, Mass.: Lexington Books.

Allan, Graham 1989. *Friendship: Developing a Sociological Perspective.* Boulder, Colo.: Westview.

Allen, Frederick Lewis. 1952. *The Big Change.* New York: Harper & Row.

Anderson, Dewey, and Percy Davidson. 1943. *Ballots and the Democratic Class Struggle.* Stanford, Calif.: Stanford University Press.

Anderson, Martin. 1978. *Welfare: The Political Economy of Welfare Reform in the United States.* Stanford, Calif.: Hoover Institution Press.

Argyle, Michael 1994. *The Social Psychology of Social Class.* New York: Routledge.

Auletta, Ken. 1982. *The Underclass.* New York: Random House.

Avery, Robert, Gregory Elliehausen, and Arthur Kennickell. 1988. "Measuring Wealth with Survey Data: An Evaluation of the 1983 Survey of Consumer Finances." *Review of Income and Wealth* 34:339–369.

Bachrach, Peter, and Morton S. Baratz. 1974. *Power and Poverty: Theory and Practice.* New York: Oxford University Press.

Bagby, Meredith. 1997. *Annual Report of the United States of America.* New York: McGraw-Hill.

Baltzell, E. Digby. 1958. *Philadelphia Gentlemen.* New York: Free Press.

Bane, Mary J. and David Ellwood 1994. *Welfare Realities: From Rhetoric to Reform.* Cambridge, Mass.: Harvard University Press.

Barnouw, Erik. 1978. *The Sponsor: Notes on a Modern Potentate.* New York: Oxford University Press.

Beeghley, Leonard. 1996. *The Structure of Social Stratification in the United States*. 2nd ed. Needham Heights, Mass.: Simon & Schuster.

Beeghley, Leonard, and John K. Cochran. 1988. "Class Identification and Gender Role Norms among Employed Married Women." *Journal of Marriage and the Family* 50:546–566.

Bell, Daniel. 1976. *The Coming of the Post-Industrial Society*. New York: Basic.

Bendix, Reinhard, and Seymour Martin Lipset, eds. 1953. *Class, Status, and Power*. 1st ed. Glencoe, Ill.: Free Press. (2nd ed., New York: Free Press, 1966).

Berg, Ivar. 1970. *Education and Jobs: The Great Training Robbery*. New York: Praeger.

Berle, Adolf A., Jr., and Gardiner C. Means. 1932. *The Modern Corporation and Private Property*. New York: Commerce Clearing House.

Berman, Paul. 1991. "A Union Man from Harvard." *New York Times Book Review*. August 11.

Bianchi, Suzanne. 1995. "Changing Economic Roles of Women and Men." In *State of the Union: America in the 1990s, Volume One: Economic Trends*, edited by Reynolds Farley. New York: Russell Sage Foundation.

Birnbach, Lisa, ed. 1980. *The Official Preppy Handbook*. New York: Workman.

Blau, Francine. 1984. "Women in the Labor Force: An Overview." In *Women: A Feminist Perspective*, edited by Jo Freeman. Palo Alto, Calif.: Mayfield.

Blau, Peter M., and Otis Dudley Duncan. 1967. *The American Occupational Structure*. New York: Wiley.

Bloch, Fred. 1977. "The Ruling Class Does Not Rule: Notes on the Marxist Theory of the State." *Socialist Revolution* 7 (1):6–28.

Blumberg, Paul M., and P.W. Paul. 1975. "Continuities and Discontinuities in Upper-Class Marriages." *Journal of Marriage and the Family* 37: 63–77.

Blumenthal, Sydney. 1986. *The Rise of the Counter Establishment*. New York: Times Books.

Blumenthal, Sydney, and Thomas Byrne Edsall, eds. 1988. *The Reagan Legacy*. New York: Pantheon.

Bonjean, Charles M., and Michael D. Grimes. 1974. "Community Power: Issues and Findings." In *Social Stratification: A Reader*, edited by Joseph Lopreato and Lionel S. Lewis. New York: Harper & Row.

Boston, Thomas D. 1988. *Race, Class and Conservatism*. Cambridge, MA.: Unwin and Hyman.

Bott, Elizabeth. 1954. "The Concept of Class as a Reference Group." *Human Relations* 7: 259–286.

————. 1964. *Family and Social Network*. London: Tavistock.

Bottomore, Tom. 1966. *Elites in Modern Society*. New York: Pantheon.

Bound, John, and Richard B. Freeman. 1992. "What Went Wrong? The Erosion of Relative Earnings and Employment Among Young Black Men in the 1980s." *Quarterly Journal of Economics* 107: 201–231.

Bourdieu, Pierre. 1984. *Distinction: A Social Critique of the Judgment of Taste*. Cambridge, Mass.: Harvard University Press.

Bowles, Samuel, and Herbert Gintis. 1976. *Schooling in Capitalist America*. New York: Basic.

Boyer, Richard, and Herbert Morais. 1975. *Labor's Untold Story*. 3rd ed. New York: United Electrical Workers.

Bradburn, Norman. 1969. *The Structure of Psychological Well-Being*. Chicago: Aldine.

Braverman, Harry. 1974. *Labor and Monopoly Capital*. New York: Monthly Review Press.

Brody, David. 1980. *Workers in Industrial America: Essays on the 20th Century Struggle*. New York: Oxford University Press.

Bronfenbrenner, Urie. 1966. "Socialization through Time and Space." In *Class, Status, and Power*, edited by Reinhard Bendix and Seymour Martin Lipset. 2nd ed. New York: Free Press.

Brooks, Thomas. 1971. *Toil and Trouble: A History of American Labor*. 2nd ed. New York: Dell.

Brown, Clifford, et al. 1995. *Serious Money: Fundraising and Contributing in Presidential Nominational Campaigns*. New York: Cambridge University Press.

Burch, Philip H., Jr. 1980. *Elites in American History: The New Deal to the Carter Administration*. New York: Holmes and Meier.

Burnham, James. 1941. *The Managerial Revolution*. New York: John Day.

Burnight, Robert G., and Parker G. Marden. 1967. "Social Correlates of Weight in an Aging Population." *Milbank Memorial Fund Quarterly* 45:75–92.

Burtless, Gary. 1987. "Inequality in America: Where Do We Stand?" *Brookings Review* 5 (Summer): 9–16.

_____. 1990. *A Future of Lousy Jobs*. Washington, D.C.: Brookings Institution.

_____. 1995. "International Trade and the Rise in Earnings Inequality." *Journal of Economic Literature*. 32: 800–816.

Burtless, Gary, R. Kent Weaver and Joshua Wiener. 1997. "The Future of the Safety Net." In *Setting National Priorities: Budget Choices for the Next Century*, edited by Robert Reishauer. Washington, D.C.: Brookings Institution.

Business Week 1996. "Special Report: How High Can CEO Pay Go?" April 22.

Business Week 1997. "Executive Pay: Special Report." April 21.

Cameron, Juan. 1978. "Small Business Trips Big Labor." *Fortune* 98 (July): 80–82.

Campbell, Angus, Gerald Gurin, and Warren E. Miller. 1954. *The Voter Decides*. Evanston, Ill.: Row, Peterson.

Cancian, Maria, et al. 1993. "Working Wives and Family Income Inequality among Married Couples." In *Uneven Tides: Rising Inequality in America*, edited by Sheldon Danziger and Peter Gottshalk. New York: Russell Sage Foundation.

Cantril, Hadley. 1951. *Public Opinion*. Princeton, N.J.: Princeton University Press.

Caplow, Theodore. 1980. "Middletown Fifty Years After." *Contemporary Sociology* 9: 46–50.

Caplow, Theodore, and Bruce Chadwick. 1979. "Inequality and Life Styles in Middletown, 1920–1978." *Social Science Quarterly* 60: 366–368.

Card, David, and Alan B. Krueger. 1992. "School Quality and Black-White Relative Earnings: A Direct Assessment." *Quarterly Journal of Economics* 107: 151-200.

Center for Political Studies. "American National Election Study, 1984: Codebook."

Centers, Richard. 1949. *The Psychology of Social Classes: A Study of Class Consciousness*. Princeton, N.J.: Princeton University Press.

_____. 1953. "Social Class, Occupation and Imputed Belief." *American Journal of Sociology* 58: 546.

Charlot, Monica. 1985. "The Ethnic Minorities Vote." In *Britain at the Polls*, edited by Austin Ranney. Washington, D.C.: American Enterprise Institute.

Citizens for Tax Justice and the Institute on Taxation and Economic Policy 1996. *Who Pays? A Distributional Analysis of the Tax Systems in All 50 States*. Washington, D.C.

Clark, Terry. 1971. "Community Structure, Decision-Making, Budget Expenditure, and Urban Renewal in 51 American Communities." In *Community Politics*, edited by Charles Bonjean. New York: Free Press.

Cohen, Jere. 1979. "Socio-Economic Status and High School Friendship Choice: Elmtown's Youth Revisited." *Social Networks* 2: 65–74.

Colclough, Glenna, and E.M. Beck. 1986. "The American Educational Structure and the Reproduction of Social Class." *Sociological Inquiry* 56: 456–476.

Coleman, James S., et al. 1966. *Equality of Educational Opportunity*. Washington, D.C.: U.S. Government Printing Office.

Coleman, Richard P., and Lee Rainwater, with Kent A. McClelland. 1978. *Social Standing in America: New Dimensions of Class*. New York: Basic.

Collier, Peter, and David Horowitz. 1976. *The Rockefellers: An American Dynasty*. New York: Holt, Rinehart & Winston.

Compton, Rosemary. 1993. *Class and Stratification: An Introduction to Current Debates*. Cambridge, Mass.: Polity.

Congressional Quarterly Press. 1976. *Guide to Congress*. 2nd ed.

Congressional Quarterly Press. 1996. "Welfare Overhaul Law." (Sept. 21): 2696–2705.

_____. 1980. "Democrats May Lose Edge in Contributions from PACs." 38: 3405–3409.

Cookson, Peter, and Caroline Hodges Persell. 1985. *Preparing for Power: America's Elite Boarding Schools*. New York: Basic.

Corcoran, M. 1995. "Rags to Rags: Poverty and Mobility in the United States." *Annual Review of Sociology*. 21: 237–267.

Corey, Lewis. 1935. *The Crisis of the Middle Class*. New York: Covici-Friede.

_____. 1953. "Problems of the Peace: The Middle Class." In *Class, Status, and Power*, edited by Reinhard Bendix and Seymour Martin Lipset. Glencoe, Ill.: Free Press.

Coser, Lewis. 1978. *Masters of Sociological Thought*. 2nd ed. New York: Harcourt Brace Jovanovich.

Coverman, Shelly. 1988. "Sociological Explanations of the Male-Female Wage Gap." in *Women Working: Theories and Facts in Perspective*, edited by Ann Stromberg and Shirley Harkess. 2nd ed. Mountain View, Calif.: Mayfield.

Coxon, Anthony, and Charles Jones. 1978. *The Images of Occupational Prestige*. New York: St. Martin's.

Croteau, David 1995. *Politics and the Class Divide: Working People and the Middle-Class Left*. Philadelphia: Temple University Press.

Cruciano, Therese 1996. "Individual Tax Returns: Preliminary Data, 1994." *SOI Bulletin: A Quarterly Statistics of Income Report* 15 (Spring): 18–24.

Current Biography. 1982–1988. New York: W.W. Wilson.

Curtis, Richard F., and Elton F. Jackson. 1977. *Inequality in American Communities*. New York: Academic.

Daalder, Hans, and Ruud Koole. 1988. "Liberal Parties in the Netherlands." In *Liberal Parties in Western Europe*, edited by Emil Kirchner. Cambridge: Cambridge University Press; pp. 151–177.

Dahl, Robert A. 1961. *Who Governs? Democracy and Power in an American City*. New Haven, Conn.: Yale University Press.

_____. 1967. *Pluralist Democracy in the United States*. Chicago: Rand McNally.

Danziger, Sheldon and Peter Gottschalk. 1995. *America Unequal*. New York: Russell Sage Foundation.

Davis, Allison, Burleigh B. Gardner, and Mary R. Gardner. 1941. *Deep South: A Social-Anthropological Study of Caste and Class*. Chicago: University of Chicago Press.

Davis, James Allen, and Tom Smith. 1990. *General Social Surveys, 1972–1990* (machine-readable data file). Chicago: National Opinion Research Center.

Davis, Mike. 1980. "The Barren Marriage of American Labour and the Democratic Party." *New Left Review*, 124: 45–84.

Demerath, N.J., III. 1965. *Social Class in American Protestantism*. Chicago: Rand McNally.

Dent, David. 1992. "The New Black Suburbs." *New York Times Magazines*. June 14.

DeParle, Jason 1996. "Welfare: Progress Hijacked." *New York Times Magazine*. December 8.

DeStefano, Linda. 1990. "Pressures of Modern Life Bring Increased Importance to Friendship." *The Gallup Poll Monthly* (March): 24–33.

Dinitz, S., F. Banks, and B. Pasamanick. 1960. "Mate Selection and Social Class: Change in the Past Quarter Century." *Marriage and Family Living* 22: 348–351.

Domhoff, G. William. 1967. *Who Rules America?* Englewood Cliffs, N.J.: Prentice-Hall.

_____. 1970. *The Higher Circles: The Governing Class in America*. Englewood Cliffs, N.J.: Prentice-Hall.

_____. 1974. *The Bohemian Grove and Other Retreats*. New York: Harper & Row.

_____. 1975. "Social Clubs, Policy Planning Groups, and Corporations." *Insurgent Sociologist* 5 (3):173–195.

_____. 1978. *Who Really Rules? New Haven and Community Power Reexamined*. New Brunswick, N.J.: Transaction.

Domhoff, G. William, and Hoyt B. Ballard, eds. 1968. *C. Wright Mills and the Power Elite*. Boston: Beacon.

Dotson, Floyd. 1950. "The Associations of Urban Workers." Unpublished Ph.D. thesis, Yale University.

Dubofsky, Melvyn. 1980. "The Legacy of the New Deal." *Executive* 6 (Spring): 8–10.

Duncan, Gregg J., et al. 1984. *Years of Poverty, Years of Plenty: The Changing Economic Fortunes of American Workers and Families.* Ann Arbor: Institute for Social Research, University of Michigan.

Duncan, Gregg, and Willard Rodgers. 1989. "Has Poverty Become More Persistent?" Institute for Social Research, University of Michigan.

Duncan, Otis Dudley. 1961. "A Socio-Economic Index for All Occupations," and "Properties and Characteristics of the Socioeconomic Index." In *Occupations and Social Status,* edited by Albert Reiss. Glencoe, Ill.: Free Press.

————. 1966. "Methodological Issues in the Analysis of Social Mobility." In *Social Structure and Mobility in Economic Development,* edited by Neil Smelser and Seymour Martin Lipset. Chicago: Aldine.

Duncan, Otis Dudley, Archibald O. Haller, and Alejandro Portes. 1968. "Peer Influences on Aspirations: A Reinterpretation." *American Journal of Sociology* 74: 119–137.

Dye, Thomas R. 1979. *Who's Running America? The Carter Years.* 2nd ed. Englewood Cliffs, N.J.: Prentice-Hall.

————. 1995. *Who's Running America: The Clinton Years.* 6th edition. Englewood Cliffs, N.J.: Prentice-Hall.

Edelman, Peter 1997. "The Worst Thing Bill Clinton Has Done." *Atlantic Magazine.* March.

Edsall, Thomas B. 1984. *The New Politics of Inequality.* New York: Norton.

Edsall, Thomas, and Mary Edsall. 1991. *Chain Reaction: The Impact of Race, Rights and Taxes in American Politics.* New York: Norton.

Edwards, Alba M., and U.S. Bureau of the Census. 1943. *U.S. Census of Population 1940: Comparative Occupational Statistics, 1870–1940.* Washington, D.C.: U.S. Government Printing Office.

Ehrenreich, Barbara. 1989. *Fear of Falling: The Inner Life of the Middle Class.* New York: Pantheon.

Eismeier, Theodore and Philip Pollock 1996. "Money in the 1994 Elections and Beyond." In *Midterm: The Elections of 1994 in Context,* edited by Philip Klinkner. Boulder, Colo.: Westview.

Ellwood, David T. 1988. *Poor Support: Poverty in the American Family.* New York: Basic.

Ellwood, David, and Mary J. Bane. 1985. "The Impact of AFDC on Family Structure and Living Arrangements." In *Research in Labor Economics* 7, edited by Ronald Ehrenberg. Greenwich, Conn.: JAI.

————. 1994. *Welfare Realities: From Rhetoric to Reform.* Cambridge, Mass.: Harvard University Press.

Erikson, Robert, and John Goldthorpe. 1992. *The Constant Flux: A Study of Class Mobility in Industrial Societies.* New York: Oxford University Press.

Farley, Reynolds. 1984. *Blacks and Whites: Narrowing the Gap?* Cambridge, Mass.: Harvard University Press.

Farley, Reynolds, ed. 1995a. *State of the Union: America in the 1990s; Volume One: Economic Trends.* New York: Russell Sage Foundation.

Farley, Reynolds, ed. 1995b. *State of the Union: America in the 1990s; Volume Two: Social Trends*. New York: Russell Sage Foundation.

Featherman, David 1979. "Opportunities are Expanding." *Society* March/April, 4, 6–11.

Featherman, David L., and Robert M. Hauser. 1978. *Opportunity and Change*. New York: Academic.

Ferguson, Thomas 1995. *Golden Rule: The Investment Theory of Party Competition and the Logic of Money Driven Political Systems*. Chicago: University of Chicago Press.

Fisher, Claude, et al. 1996. *Inequality by Design: Cracking the Bell Curve Myth*. Princeton, N.J.: Princeton University Press.

Forbes. 1990. "The Forbes 400." October 22.

_____. 1991. "What 800 Companies Paid their Bosses." May 27.

_____. 1996. "The Forbes 400." October 14.

Fortune. 1937. "The Industrial War" 14 (November): 105–110, 156, 160, 166.

Frank, Robert H., and Philip J. Cook. 1995. *The Winner-Take-All Society*. New York: Simon & Schuster.

Fraser, Douglas. 1978. "UAW President Fraser Resigns from Labor-Management Group." *Radical History Review* 18 (Fall): 117–121.

Frears, John. 1988. "Liberalism in France." In *Liberal Parties in Western Europe*, edited by Emil Kirchner. Cambridge: Cambridge University Press, pp. 124–150.

Freeman, Richard B. 1996. "Labor Market Institutions and Earnings Inequality." *New England Economic Review*. Special Issue (May/June): 157–181.

Fussell, Paul. 1983. *Class: A Guide Through the American Status System*. New York: Ballantine.

Galbraith, John Kenneth. 1958. *The Affluent Society*. Boston: Houghton Mifflin.

_____. 1967. *The New Industrial State*. Boston: Houghton Mifflin.

Gallup, George, Jr. 1988. *The Gallup Poll: Public Opinion 1987*. Wilmington, Del.: Scholarly Resources.

_____. 1989. *The Gallup Poll: Public Opinion 1988*. Wilmington, Del.: Scholarly Resources.

Garfinkel, Irwin, and Sara McLanahan. 1986. *Single Mothers and Their Children*. Washington, D.C.: Urban Institute.

Gecas, Viktor. 1979. "The Influence of Social Class on Socialization." In *Contemporary Theories about the Family*, edited by W.R. Burr et al. Vol. I. New York: Free Press.

Geoghegan, Thomas. 1991. *Which Side Are You on: Trying to Be for Labor When It's Flat on Its Back*. New York: Farrar, Strauss & Giroux.

Giddens, Anthony. 1973. *The Class Structure of the Advanced Societies*. New York: Harper & Row.

Gilbert, Dennis. 1981. "Cognatic Descent Groups in Upper-Class Lima (Peru)." *American Ethnologist* 8: 739–757.

Ginsberg, Benjamin, and Martin Shefter. 1990. *Politics by Other Means: The Declining Importance of Elections in America*. New York: Basic.

Gittleman, Maury. 1994. "Earnings in the 1980s: An Occupational Perspective." *Monthly Labor Review*. 117: 16-27.

Glenn, Norval D. 1975. "The Contribution of White Collars to Occupational Prestige." *Sociological Quarterly* 16: 184–189.

Glenn, Norval D., and Jon P. Alston. 1968. "Cultural Distances Among Occupational Categories." *American Sociological Review* 33: 365–382.

Goldstein, Marshall. 1962. "Absentee Ownership and Monolithic Power Structures." In *Trends in Comparative Community Studies*, edited by Bert E. Swanson. Kansas City, Mo.: Community Studies.

Goldthorpe, John H., and Keith Hope. 1972. "Occupational Grading and Occupational Prestige." In *The Analysis of Social Mobility: Methods and Approaches*, edited by Keith Hope. Oxford, England: Clarendon.

Green, Mark. 1979. *Who Runs Congress?* 3rd ed. New York: Bantam.

Greene, Bert. 1978. *Pity the Poor Rich.* Chicago: Contemporary.

Greenstone, J. David. 1977. *Labor in American Politics.* Chicago: University of Chicago Press.

Grusky, David, ed. 1994. *Social Stratification: Class, Race and Gender in Sociological Perspective.* Boulder, Colo.: Westview.

Hacker, Andrew 1997. *Money: Who Has How Much and Why?* New York: Scribners.

Hacker, Louis. 1970. *The Course of American Economic Growth and Development.* New York: Wiley.

Halberstam, David. 1972. *The Best and the Brightest.* New York: Random House.

Halle, David 1984. *America's Working Man: Work, Home and Politics among Blue-Collar Property Owners.* Chicago: University of Chicago Press.

Hamilton, Alexander. 1780. Quoted in Arthur J. Schlesinger, Jr., *The Age of Jackson.* Boston: Little Brown (1945), p. 10.

Hamilton, Richard. 1966. "The Marginal Middle Class: A Reconsideration." *American Sociological Review* 31: 192–199.

————. 1972. *Class and Politics in the United States.* New York: Wiley.

————. 1975. *Restraining Myths: Critical Studies of United States' Social Structure and Politics.* Beverly Hills, Calif.: Sage.

Harrington, Michael. 1962. *The Other America: Poverty in the United States.* New York: Macmillan.

Harrison, Bennett, and Barry Bluestone. 1988. *The Great U-Turn: Corporate Restructuring and the Polarizing of America.* New York: Basic.

Haveman, Robert, and Barbara Wolf 1994. *Succeeding Generations: On the Effects of Investments in Children.* New York: Russell Sage Foundation.

Herman, Edward S. 1981. *Corporate Control, Corporate Power.* Cambridge, England: Cambridge University Press.

Herrstein, Richard, and Charles Murray 1994. *The Bell Curve: Intelligence and Class Structure in American Life.* New York: Free Press.

Higham, John. 1963. *Strangers in the Land.* New York: Atheneum.

Hodge, Robert W., Donald J. Treiman, and Peter H. Rossi. 1966. "A Comparative Study of Occupational Prestige." In *Class, Status and Power,* edited by Reinhard Bendix and Seymour Martin Lipset. 2nd ed. New York: Free Press.

Hodge, Robert W., and Donald Treiman. 1968. "Class Identification in the United States." *American Journal of Sociology* 73: 535–547.

Hodges, Harold M. 1964. *Social Stratification: Class in America.* Cambridge, Mass.: Schenkman.

Hoffman, Saul. 1977. "Marital Instability and the Economic Status of Women." *Demography* 14: 67–76.

Hollingshead, August B. 1949. *Elmtown's Youth.* New York: Wiley.

————. 1950. "Cultural Factors in the Selection of Marriage Mates." *American Sociological Review* 15: 619–627.

Hollingshead, August B., and Frederick Redlich. 1958. *Social Class and Mental Illness: A Community Study.* New York: Wiley.

Horan, Patrick M. 1978. "Is Status Attainment Research Atheoretical?" *American Sociological Review* 43: 534–541.

Hout, Michael. 1988. "More Universalism, Less Structural Mobility: The American Occupational Structure in the 1980s." *American Journal of Sociology* 93: 1358–1400.

Hout, Michael, et al. 1995. "Class Voting in the U.S. 1948–92." *American Sociological Review* 60:802–828.

Howe, Louise. 1977. *Pink Collar Worker: Inside the World of Woman's Work.* New York: Putnam.

Hunter, Floyd. 1953. *Community Power Structure: A Study of Decision Makers.* Chapel Hill: University of North Carolina Press.

Inkeles, Alex, and Peter H. Rossi. 1956. "National Comparisons of Occupational Prestige." *American Journal of Sociology* 61: 329–339.

Jackman, Mary. 1979. "The Subjective Meaning of Social Class Identification in the United States." *Public Opinion Quarterly* 43: 443–462.

Jackson, Elton F., and Richard F. Curtis. 1968. "Conceptualization and Measurement in the Study of Social Stratification." In *Methodology in Social Research,* edited by Herbert M. Blalock, Jr., and Ann B. Blalock. New York: McGraw-Hill.

Jargowsky, Paul 1996. "Take the Money and Run: Economic Segregation in U.S. Metropolitan Areas." *American Sociological Review.* 61: 984-998.

Jaynes, Gerald David, and Robin M. Williams, Jr., eds. 1989. *A Common Destiny: Blacks and American Society.* Washington, D.C.: National Academy Press.

Jefferson, Thomas. 1821. "Autobiography." In *The Life and Selected Writings of Thomas Jefferson,* edited by Adrienne Koch and William Peden. New York: Modern Library, 1944.

Jencks, Christopher, et al. 1972. *Inequality: A Reassessment of the Effect of Family and Schooling in America.* New York: Basic.

————. 1979. *Who Gets Ahead?* New York: Basic.

————. 1991. "Is the American Underclass Growing?" In *The Urban Underclass,* edited by Christopher Jencks and Paul Peterson. Washington, D.C.: Brookings Institution.

Jencks, Christopher, and Paul Peterson. 1991. *The Urban Underclass.* Washington, D.C.: Brookings Institution.

Johnson, Haynes and David Broder 1996. *The System: The American Way of Politics at the Breaking Point.* Boston: Little Brown.

Judis, John B. 1991. "Twilight of the Gods." *Wilson Quarterly* 5 (Autumn): 43–57.

Kahl, Joseph A. 1953. "Educational and Occupational Aspirations of Common Man Boys." *Harvard Educational Review* 23: 186–203.

_____. 1957. *The American Class Structure.* 1st ed. New York: Rinehart.

Kahl, Joseph A., and James A. Davis. 1955. "A Comparison of Indexes of Socio-Economic Status." *American Sociological Review* 20: 317–325.

Kanter, Rosabeth. 1977. *Men and Women of the Corporation.* New York: Basic.

Karabel, Jerome, and A.H. Halsey, eds. 1977. *Power and Ideology in Education.* New York: Oxford University Press.

Katz, Lawrence F., and Kevin M. Murphy. 1992. "Changes in Relative Wages, 1963–1987: Supply and Demand Factors." *Quarterly Journal of Economics.* 107: 35–78.

Kassalow, Everett. 1978. "How Some European Nations Avoid U.S. Levels of Industrial Conflict." *Monthly Labor Review* 101 (April): 97.

Kaus, Mickey 1992. *The End of Inequality.* New York: Basic.

Kaysen, Carl. 1957. "The Social Significance of the Modern Corporation." *American Economic Review* 47: 311–319.

Kennickell, Arthur B.; Douglas A. McManus and R. Louise Woodburn. 1996. "Weighting Design for the 1992 Survey of Consumer Finances." Unpublished paper.

Kerr, Clark, and Abraham Seigal. 1954. "The Interindustry Propensity to Strike — An International Comparison." In *Industrial Conflict,* edited by Arthur Kornhauser, et al. New York: McGraw-Hill.

Kichen, Steve, et al. 1996. "The Private 500." *Forbes.* December 2.

Klinkner, Philip, ed, 1996. *Midterm: The Elections of 1994 in Context.* Boulder, Colo.: Westview.

Kodrzycki, Yolanda K. 1996. "Labor Market and Earnings Inequality: A Status Report." *New England Economic Review.* May/June 1996: 11–25.

Koenig, Thomas. 1980. "Corporate Support for Political Contribution Disclosure." Unpublished paper presented at the American Sociological Association, New York.

Kohn, Melvin L. 1969. *Class and Conformity: A Study in Values.* Homewood, Ill.: Dorsey.

_____. 1976. Social Class and Parental Values: Another Conformation of the Relationship. *American Sociological Review* 41: 538–545.

_____. 1977. *Class and Conformity.* 2nd ed. Chicago: University of Chicago Press.

Kohn, Melvin L., and Carmi Schooler. 1983. *Work and Personality: An Inquiry into the Impact of Social Stratification.* Norwood, N.J.: Ablex.

Komarovsky, Mirra. 1946. "The Voluntary Associations of Urban Dwellers." *American Sociological Review* 11: 689–698.

_____. 1962. *Blue Collar Marriage.* New York: Vintage.

Krugman, Paul and Robert Lawrence 1994. "Trade, Jobs, and Wages." *Scientific American* April: 44–49.

Lamont, Michele 1992. *Money, Morals and Manners: The Culture of the French and the American Upper-Middle Class.* Chicago: University of Chicago Press.

Landecker, Werner S. 1981. *Class Crystallization.* New Brunswick, N.J.: Rutgers University Press.

Langerfeld, Steven. 1981. "To Break a Union." *Harpers* 262 (May): 16–21.

Langman, Lauren 1987. "Social Stratification." In *Handbook of Marriage and the Family,* edited by Marvin Sussman and Suzanne Steinmetz. New York: Plenum.

Laumann, Edward O. 1966. *Prestige and Association in an Urban Community.* Indianapolis: Bobbs-Merrill.

_____. 1973. *Bonds of Pluralism: The Form and Substance of Urban Social Networks.* New York: Wiley.

Lazerow, Michael 1995. "Millionaires Now 14% of House." *Roll Call,* July 10.

Leahy, Robert. 1981. "The Development of the Conception of Economic Inequality." *Child Development* 52: 523–532.

_____. 1983. *The Child's Construction of Social Inequality.* New York: Academic.

LeMasters, E.E. 1975. *Blue-Collar Aristocrats: Life-Styles at a Working-Class Tavern.* Madison: University of Wisconsin Press.

Lenski, Gerhard. 1954. "Status Crystallization: A Non-Vertical Dimension of Social Status." *American Sociological Review* 19: 405–413.

_____. 1966. *Power and Privilege: A Theory of Social Stratification.* New York: McGraw-Hill.

Lerner, Robert, Althea Nagai, and Stanley Rothman. 1996. *American Elites.* New Haven, Conn.: Yale University Press.

Levine, Donald M., and Mary Jo Bane, eds. 1975. *The "Inequality" Controversy: Schooling and Distributive Justice.* New York: Basic.

Levitan, Sar A. 1990. *Programs in Aid of the Poor.* 6th ed. Baltimore: Johns Hopkins University Press.

Levitan, Sar, and Isaac Shapiro. 1987. *Working But Poor: America's Contradiction.* Baltimore: Johns Hopkins University Press.

Levitan, Sar A., and Robert Taggart. 1976. *The Promise of Greatness.* Cambridge, Mass.: Harvard University Press.

Levy, Frank, and Richard J. Murnane. 1992. "U.S. Earnings Levels and Earnings Inequality: A Review of Recent Trends and Proposed Explanations." *Journal of Economic Literature.* 30: 1333–1381.

Lewis, Neil. 1996. "This Mr. Smith Gets his Way in Washington." *New York Times,* Oct. 12.

Lewis, Sinclair. 1922. *Babbitt.* New York: Harcourt Brace Jovanovich.

Link, Arthur S., and William Cotton. 1973. *American Epoch.* Vol. I. 4th ed. New York: Knopf.

Lipset, Seymour Martin. 1960. *Political Man.* New York: Doubleday.

_____. 1981. *Political Man.* Expanded edition. Baltimore: Johns Hopkins University Press.

Litwack, Leon, ed. 1962. *The American Labor Movement.* Englewood Cliffs, N.J.: Prentice-Hall.

Lopata, Helena Z., et al. 1980. "Spouses' Contributions to Each Other's Roles." In *Dual-Career Couples*, edited by Fran Pepitone-Rockwell. Beverly Hills, Calif.: Sage.

Lord, Walter. 1955. *A Night to Remember*. New York: Henry Holt.

Lorwin, Lewis L. 1933. *The American Federation of Labor*. Washington, D.C.: Brookings Institution.

LTV Corporation. 1990. *A Guide to the 102nd Congress: 1st Session Datebook/Calendar*. Washington, D.C.: Author.

Lubell, Samuel. 1956. *The Future of American Politics*. 2nd ed. Garden City, N.Y.: Doubleday Anchor.

Lundberg, Ferdinand. 1968. *The Rich and the Super-Rich*. New York: Lyle Stuart.

———. 1975. *The Rockefeller Syndrome*. Secaucus, N.J.: Lyle Stuart.

Lynd, Robert S., and Helen Merrell Lynd. 1929. *Middletown*. New York: Harcourt Brace Jovanovich.

———. 1937. *Middletown in Transition*. New York: Harcourt Brace Jovanovich.

Mackenzie, Gavin 1973. *The Aristocracy of Labor: The Position of Skilled Craftsmen in the American Class Structure*. New York: Cambridge University Press.

Madison, James (with Alexander Hamilton and John Jay). 1787. *The Federalist Papers*. New York: New American Library (1961).

Makinson, Larry. 1990. *Open Secrets: The Dollar Power of PACs in Congress*. Washington D.C.: Congressional Quarterly.

Makinson, Larry and Joshua Goldstein 1996. *Open Secrets: the Encyclopedia of Congressional Money and Politics*. 4th edition. Washington, D.C.: Congressional Quarterly.

Marcus, Ruth and Charles Babcock 1997. "The System Cracks Under the Weight of Cash." *Washington Post*. February 9.

Marx, Karl. 1979. *The Marx-Engels Reader*, edited by Robert C. Tucker. 2nd ed. New York: Norton.

Massey, Douglas 1996. "The Age of Extremes: Concentrated Affluence and Poverty in the Twenty-First Century." *Demography*. 33: 395–412.

McLeod, Jay. 1987. *Ain't No Makin' It: Leveled Aspirations in a Low-Income Neighborhood*. Boulder, Colo.: Westview.

Miller, Herman P. 1971. *Rich Man, Poor Man*, rev. ed. New York: Thomas Y. Crowell.

Mills, C. Wright. 1951. *White Collar*. New York: Oxford University Press.

———. 1956. *The Power Elite*. New York: Oxford University Press.

———. 1968. "Comment on Criticism." In *C. Wright Mills and the Power Elite*, edited by G. William Domhoff and Hoyt B. Ballard. Boston: Beacon.

Mintz, Beth. 1975. "The President's Cabinet, 1897–1972." *Insurgent Sociologist* 5(3): 131–149.

Mortenson, Thomas G. 1991. *Equity of Higher Educational Opportunity for Women, Black, Hispanic and Low Income Students*. Iowa City, Iowa: American College Testing Program.

Mortenson, Thomas G., and Zhijun Wu. 1990. *High School Graduation and College Participation of Young Adults by Family Income Backgrounds: 1970–1989*. Iowa City, Iowa: American College Testing Program.

Murray, Charles A. 1984. *Losing Ground: American Social Policy*. New York: Basic.

Nagle, John. 1977. *System and Succession: The Social Bases of Political Elite Recruitment*. Austin: University of Texas Press.

Naimark, Hedwin. 1981. *The Development of the Understanding of Social Class*. Doctoral dissertation, New York University.

Nakao, Keiko, and Judith Treas. 1990. "Revised Prestige Scores for All Occupations." Chicago: National Opinion Research Center (unpublished paper).

National Commission on Children. 1991. *Beyond Rhetoric: A New American Agenda for Children and Families*. Washington, D.C.: U.S. Government Printing Office.

Newman, Katherine S. 1988. *Falling from Grace: The Experience of Downward Mobility in the American Middle Class*. New York: Free Press.

Nock, Steven L., and Peter H. Rossi. 1978. "Ascription versus Achievement in the Attribution of Family Social Status." *American Journal of Sociology* 84: 565–590.

NORC (National Opinion Research Center). 1953. "Jobs and Occupations: A Popular Evaluation." In *Class, Status, and Power*, edited by Reinhard Bendix and Seymour M. Lipset. Glencoe, Ill.: Free Press.

Oakes, Jeannie. 1985. *Keeping Track: How High Schools Structure Inequality*. New Haven, Conn.: Yale University Press.

Oliver, Melvin and Thomas Shapiro 1995. *Black Wealth and White Wealth: A New Perspective on Racial Inequality*. New York: Routledge.

Ornati, Oscar. 1966. *Poverty Amidst Affluence*. New York: Twentieth Century Fund.

Orshansky, Mollie. 1974. "How Poverty Is Measured." In *Sociology of American Poverty*, edited by Joan Huber, et. al. Cambridge, Mass.: Schenkman.

Ossowski, Stanislaw. 1963. *Class Structure in the Social Consciousness*. New York: Free Press.

Osterman, Paul. 1991. "Gains from Growth? The Impact of Full Employment on Poverty in Boston." In *The Urban Underclass*, edited by Christopher Jencks and Paul Peterson. Washington, D.C.: Brookings Institution.

Ostrander, Susan. 1984. *Women of the Upper Class*. Philadelphia: Temple University Press.

Pakulski, Jan, and Malcolm Waters 1996. *The Death of Class*. Thousand Oaks, Calif.: Sage.

Pen, Jan. 1971. *Income Distribution*. London, England: Allen Lane.

Persell, Caroline Hodges. 1977. *Education and Inequality*. New York: Free Press.

Phillips, Kevin. 1990. *The Politics of Rich and Poor: Wealth and the American Electorate in the Reagan Aftermath*. New York: Random House.

Pianin, Eric 1997. "How Business Found Benefits in Wage Bill." *Washington Post*, February 11.

Piven, Frances Fox, and Richard A. Cloward. 1971. *Regulating the Poor: The Functions of Public Welfare*. New York: Pantheon.

Polsby, Nelson. 1970. "How to Study Community Power: The Pluralist Alternative." In *The Structure of Community Power*, edited by Michael Aiken and Paul E. Mott. New York: Random House.

Pomper, Gerald M. 1989. "The Presidential Power." In *The Election of 1988*, edited by Gerald M. Pomper. Chatham, N.J.: Chatham House, pp. 153–176.

Rainwater, Lee. 1965. *Family Design: Marital Sexuality, Family Size, and Contraception*. Chicago: Aldine.

_____. 1974. *What Money Buys*. New York: Basic.

Ramsey, Patricia. 1991. "Young Children's Awareness and Understanding of Social Class Differences." *Journal of Genetic Psychology* 152: 71–82.

Reiss, Albert. 1961. *Occupations and Social Status*. Glencoe, Ill.: Free Press.

Reissman, Leonard. 1954. "Class, Leisure and Participation." *American Sociological Review* 19: 74–84.

Reischauer, Robert 1997. *Setting National Priorities: Budget Choices for the Next Century*. Washington, D.C.: Brookings Institution.

Rieder, Jonathan. 1985. *Canarsie: The Jews and Italians of Brooklyn against Liberalism*. Cambridge, Mass.: Harvard University Press.

Riesman, David. 1953. *The Lonely Crowd: A Study of the Changing American Character*. New Haven, Conn.: Yale University Press.

Ritter, Kathleen, and Lowell Hargens. 1975. "Occupational Positions and Class Identifications of Married Women." *American Journal of Sociology* 80: 934–948.

Ritzer, George. 1996. *The McDonaldization of Society: an investigation into the changing character of contemporary social life*. Thousand Oaks, Calif.: Pine Forge.

Rose, Stephen J. 1992. *Social Stratification in the United States: The American Profile Poster Revised and Expanded*. New York: New Press.

_____. 1993. "Declining Family Incomes in the 1980s: New Evidence from Longitudinal Data." *Challenge*. November-December, 1993: 29-36.

_____. 1994. *On Shaky Ground: Rising Fears About Income and Earnings*. Washington, D.C.: National Commission for Employment Policy.

Rosen, Ellen Israel 1987. *Bitter Choices: Blue-Collar Women in and out of Work*. Chicago: University of Chicago Press.

Ross, Howard, 1968. "Economic Growth and Change in the United States under Laissez-Faire: 1870–1929." In *The Age of Industrialization in America*, edited by Frederic Cople Jaher. New York: Free Press.

Rossi, Peter H. 1989. *Down and Out in America: The Origins of Homelessness*. Chicago: University of Chicago Press.

Rubin, Lillian Breslow. 1976. *Worlds of Pain: Life in the Working-Class Family*. New York: Basic.

_____. 1994. *Families on the Faultline: America's Working Class Speaks About the Family, the Economy, Race, and Ethnicity*. New York: HarperCollins.

Rubin, Z. 1968. "Do American Women Marry Up?" *American Sociological Review* 5: 750–760.

Ryscavage, Paul, et al. 1992. "The Impact of Demographic, Social and Economic Change on the Distribution of Income." In U.S. Census, *Studies in the Distribution of Income*.

Sachs, Jeffrey and Howard Shatz. 1994. "Trade and Jobs in U.S. Manufacturing." *Brookings Papers on Economic Activity* Number 1.

Sawhill, Isabel. 1989. "The Underclass: An Overview." *The Public Interest* 96 (Summer): 3–15.

Schiller, Bradley. 1989. *The Economics of Poverty and Discrimination.* 5th ed. Englewood Cliffs, N.J.: Prentice-Hall.

Schreiber, E.M., and G.T. Nygreen. 1970. "Subjective Social Class in America: 1945–1968." *Social Forces* 45: 348–356.

Schwartz, John E. 1983. *America's Hidden Success: A Reassessment of Twenty Years of Public Policy.* New York: Norton.

Schwartz, Marvin, and Barry Johnson. 1990. "Estimates of Personal Wealth, 1986." *Statistics of Income Bulletin* 9 (Spring):63–78.

Schwartz, Michael, ed. 1987. *The Structure of Power in America: The Corporate Elite as a Ruling Class.* New York: Holmes & Meier.

Sernau, Scott, ed. 1996. *Social Stratification Courses: Syllabi and Instructional Materials.* 3rd ed. Washington, D.C.: American Sociological Association.

Sewell, William H., and Robert M. Hauser. 1975. *Education, Occupation and Earnings.* New York: Academic.

Sewell, William H., and Vimal P. Shah. 1977. "Socioeconomic Status, Intelligence, and the Attainment of Higher Education. In *Power and Ideology in Education,* edited by Jerome Karabel and A.H. Halsey. New York: Oxford University Press.

Shostak, Arthur and William Gomberg, eds. 1964. *The Blue Collar World: Studies of the American Worker.* Englewood Cliffs, N.J. : Prentice Hall.

Simmons, Robert G., and Morris Rosenberg. 1971. "Functions of Children's Perceptions of the Stratification System." *American Sociological Review* 36: 235–249.

Simkus, Albert. 1978. "Residential Segregation by Occupation and Race." *American Sociological Review* 43: 81–93.

Smith, David H. and Jacqueline Macaulay. 1980. *Participation in Social and Political Activities.* San Francisco: Jossey-Bass.

Smith, James D. 1984. "Trends in the Concentration of Personal Wealth in the United States, 1958–1976." *Review of Income and Wealth* 30: 419–428.

Smith, James P., and Finis R. Welch. 1989. "Black Economic Progress After Myrdal." *Journal of Economic Literature* 27: 519–564.

Snow, David and Leon Anderson 1993. *Down on their Luck: Homeless Street People.* Berkeley: University of California Press.

Sorauf, Frank 1992. *Inside Campaign Finance: Myths and Realities.* New Haven, Conn.: Yale University Press.

Sorensen, Annemette 1994. "Women, Family and Class, " *Annual Review of Sociology* 20: 27–47.

Stack, Carol B. 1974. *All Our Kin.* New York: Harper & Row.

Statesman's Year-Book. 1979–1990. New York: St. Martin's.

Stendler, Celia Burns. 1949. *Children of Brasstown: Their Awareness of the Symbols of Social Class.* Urbana: University of Illinois Press.

Strole, Leo, et al. 1978. *Mental Health in the Metropolis: The Midtown Manhattan Study.* Revised and enlarged ed. New York: New York University Press.

Sussman, Marvin and Suzanne Steinmetz. 1987. *Handbook of Marriage and the Family.* New York: Plenum.

Sweezy, Paul. 1968. "Power Elite or Ruling Class?" In *C. Wright Mills and the Power Elite,* edited by G. William Domhoff and Hoyt B. Ballard. Boston: Beacon.

Szymanski, Albert. 1978. *The Capitalist State and the Politics of Class.* Cambridge, Mass.: Winthrop.

Teixeira, Ruy 1992. *The Disappearing American Voter.* Washington, D.C: Brookings Institution.

Terkel, Studs. 1974. *Working: People Talk About What They Do All Day and How They Feel About It.* New York: Avon.

Thompson, E.P. 1963. *The Making of the English Working Class.* New York: Vintage.

Treas, Judith, and Ramon Torrecilha. 1995. "The Older Population." in *The State of the Union: America in the 1990s; Volume Two: Social Trends.* New York: Russell Sage Foundation.

Treiman, Donald. 1977. *Occupational Prestige in Comparative Perspective.* New York: Academic.

Treiman, Donald, and Patricia Roos. 1983. "Sex and Earnings in Industrial Society: A Nine-Nation Comparison." *American Journal of Sociology* 89: 612–650.

Tuchman, Gaye, ed. 1974. *The TV Establishment: Programming for Power and Profit.* Englewood Cliffs, N.J.: Prentice-Hall.

Tutor, Jeannette. 1991. "The Development of Class Awareness in Children." *Social Forces* 49: 470–476.

U.S. Bureau of the Census. 1973. *Statistical Abstract of the United States: 1973.*

_____. 1975. *Historical Statistics of the United States.* Bicentennial Edition. 2 Parts.

_____. 1976. *Bicentennial Statistics, Pocket Data Book, U.S.A.*

_____. 1980a. *The Social and Economic Status of the Black Population in the United States: An Historical View, 1790–1978.* Current Population Reports, Special Studies Series P-23, no. 80.

_____. 1980b. *Statistical Abstract of the United States: 1980.*

_____. 1983. *Statistical Abstract of the United States: 1984.*

_____. 1984a. *Statistical Abstract of the United States: 1985.*

_____. 1984b. *Money Income and Poverty Status of Families and Persons in the United States: 1983 — Advance Data.* Current Population Reports. Series P-60, no. 145.

_____. 1984c. *1980 Census of Population, United States Summary, Section A: United States.* Characteristics of the Population. Series PC80–1– D1-A.

_____. 1986a. *Statistical Abstract of the United States, 1987.*

_____. 1986b. *Characteristics of the Population Below the Poverty Level: 1984.* Series P-60, no. 152.

_____. 1989a. *Money Income of Households, Families, and Persons in the United States: 1987.* Current Population Reports. Series P-60, no. 162.

_____. 1990a. *Money Income and Poverty Status in the United States, 1989.* Current Population Reports. Series P-60, no. 168.

_____. 1990b. *Trends in Income, by Selected Characteristics: 1947 to 1988.* Current Population Reports. Series P-60, no. 167.

_____. 1990c. *Statistical Abstract of the United States: 1990.*

_____. 1990d. *Measuring the Effect of Benefits and Taxes on Income and Poverty: 1989.* Current Population Reports. Series P-60, no. 169-RD.

_____. 1990e. *Transition in Income and Poverty Status: 1985–86.* Series P-70, no. 18.

_____. 1991a. *Money Income of Households, Families, and Persons in the United States: 1990.* Series P-60, no. 174.

_____. 1991b. *Poverty in the United States: 1990.* Series P-60, no. 175.

_____. 1991c. *Measuring the Effect of Benefits and Taxes on Income and Poverty: 1990.* Series P-60, no. 176-RD.

_____. 1991d. *Transition in Income and Poverty Status: 1987–88.* Series P-70, no. 24.

_____. 1992. *Workers with Low Earnings: 1964–1990.*

_____. 1993. *Money Income of Households, Families, and Persons in the United States: 1992.* Series P60-184.

_____. 1994. *Dynamics of Economic Well-Being: Labor Force and Income, 1990 to 1992.* Series P70-40.

_____. 1995a. *Statistical Abstract of the United States, 1995.*

_____. 1995b. *Child Support for Custodial Mothers and Fathers: 1991.* Series P60-187.

_____. 1995c. *Asset Ownership of Households: 1993.*

_____1996a. *Income, Poverty, and Valuation of Noncash Benefits: 1994.* Series P60-189.

_____. 1996b. "A Brief Look at Postwar U.S. Income Inequality." *Current Population Reports.* Series P60-191.

_____. 1996c. *Selected Income Distribution Statistics.* Series P60-191.

_____. 1996d. *Poverty in the United States: 1995.*

_____. 1996e. *Money Income in the United States: 1995.*

_____. 1996f *Statistical Abstract of the United States: 1996.*

_____. n.d. *Asset Ownership of Households: 1993.* Series P70-47. Detailed Tables.

U.S. Department of Justice. 1990. *Sourcebook of Criminal Justice Statistics: 1989.*

U.S. Department of Labor. 1978. *Employment and Training Report to the President.*

_____. 1980. *Handbook of Labor Statistics.*

_____. 1984. *Employment and Earnings.* January.

_____. 1990a. *Outlook 2000.*

_____. 1990b. *Employment and Earnings.* January.

_____. 1991. *Employment and Earnings.* January.

_____. 1994. *Report on the American Workforce*.

_____. 1995. "BLS Projections to 2005." Monthly Labor Review. 118: 3-9. (November)

_____. 1996a. *Consumer Expenditures in 1994*. (February)

_____. 1996b. *Employment and Earnings*. (January 1996)

U.S. General Accounting Office. 1990. *Voting: Some Procedural Changes and Informational Activities Could Increase Turnout*.

U.S. House of Representatives, Committee on Ways and Means. 1990. *Tax Progressivity and Income Distribution*. March 26.

U.S. House of Representatives, Committee on Ways and Means. 1992. *Overview of Entitlement Programs: Background Material and Data on Programs within the Jurisdiction of the Committee of Ways and Means*. May 15.

_____. 1995. *The 1994-1995 Green Book*.

_____. 1991a. *Overview of the Federal Tax System*. WMCP-102–7.

_____. 1991b. *Background Material and Data on Programs within the Jurisdiction of the Committee of Ways and Means*.

U.S. Internal Revenue Service. 1988. *Statistics of Income*. Fall.

_____. 1995. *Statistics of Income*. Spring.

_____. 1996. *Statistics of Income*. Spring.

U.S. Public Health Service. 1990. *Prevention Profile, Health, United States, 1989*.

Useem, Michael 1996. *Investor Capitalism: How Money Managers are Changing the Face of Corporate America* New York: Basic.

Vanfossen, Beth Ensminger. 1977. "Sexual Stratification and Sex-Role Socialization." *Journal of Marriage and the Family* 39: 563–574.

Veblen, Thornstein. 1934. *The Theory of the Leisure Class*. New York: Modern Library. (First published 1899).

Wallace, Phyllis A., ed. 1982. *Women in the Workplace*. Boston: Auburn House.

Walton, John. 1970. "A Systematic Survey of Community Power Research." In *The Structure of Community Power*, edited by Michael Aiken and Paul E. Mott. New York: Random House.

Warner, W. Lloyd, and Paul S. Lunt. 1941. *The Social Life of a Modern Community*. New Haven, Conn.: Yale University Press.

Warner, W. Lloyd, et al. 1949a. *Democracy in Jonesville*. New York: Harper & Row.

_____. 1949b. *Social Class in America*. Chicago: Science Research Associates.

_____. 1973. *Yankee City*. Abridged Edition. New Haven, Conn.: Yale University Press.

Wattenberg, Ben J. 1974. *The Real America*. Garden City, N.Y.: Doubleday.

Weber, Max. 1946. *From Max Weber: Essays in Sociology*, edited by H.H. Gerth and C. Wright Mills. New York: Oxford University Press.

Who's Who in America. 1980. 41st ed. Chicago: Marquis.

Wetzel, James R. 1995. "Labor Force, Unemployment, and Earnings." in *State of the Union: America in the 1990s, Volume One: Economic Trends,* edited by Reynolds Farley. New York: Russell Sage Foundation.

Who's Who in American Politics. 1979. 7th ed. New York: Bowker.

Whyte, Martin King. 1990. *Dating, Mating and Marriage.* New York: Aldine de Gruyter.

Whyte, William H. 1952. *Is Anybody Listening?* New York: Simon & Schuster.

Wilson, William J. 1980. *The Declining Significance of Race.* 2nd ed. Chicago: University of Chicago Press.

————. 1987. *The Truly Disadvantaged: The Inner City, the Underclass, and Public Policy.* Chicago: University of Chicago Press.

————. 1991. "Public Policy Research and *The Truly Disadvantaged.*" In *The Urban Underclass,* edited by Christopher Jencks and Paul Peterson. Washington, D.C.: Brookings Institution.

————. 1996. *When Work Disappears: The World of the New Urban Poor.* New York: Knopf.

Wilson, William Julius, and Katheryn Nickerman. 1986. "Poverty and Family Structure: The Widening Gap Between Evidence and Public Policy Issues." In *Fighting Poverty: What Works and What Doesn't,* edited by Sheldon Danziger and Daniel H. Weinberg. Cambridge, Mass.: Harvard University Press.

Wolf, Tom 1987. *The Bonfire of the Vanities.* New York: Farrar, Straus & Giroux.

Wolff, Edward. 1993. "The Structure of Wealth Inequality: A Report to the Twentieth Century Fund." Unpublished paper.

————. 1996. *Trends in Household Wealth, 1983–1992: Report Submitted to the Department of Labor.* Unpublished paper.

Wolfinger, Raymond E. 1973. *The Politics of Progress.* Englewood Cliffs, N.J.: Prentice-Hall.

Wool, Harold. 1976. *The Labor Supply for Lower-Level Occupations.* New York: Praeger.

Wright, James. 1989. *Address Unknown: The Homeless in America.* New York: Aldine de Gruyter.

Zeitlin, Maurice. 1980. *Classes, Class Conflict and the State.* Cambridge, Mass.: Winthrop.

Name Index

Subject Index

Credits

This constitutes an extension of the copyright page. The numbers in parentheses after each entry are the page numbers on which the reprinted material appears.

Bott, Elizabeth. 1954. "The Concept of Class as a Reference Group." *Human Relations* 7: 259–286. Reprinted by permission of Plenum Publishing Corporation. (228)

Centers, Richard. 1949. *The Psychology of Social Classes: A Study of Class Consciousness.* Princeton, N.J.: Copyright © 1949, © renewed by Princeton University Press. Reprinted by permission of Princeton University Press. (222)

Coleman, Richard P., and Lee Rainwater. 1978. *Social Standing in American Society.* Copyright © 1978 by Basic Books, Inc. Reprinted by permission of Basic Books, a division of HarperCollins Publishers Inc. (34–37)

Dahl, Robert A. 1961. *Who Governs? Democracy and Power in an American City.* New Haven, Conn.: Yale University Press. Reprinted by permission of Yale University Press. (180–181)

Davis, Allison, Burleigh B. Gardner, and Mary R. Gardner, 1941. *Deep South: A Social-Anthropological Study of Caste and Class.* Chicago: University of Chicago Press. Copyright 1941 by The University of Chicago. Reprinted by permission of The University of Chicago Press. (30–31)

Featherman, David L., and Robert M. Hauser. 1978. *Opportunity and Change.* New York: Academic Press. Reprinted by permission of Academic Press and David L. Featherman. (151, 154)

Halberstam, David. 1972. *The Best and the Brightest.* New York: Random House. Reprinted by permission of Random House. (205)

Jackman, Mary. 1979. "The Subjective Meaning of Social Class Identification in the United States." *Public Opinion Quarterly* 43: 443–462. Reprinted by permission of Elsevier Science Publishing Co., Inc. (42)

Laumann, Edward O. 1966. *Prestige and Association in an Urban Community.* Indianapolis: Bobbs-Merrill. Reprinted by permission of Bobbs-Merrill Co., Inc. (123, 133)

LeMasters, E. E. 1975. *Blue-Collar Aristocrats: Life Styles at a Working-Class Tavern.* University of Wisconsin Press. Reprinted by permission of University of Wisconsin Press. (127, 128)

Lewis, Sinclair. 1992. *Babbitt.* New York: Harcourt Brace Jovanovich. Reprinted by permission of Harcourt Brace Jovanovich. (23)

Lynd, Robert S., and Helen Merrell Lynd. 1929. *Middletown.* New York: Harcourt Brace Jovanovich. Copyright 1929 Harcourt Brace Jovanovich, Inc.; copyright 1957 Robert S. and Helen M. Lynd. Reprinted by permission of Harcourt Brace Jovanovich. (59)

Marx, Karl. 1979. *The Marx-Engels Reader,* edited by Robert C. Tucker. 2nd ed. New York: W.W. Norton. Reprinted by permission of W.W. Norton & Company, Inc. (6)

Mills, C. Wright. 1956. *The Power Elite.* New York: Oxford University Press. Reprinted by permission of Oxford University Press. (30, 187)

Mortenson, Thomas G. 1991. *Equity of Higher Education Opportunity for Women, Black, Hispanic, and Low Income Students.* Iowa City: Iowa: American College Testing Service. Reprinted by permission of American College Testing Service. (170, 172)

Rainwater, Lee. 1965. *Family Design: Marital Sexuality, Family Size, and Contraception.* Chicago: Aldine. Reprinted by permission of Lee Rainwater. (125)

Rubin, Lillian Breslow. 1976. *Worlds of Pain: Life in the Working-Class Family.* New York: Basic Books. Copyright © 1976 by Lillian Breslow Rubin. Reprinted by permission of Basic Books, a division of HarperCollins Publishers Inc. (129)

Warner, W. Lloyd, and Paul S. Lunt. 1941. *The Social Life of a Modern Community.* New Haven, Conn.: Yale University Press. Reprinted by permission of Yale University Press. (25–27)

Warner, W. Lloyd, et al. 1949. *Social Class in America.* New York: Harper & Row. Reprinted by permission of Harper & Row Publishers, Inc. (28)